AF361743

Fiber Optic Sensors for Structural and Geotechnical Monitoring

Fiber Optic Sensors for Structural and Geotechnical Monitoring

Special Issue Editor

Michele Arturo Caponero

MDPI • Basel • Beijing • Wuhan • Barcelona • Belgrade • Manchester • Tokyo • Cluj • Tianjin

Special Issue Editor
Michele Arturo Caponero
ENEA — Research Centre of Frascati - Photonics Micro- and Nanostructures Laboratory
Italy

Editorial Office
MDPI
St. Alban-Anlage 66
4052 Basel, Switzerland

This is a reprint of articles from the Special Issue published online in the open access journal *Sensors* (ISSN 1424-8220) (available at: https://www.mdpi.com/journal/sensors/special_issues/ FOS_Structural_Geotechnical_Monitoring).

For citation purposes, cite each article independently as indicated on the article page online and as indicated below:

LastName, A.A.; LastName, B.B.; LastName, C.C. Article Title. *Journal Name* **Year**, *Article Number*, Page Range.

ISBN 978-3-03936-032-1 (Hbk)
ISBN 978-3-03936-033-8 (PDF)

Contents

About the Special Issue Editor

Michele Arturo Caponero is a senior researcher at ENEA, the Italian National Agency for New Technologies, Energy and Economic Sustainable Development. He is a member of the Working Group TC86C Fibre Optic Sensors of IEC, the International Electrotechnical Commission. His research activity is mainly focused on the development of sensing systems based on fiber optic technology for applications in various fields: civil and geotechnical engineering, nuclear physics, cultural heritage conservation, biomedicine. He has been the reference person for ENEA in many publicly and privately funded research projects focused on industrial applications of fiber optic sensors. He holds various patents and has authored over 100 research papers.

Editorial

Special Issue "Fibre Optic Sensors for Structural and Geotechnical Monitoring"

Michele Arturo Caponero

Photonics Micro and Nanostructures Laboratory, ENEA Research Centre of Frascati, Via Enrico Fermi 45, Frascati, 00044 Rome, Italy; michele.caponero@enea.it

Received: 21 April 2020; Accepted: 22 April 2020; Published: 24 April 2020

Abstract: In this editorial on the special issue "Fibre Optic Sensors for Structural and Geotechnical Monitoring" a review of the contribution papers selected for publication is given. Each paper is briefly summarized, presenting its objective and methods, then a comment is given about the relevance of the work with respect to the advance and the spreading of the fibre optic technology for monitoring applications.

1. Introduction

The use of sensors based on fibre optic technology allows a broad range of applications in the fields of structural and geotechnical monitoring, which can effectively improve maintenance of infrastructures and safety of communities. Thanks to its valuable features, such as distributed monitoring, easy and endurance of cabling, long term stability, reliable response in both static and dynamic regime, fibre optic technology has already provided innovative and efficient solutions to quite difficult monitoring problems. The worldwide increasing attention to infrastructures and communities with resilience capabilities against natural disasters has opened up new and challenging prospective applications for the use of fibre optic technology for structural and geotechnical monitoring.

This Special Issue collects contributions in the development and application of monitoring solutions based on fibre optic technology for structural and geotechnical engineering works and issues. In the following, the content of the contributions is reviewed, providing a brief introduction to the work presented in the paper and commenting the relevance of the work with respect to the advance and the spreading of the fibre optic technology for monitoring applications. All contributions, following the recommendation in the invitation to this Special Issue, provide a comprehensive discussion and report a rich bibliography on the current trends and the issues relative to the work presented in the paper.

2. Contributions

Paper [1] reports the monitoring activity successfully tested in a marble quarry, by Brillouin sensing combined with drone photogrammetry and geotechnical survey. The monitoring activity is intended to both raise the safety at work and assist the extraction planning. Beyond the specific application, the paper confirms the effective capability of fibre optic distributed sensing techniques to monitor complex geomorphological sites. Moreover, the paper proposes and demonstrates an efficient procedure to integrate fibre optic distributed sensors in existing monitoring systems based on an array of spatially localized geotechnical sensors (extensometers, crackmeters, inclinometers, topographic markers, etc.). That is a procedure of relevant interest, as it can facilitate the spreading of fibre optic distributed sensing, which can thus be proposed as an efficient tool to upgrade monitoring installations based on traditional technologies.

Paper [2] reports the successful demonstration, on a small mockup, of the use of fibre optic distributed sensing to characterize plate foundation support. As the execution of distributed strain measurements with very high-resolution is critical to the demonstration on the small mockup, a suitable technique is adopted, which is based on a proprietary implementation of the optical frequency domain reflectometry. In the perspective of future demonstration on a true size mockup and with the envision of engineering application on real works, since very high-resolution will not be necessary, the authors preview the use of standard fibre optic distributed sensing techniques. Beyond the specific application, the paper confirms the high versatility of the fibre optic distributed sensing technologies, that prove themselves as a major candidate for the development of self-monitoring extended engineering structures.

Paper [3] reports the use of fibre optic distributed monitoring to investigate properties of geomaterial samples according to the laboratory procedure named direct simple shear test. A new procedure is proposed to evaluate some component of the stresses in the sample, taking advantage of the very high-resolution measurements that can be done by use of a proprietary implementation of the optical frequency domain reflectometry. The paper presents the production of an upgraded sample holder for direct simple shear test, the execution of the test, and a discussion of the results. The sample holder is made by 3D printing, with grooves in which the optical fibre distributed sensor is installed. The paper finely demonstrates the use of high-resolution fibre optic distributed sensing for the direct simple shear test. The paper, with a new application of a fibre optic sensing technique to laboratory tests, assesses the broad range of potential applications of the fibre optic sensors in the geothechnical field, so far mostly proved for large-scale in-field applications.

Paper [4] reports a novel method of data processing for distributed acoustic sensing by phase-sensitive optical time domain reflectometry. The method, based on spatial kurtosis, is intended to improve the signal detection in noisy systems and in presence of environment perturbations (high signal to noise ratio). The method is validated in laboratory and outdoor experimental tests, simulating structural cracking and malicious digging. Results provide a valid perspective of the broadening of use of distributed acoustic sensing for applications in structural health monitoring and intrusion surveillance. In fact, the method for basic (no proprietary) optoelectronic hardware meets requirements for real-time data analysis and does not lower the spatial resolution provided by other assessed signal processing methods.

Paper [5] reports an application of distributed temperature sensing to monitor the final construction phase and commissioning of wells serving a CO_2 geological storage. In the paper, offline data analysis is done adopting different temperature calibration methods and results are compared. Beyond the specific application, the paper addresses practical procedures to face the impossibility to properly perform calibration of cables before installation, which is an often-recurring situation when working in a real construction yard. The paper provides a relevant example of an application in which the same fibre optic distributed monitoring installation can be first used to monitor first the correct production of the infrastructure, and can be later used to monitor it along its working life. Moreover, the installed sensing line is shown to have fibres for distributed acoustic sensing too, which makes the wells a relevant potential demonstration site for multifunctional monitoring by fibre optic distributed monitoring.

Paper [6] reports an application of structural health monitoring of a railway tunnel. Monitoring is done by strain gauges and pressure cells based on the fibre Bragg grating technology. Monitoring is focused on the interaction soil-tunnel, the tunnel being built in open space and later buried. The work done shows an impressive engineering effort, with the adoption of rugged solutions to protect the installation from the severe working yard conditions. Moreover, a valuable system for remote control and automatic data collection is at service of the installation. Beyond the specific application, the paper shows the high maturity of the technological solutions available to develop monitoring systems based on the the fibre Bragg grating sensors which offer an unrivalled versatility in the production of customised and multifunctional chain of sensors.

Paper [7] reports use of fibre Bragg grating sensors in a real size test on prefabricated concrete hollow piles. The test is intended to characterize the performance of the piles during the installation and the subsequent loading test. Chains of sensors were were installed on the external surface of the piles along diametrically opposite longitudinal grooves. Piles were installed in clay subsoil by a clamping-and-jacking machine, with a step-by-step pushing action. The loading test was done 17 days later, with an hydraulic jack inserted between the pile upper end and loading platform loaded with concrete blocks. Results provide an effective characterization of the performance of the piles. Beyond the specific application, the work points out the potential of the fibre Bragg sensor technology in solving problems related to the characterization and monitoring of the civil and geotechnical engineering infrastructures. In fact, the technology offers both fast response to dynamic or transitory events and long term stability to static loads, which are paramount features whenever measurements shall be done before and after the production, with some curing/settlement time to wait for.

Paper [8] presents a displacement sensor developed for structural health monitoring in the civil engineering field, with specific potential applications to monitor spring dampers for floating slab installation. The displacement sensor design is based on a sliding block that affects the deflection of a cantilever; the cantilever deflection is monitored by two fibre Bragg grating sensors. The paper presents a prototype sensor dimensioned and produced for an intended application on floating slabs of a subway line in Beijing. The performance of the prototype is investigated for sensitivity in the full displacement range, repeatability, temperature compensation capability and creep. The paper shows the high versatility and potentiality of fibre Bragg grating sensors in being used to equip machinery transducers for mechanical measurements. The use of such sensors for transducers devoted to the structural health monitoring of civil engineering infrastructures can effectively benefit of their peculiar features, as for instance the data taking in wavelength division multiplexing which can greatly lower the cost of installations with many sensors to be deployed with long routing.

Paper [9] investigates the structural health capabilities of fibre Bragg grating sensors embedded in components made by ultrasonic additive manufacturing. Sensors are embedded during the manufacturing process of standard specimens for laboratory crack tests. During the tests, the effectiveness of the sensors in detecting the crack and its evolution is characterised. Sensors are also thermally tested to investigate the upper temperature operational value in the embedded condition. The paper contribute to assess the versatility of the fibre Bragg grating sensors as embeddable sensors in components made by special metallurgic process. Thus, being already assessed since long the possibility of their embedding in carbon and glass composite materials, fibre Bragg grating sensors demonstrate their unrivalled potential to provide prognostic capabilities to high technology mechanical components.

Paper [10] reports the use of fibre Bragg grating sensors to characterize an innovative composite component of aircraft landing gear. Several sensors are used to monitor strain while performing mechanical tests to simulate in laboratory the expected operating stress condition. Sensors are glued on the surface of the component, and preliminary tests are done to select the best adhesive for the specific application. Preliminary tests also confirm that no spectral distortion of the signal of the sensors occurs in the full range of the load test, which guarantee both the correct working condition of the sensors the correct decoding of their spectroscopic signal. Beyond the specific application, the paper confirms the potential of fibre Bragg grating sensors to be used for fine mechanical measurements, with easy production of minimally invasive and easy cabling multifunctional chain of sensors. Moreover, the spectral signature of the sensors provides a reliable feature to control their correct functioning.

Paper [11] reports a demonstration of distributed dynamic sensing by use of ultra-weak fibre Bragg grating technology. The test site is a subway tunnel in regular service, the demonstration is for intrusion alarm and structural heath monitoring. Distribute sensing is tested for a length of 5 km, with spatial resolution of 5 m. Sensing is done by optical time domain reflectometry to evaluate vibration position and by phase demodulation to evaluate frequency and amplitude of the vibration. Beyond the specific application, the paper confirms the availability of emerging and promising techniques for distributed acoustic sensing that can go beyond the limits of the traditional ones and can manage the demanding specifications often required for retrofitting interventions on existing large civil infrastructure, as for instance high sensitivity, real-time response, operatibility with high signal-to-noise ratio.

Paper [12] proposes and tests in the laboratory a method for corrosion detection and evaluation of concrete rebars. The method is applied by use of an array of long gage sensors, each sensor being made by a fibre Bragg grating installed in a tubular housing whose length defines the sensor length gauge. The array of sensors is used to perform strain modal identification, in turn used to perform corrosion detection and evaluation, working out both location and quantification. The proposed method is tested and validated on a reinforced concrete beam subjected to controlled accelerated corrosion procedure. The paper is a relevant example of innovation procedure for structural health monitoring of reinforced concrete infrastructures, whose implementation is facilitated by the unrivalled property of the fibre Bragg grating technology in providing both dynamic and stable static strain monitoring.

Paper [13] proposes a strain measurement procedure based on the bending loss variation of an optical fibre specially arranged on the component to be monitored. The special arrangement consists of a short length of the optical fibre stretched and twisted around a few small cylinders. Sensing is encoded in the intensity of the light transmitted trough the fibre. The cylinders are attached to the component to be monitored, with the correct pattern according to the strain measurement to be done. As the pattern of the cylinders changes according to the deformation experienced by the component, the bending of the fibre changes accordingly and in turn a variation of the bending loss occurs. The paper presents the results obtained with composite specimens subject to tensile and bending tests. The paper is an interesting example of the use of basic properties of the optical fibre and simple setup to perform strain measurements with the possibility to easily adjust both sensitivity and gauge length.

Paper [14] proposes a deflection sensor based on the bending loss of machined plastic optical fibre. The sensor is intended for application in structural health monitoring of civil and geotechnical infrastructures. In particular, its use is proposed as a permanent monitoring device to be used with scheduled surveys by topographic techniques. The sensor is in the form of a bar with four plastic fibres running longitudinally on its surface, equally spaced along its circumference. The sensing element is a short segment of the plastic fibre, machined to have saw-teeth-shaped groves on its cladding in order to maximize bending losses. The four sensing elements monitor the bending of the bar, which has to be stuck on the structure to be monitored for deflection. The paper is an interesting example of production of a modular and cost-effective monitoring tool by use of plastic fibres and exploitation of the basic properties of guided light transmission.

Paper [15] proposes and studies a bending sensor intended for structural health monitoring of underwater civil engineering structures. The sensor is based on the use of a plastic fibre and specific pulsed light-emitting and power-measuring circuitry. Along the plastic fibre, a short sensing segment is produced by making a machined cut which acts as a light leakage zone. Leakage depends on the local bending of the fibre, thus power loss encodes the measurement of the local bending. The paper presents and verifies a model of the bending loss mechanism having parameters which related to the geometry of the leakage zone. Laboratory tests validate a prototype of the proposed sensor and provide its experimental characterisation. The paper is a relevant example of the application oriented development of a sensor based on a basic property, namely the bending loss, whose metrological features are upgraded with respect of the state of the art thanks to simple but effective encoding/decoding signal solutions.

Funding: This research received no external funding.

Conflicts of Interest: The author declares no conflict of interest.

References

1. Lanciano, C.; Salvini, R. Monitoring of Strain and Temperature in an Open Pit Using Brillouin Distributed Optical Fiber Sensors. *Sensors* **2020**, *20*, 1924. [CrossRef] [PubMed]
2. Skar, A.; Klar, A.; Levenberg, E. Load-Independent Characterization of Plate Foundation Support Using High-Resolution Distributed Fiber-Optic Sensing. *Sensors* **2019**, *19*, 3518. [CrossRef] [PubMed]
3. Klar, A.; Roed, M.; Rocchi, I.; Paegle, I. Evaluation of Horizontal Stresses in Soil during Direct Simple Shear by High-Resolution Distributed Fiber Optic Sensing. *Sensors* **2019**, *19*, 3684. [CrossRef] [PubMed]

4. Jiang, F.; Li, H.; Zhang, Z.; Zhang, Y.; Zhang, X. Localization and Discrimination of the Perturbation Signals in Fiber Distributed Acoustic Sensing Systems Using Spatial Average Kurtosis. *Sensors* **2018**, *18*, 2839. [CrossRef] [PubMed]
5. Lee, D.S.; Park, K.G.; Lee, C.; Choi, S.-J. Distributed Temperature Sensing Monitoring of Well Completion Processes in a CO_2 Geological Storage Demonstration Site. *Sensors* **2018**, *18*, 4239. [CrossRef] [PubMed]
6. Xu, T.; Wang, M.; Yu, L.; Lv, C.; Dong, Y.; Tian, Y. Research on the Earth Pressure and Internal Force of a High-Fill Open-Cut Tunnel Using a Bilayer Lining Design: A Field Test Using an FBG Automatic Data Acquisition System. *Sensors* **2019**, *19*, 1487. [CrossRef] [PubMed]
7. Kou, H.-L.; Diao, W.-Z.; Liu, T.; Yang, D.-L.; Horpibulsuk, S. Field Performance of Open-Ended Prestressed High-Strength Concrete Pipe Piles Jacked into Clay. *Sensors* **2018**, *18*, 4216. [CrossRef] [PubMed]
8. Guo, Y.; Liu, W.; Xiong, L.; Kuang, Y.; Wu, H.; Liu, H. Fiber Bragg Grating Displacement Sensor with High Abrasion Resistance for a Steel Spring Floating Slab Damping Track. *Sensors* **2018**, *18*, 1899. [CrossRef] [PubMed]
9. Chilelli, S.K.; Schomer, J.J.; Dapino, M.J. Detection of Crack Initiation and Growth Using Fiber Bragg Grating Sensors Embedded into Metal Structures through Ultrasonic Additive Manufacturing. *Sensors* **2019**, *19*, 4917. [CrossRef]
10. Iadicicco, A.; Natale, D.; Di Palma, P.; Spinaci, F.; Apicella, A.; Campopiano, S. Strain Monitoring of a Composite Drag Strut in Aircraft Landing Gear by Fiber Bragg Grating Sensors. *Sensors* **2019**, *19*, 2239. [CrossRef] [PubMed]
11. Gan, W.; Li, S.; Li, Z.; Sun, L. Identification of Ground Intrusion in Underground Structures Based on Distributed Structural Vibration Detected by Ultra-Weak FBG Sensing Technology. *Sensors* **2019**, *19*, 2160. [CrossRef] [PubMed]
12. Cheng, Y.; Zhao, C.; Zhang, J.; Wu, Z. Application of a Novel Long-Gauge Fiber Bragg Grating Sensor for Corrosion Detection via a Two-level Strategy. *Sensors* **2019**, *19*, 954. [CrossRef] [PubMed]
13. Choi, S.-J.; Jeong, S.-Y.; Lee, C.; Park, K.G.; Pan, J.-K. Twisted Dual-Cycle Fiber Optic Bending Loss Characteristics for Strain Measurement. *Sensors* **2018**, *18*, 4009. [CrossRef] [PubMed]
14. Marković, M.Z.; Bajić, J.S.; Batilović, M.; Sušić, Z.; Joža, A.; Stojanović, G.M. Comparative Analysis of Deformation Determination by Applying Fiber-optic 2D Deflection Sensors and Geodetic Measurements. *Sensors* **2019**, *19*, 844. [CrossRef] [PubMed]
15. Chen, J.; Cao, C.; Huang, Y.; Zhang, Y.; Ge, Y. Research on Optical Fiber Sensor Based on Underwater Deformation Measurement. *Sensors* **2019**, *19*, 1115. [CrossRef] [PubMed]

 *sensors*

Article

Monitoring of Strain and Temperature in an Open Pit Using Brillouin Distributed Optical Fiber Sensors

Chiara Lanciano and Riccardo Salvini *

Earth and Physical Sciences and Centre for Geotechnologies CGT, Department of Environment, University of Siena, Via Vetri Vecchi 34, 52027 San Giovanni Valdarno (AR), Italy; chiara.lanciano@student.unisi.it
* Correspondence: riccardo.salvini@unisi.it; Tel.: +39-055-9119441

Received: 31 December 2019; Accepted: 25 March 2020; Published: 30 March 2020

Abstract: Marble quarries are quite dangerous environments in which rock falls may occur. As many workers operate in these sites, it is necessary to deal with the matter of safety at work, checking and monitoring the stability conditions of the rock mass. In this paper, some results of an innovative analysis method are shown. It is based on the combination of Distributed Optical Fiber Sensors (DOFS), digital photogrammetry through Unmanned Aerial Vehicle (UAV), topographic, and geotechnical monitoring systems. Although DOFS are currently widely used for studying infrastructures, buildings and landslides, their use in rock marble quarries represents an element of peculiarity. The complex morphologies and the intense temperature range that characterize this environment make this application original. The selected test site is the Lorano open pit which is located in the Apuan Alps (Italy); here, a monitoring system consisting of extensometers, crackmeters, clinometers and a Robotic Total Station has been operating since 2012. From DOFS measurements, strain and temperature values were obtained and validated with displacement data from topographic and geotechnical instruments. These results may provide useful fundamental indications about the rock mass stability for the safety at work and the long-term planning of mining activities.

Keywords: marble quarry; distributed optical fiber sensors; brillouin shift frequency; strain; temperature; unmanned aerial vehicle; robotic total station; geotechnical monitoring system

---

1. Introduction

Due to its geological characteristics, in Italy, extraction sites are widely diffused in all regions [1]. In particular, the Tuscany Region is characterized by the presence of marble quarries that represent the strong economic activity for the area. The only province of Massa-Carrara in the year 2016 extracted 844,000 tons of marble blocks [2].

Despite their economic importance, marble quarries are quite dangerous environments in which rock falls and accidents may occur. As many workers operate in these sites, it is necessary to deal with the issue of safety at work. The use of innovative technologies and the analysis of the stability of quarry slopes can contribute to improving the conditions of safety in the workplace.

To obtain information about the status of the rock mass and the slope stability, it is necessary to measure data that characterize the observed site as the properties of rock mass and joint systems. There are several methods to acquire this type of data; the most widespread include classical structural and engineering-geological surveys, geotechnical sensors (ex. extensometers, crackmeters), ground-based radar interferometry, GNSS (Global Navigation Satellite System) and Robotic Total Station. In this paper, an innovative monitoring system developed as part of the R&D project POR FESR 2014-2020, named "Real-time monitoring of quarry walls using fiber optic sensors" [3,4], is shown. The project was led by the "Cooperativa Cavatori Lorano" (Carrara, Italy) and came about from the collaboration with the "Centre of Geotechnologies" (CGT) of Siena University (Italy) and "Geo Explorer S.r.l", a start-up society of Siena University, which is part of the Regional Technological Marble District. The project partnership also consisted of the "Cooperativa Apuana Vagli di Sopra" (Lucca, Italy) and the "Cooperativa Levigliani" (Lucca), managers of another open pit and of an underground marble quarry, respectively, where the use of DOFS was also experimented. The project planned activities concerned the implementation of a monitoring system of the potentially unstable marble quarry fronts using DOFS. The aim of this study was to develop a more efficient and well spatially distributed system, compared to traditional surveillance techniques, capable of advising movements of the rock mass as indicators of changed stability condition and sources of potential risk.

The recent literature reports on application of fiber optic networks in mines to monitor the structural integrity, environmental safety and production parameters. Fiber optic is capable of seismic event and mine pressure detection, methane gas monitoring, temperature monitoring and water pressure monitoring in a way to provide information for accident prediction and early warning. Liu et al. described an all fiber optic comprehensive underground coal mine safety monitoring system in China [5]. Already in 2007, Naruse et al. presented a work on the installation of an underground mine monitoring system based on a fiber optic system in the El Teniente mine (Chile) aiming to monitor the deformation due to mining activities. The monitoring system consisted of optical fiber sensors attached to the tunnel ceiling and sidewalls using rock bolts [6]. Bin and Hua presented a paper on Brillouin Optical Time Domain Reflectometry (BOTDR) using sensors installed in boreholes to monitor rock deformation within an excavated roadway in Zhangji coal mine, China [7]. Zhao et al. proposed a displacement monitoring methodology of rock layers overlying a coal seam in Zhu Xian-zhuang mine (China) based on fiber Bragg grating displacement sensors [8]. Cheng et al. measured the deformation of overlying rock layers of a coal seam by employing a BOTDR-based monitoring method [9]. Wang and Luan built a fiber mesh structure on mine roof and conducted BOTDR strain measurement [10]. Zhigang et al. conducted an experimental study using fiber optic sensing on the monitoring of deformation in the shallow layers of waste rock from the mining process in the Chinese Nanfen open pit iron mine [11]. Arzu et al. built up a laboratory experiment set-up containing an optical fiber system to simulate landslide phenomenon and to record movements [12].

In Italy, Matano et al. reported the implementation of an integrated system aimed at controlling the rock slope stability in the Coroglio tuff cliff, located in the highly urbanized coastal area of Naples at the border of the active volcanic caldera of Campi Flegrei [13]. Schenato et al. used a distributed optical fiber sensing system to measure landslide-induced strains on an optical fiber buried in a large-scale physical model of a slope [14].

In this work, among the three available sites of the R&D project, the Lorano "I" N° 22 quarry (Pradetto site—Figure 1) was selected to describe DOFS monitoring system results. In this quarry, an integrated topographic-geotechnical monitoring system, constituted by a Robotic Total Station (RTS), measuring every day several prisms, extensometers, crackmeters and clinometers, has been active since 2012 [15]. Therefore, thanks to this configuration, it was theoretically possible to compare new DOFS results with data acquired by the other techniques already widely discussed in [15]. The data presented in this paper refer to the year 2018.

Figure 1. (**a**) Location of the Lorano marble quarry in Italy. (**b**) Panoramic image of the investigated Pradetto cut site.

The test site, as already said, is located in the Apuan Alps, a mountain range in northern Tuscany (Italy, Figure 1a) delimited by the following natural boundaries: the Serchio River to the NE and SE, the Aulella River to the N, the Magra River to the NW and the Versilia coastal plain between the Magra and Serchio rivers to the SW. The name "Alps" is due to a very typical alpine appearance consisting of high peaks, narrow ridges and deep-cut valleys. These mountains represent the most important tectonic window of the Apennine chain, a fold and thrust belt produced by the convergence of the African plate towards the European one [16–22]. First described by [23], the Apuan Alps complex, composed by the Massa and Apuan tectono-metamorphic units, is interpreted as result of two main tectonic phases known as "D1" and "D2" [15,24–28]. The first is a ductile compressional event (late Oligocene-very early Miocene), which originated from a progressive deformation with intense foliation [29]. The second, a ductile extensional occurrence dating back to the early Miocene, produced both folds and high-strain shear zones [29].

The Lorano open pit, which falls within the Apuan Unit, where the Upper Triassic–Oligocene metasedimentary sequence overlaps the Palaeozoic basement [15,24,25], is located in the normal limb of the "Pianza anticline". The latter and the "Vallini syncline" form an antiform–synform pair marked by a core of Jurassic marbles and cherty meta-limestones; these are minor folds (hectometer-scale) between the "Carrara syncline" and the "Vinca anticline", structures that can be referred to as the D1 phase [15,30].

The study area belongs to the Torano marble extractive basin, where there are several active open pits with quarry walls that can reach hundreds of meters in height: the landscape is therefore characterized by natural and anthropic slopes giving a very complex morphology.

The dominant variety of marble in the Lorano quarry is the "White Marble" (about 100–200 μm grain size), with colors varying from white to ivory–white and pearl–white to light grey [31]. Moreover, the "Ordinary Marble" (about 200 μm grain size) characterized by colors from pearl–white to light grey [32] and two subordinate categories, the "Veined Grey Marble" and the "Breached Marble" [15,33], outcrop.

A typical Mediterranean climate, with hot dry summers and cold wet winters, affects the quarry area; the copious rainfall (over 3000 mm yr−1) shows a primary maximum value in the autumn season and two secondary peaks in winter and spring [15,34].

A very important element of the Lorano open pit is the Pradetto cut site (Figure 1b), a marble buttress (about 120 m high, 30 m wide and 40 m deep) derived from previous mining activities and object of monitoring activities as described in [35]. Accessible from three sides, at its base, the excavations keep on with a downward feed. While the "Ordinary marble" outcrops in the buttress, the "Veined Grey Marble" characterizes the mountain above. Regarding the structural and engineering-geological analysis, previous studies [15,35,36] show four high angle sets of discontinuities and, despite a good quality of the rock mass (from the Basic Rock Mass Rating RMR$_b$ [37]), potentially unstable joint systems along the three different slopes of the buttress were highlighted from kinematic stability analyses [15].

2. Materials and Methods

2.1. UAV Photogrammetry

Under the guidance of previous results on slopes stability, DOFS were installed on the buttress by specialized climbing workers. With the scope of determining and georeferencing the DOFS exact position and facilitating comparisons with the topographic-geotechnical monitoring system, an aerial photogrammetric survey was carried out through an Unmanned Aerial Vehicle (UAV). The photogrammetric survey was carried out using the Aibotix^{TM} X6 V1 multirotor drone (Figure 2) which, with an autonomy of about 15 minutes, can operate in the visible range (400-700 nm) of the electromagnetic spectrum using a Nikon Coolpix A type camera. The equipment also consists of i) the Inertial Navigation System (INS) with GNSS, accelerometers and gyroscopes, ii) a video camera for remote inspection and iii) the flight management software. The flight was designed defining a Ground Sampling Distance (GSD) of about 1.4 cm/pix and an average flight distance from the slopes of about 50 m. As Ground Control Points (GCPs), which are necessary to improve the accuracy of the exterior orientation of the photographs, and Check Points (CPs), whose function is positional accuracy assessment, natural and artificial targets of known coordinates taken from previous works [35] (i.e., aerial photogrammetric surveys and terrestrial laser scanning) were used. For this reason, it was not necessary to perform a new topographic survey.

Photogrammetric data processing was performed using the code Agisoft^{TM} Metashape Professional which is based on the "Structure from Motion" (SfM) technique. This is a "range imaging" methodology, belonging to the "computer vision" and the visual perception [38–42], aimed at the reconstruction of three-dimensional structures starting from sequences of two-dimensional images. The software uses robust algorithms which allow to adjust the orientation of the frames and generate three-dimensional georeferenced and scaled point clouds, Digital Elevation Models (DEMs), three-dimensional mesh-like models and orthophotos of the area of interest.

For this case study, the image processing involved 548 digital photographs and the positioning of 49 GCPs and 8 CPs. The final Root Mean Square Error (RMSE) of the exterior orientation phase is around 8 cm for GCPs and 10 cm for CPs.

Figure 2. UAV survey at the Lorano marble quarry.

2.2. Topographic Monitoring System

The topographic monitoring system installed at the open pit consists of a Leica^{TM} TCA2003 RTS (Figure 3a), which is a tool for topographic survey that combines a laser distancemeter, an electronic theodolite and a computer on a single device. More details about the instrument can be found in [15].

At the Lorano site, the RTS was placed on top of a stable marble block [15], at an approximately 300 m distance from the buttress and it was protected by a metallic box with anti-aberration glasses (Figure 3a). The RTS, starting from December 1st 2012, automatically detects, every 6 h (at 0.00, 06.00, 12.00 and 18.00 h), the 3D distance measurement of 24 prisms (an example in Figure 3b) positioned both on the pillar (20 measurement points) and outside it (4 reference points), so as to be able to discriminate between the relative movements due to local fracturing from the absolute ones, i.e., those of the entire pillar. The reference points (Figure 4), moreover, allow to have additional control in external stable areas, useful, above all, in the calibration phase of the monitoring system.

Figure 3. (**a**) RTS installed at the Lorano marble quarry. (**b**) Detail of a prism placed on the excavation site walls.

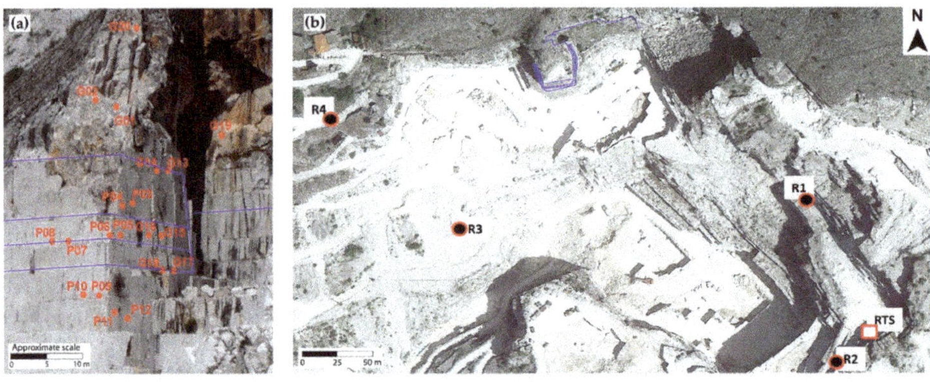

Figure 4. (**a**) Perspective photograph of the marble buttress with RTS measurement prisms (red points) and the DOFS trace (blue lines). (**b**) UAV-orthophoto showing the position of RTS, reference prisms (black and red points) and fiber optics (blue lines).

The transfer of the acquired data takes place in real-time through a telephone line that links the RTS and the CGT, where a PC controls the whole system functioning and data storage and processing. Commercial software packages (i.e., GeoMoS Monitor from Leica™, Analysis from Geodesia™ and System Anywhere from Geodesia™) control the RTS, process the data and produce instantaneous time-displacement graphs.

2.3. Geotechnical Monitoring System

The geotechnical monitoring system of the marble buttress consists of 12 monoaxial mechanical crackmeters (FS_n in Figure 5), 1 three-directional crackmeter (FS3D), 2 biaxial clinometers (CL_n) and 4 multipoint borehole extensometers (ES_n), the deepest of which was placed at a depth of 30 m [15,43,44]. This system has been operative since July 27, 2012, providing high-temporal-frequency deformation trends to be compared with seasonal variations in the climatological data (rainfall and temperature) and data from the RTS [15].

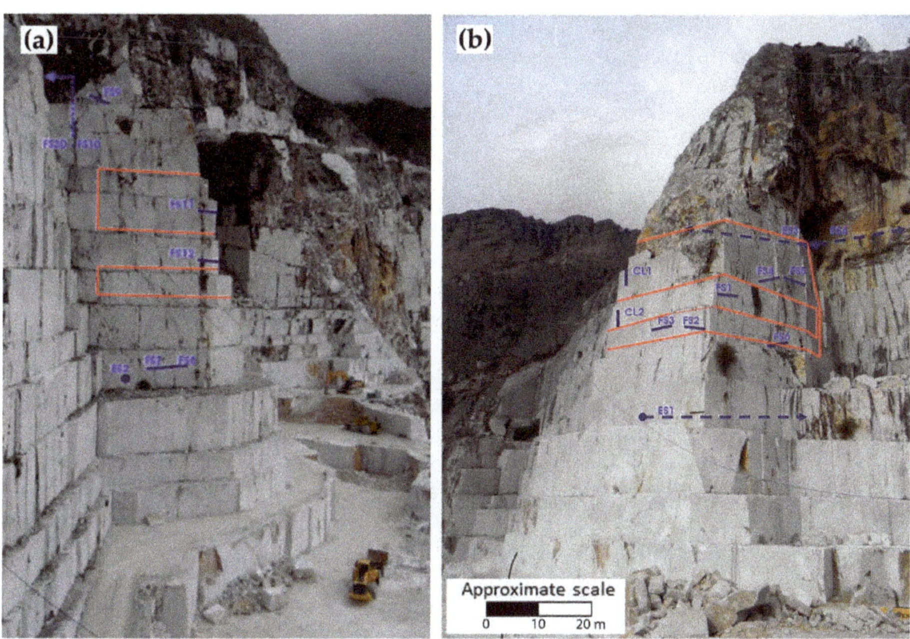

Figure 5. Geotechnical sensors (in blue) and fiber optics cable (in red) on the western (**a**) and southern-eastern (**b**) sides of the buttress.

2.4. DOFS Monitoring System

Sensors based on optical fibers [45–58] are currently widely used for monitoring infrastructures, bridges, dams, buildings and landslides with many inherent advantages [45] including i) resistance to electromagnetic interference, ii) light weight, iii) small size, iv) high sensitivity, v) high-temperature performance, vi) immunity to corrosion and vii) large bandwidth [46].

DOFS are cables of optical fiber which offer the possibility of monitoring variations of one-dimensional structural physical fields along the entire optical fiber in a truly distributed way [51,52].

Among the different types of DOFS, Brillouin scattering-based sensors allow to obtain distributed strain and temperature measurements by detecting a frequency shift [50]. Specifically, Brillouin scattering is a process of diffusion of acoustic light thermally generated in the optical medium; due to the Doppler effect, the diffused light presents, with respect to the incident light wave, a frequency shift v_B named "Brillouin Frequency Shift" (BFS):

$$v_B = \frac{2nV_a}{\lambda_0} \tag{1}$$

where n is the effective refractive index, λ_0 the wavelength of the incident light in the void and V_a the velocity of the acoustic wave [53]. The BFS is a parameter related to the optical and elastic properties of the fiber [54] and it depends on both material temperature and density; moreover, the elastic properties of silica (glass fibers) are such that an induced strain causes a change in volume and therefore, a variation in the material density. Then, any local variation of temperature and/or fiber strain, acting on the acoustic speed, produces a variation in the local value of the BFS [55]. Experimental measurements show an excellent linearity in Brillouin shift dependence on strain and temperature [54]. The relationship between the variations of strain ($\Delta\varepsilon$), temperature (ΔT) and BFS (Δv_B) is described as [53]

$$\Delta v_B(\varepsilon/T) = C_T \Delta T + C_\varepsilon \Delta\varepsilon \tag{2}$$

where C_T = (1.26 MHz/°C) and C_ε = (0.06 MHz/µε) are, respectively, the temperature and strain coefficients at 1550 nm for a single mode silica fiber [53] (the coefficients values change slightly for different types of single mode fibers). In particular, the fiber used at the Lorano site was calibrated in laboratory in order to obtain the calibration coefficients; the following temperature and strain coefficients were obtained: C_T = 1 MHz/°C and C_ε = 0.05 MHz/µε.

Sensors based on Brillouin scattering can be classified into two main types: sensors based on spontaneous Brillouin scattering, in which only the incident light is launched into the optical fiber, and sensors based on stimulated Brillouin scattering (SBS), characterized by an additional stimulation on the generation of phonons [56]. Among the techniques based on SBS, the Brillouin Optical Time Domain Analyse (BOTDA) methodology [46,52,54,56], due to its powerful signal and spatial resolution, was considered suitable and, therefore, adopted for monitoring the buttress at the Lorano marble quarry. Data were obtained using the OSD-1 system [58] (Figure 6a) provided by the company Optosensing S.r.l. The OSD-1 system, which was periodically checked both on site and remotely via PC, is a control unit able to send the logs directly to a centralized enterprise-type database. Logs provide distance from the source and microstrain values that are archived into the database in a georeferenced mode.

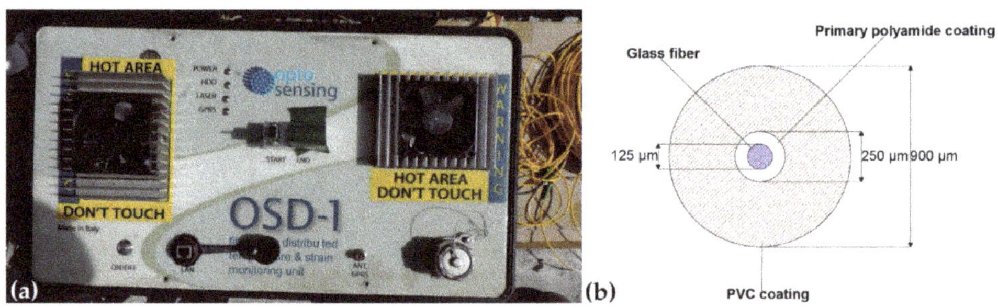

Figure 6. (**a**) Photograph of the OSD-1 measurement unit. (**b**) Cross section of the installed optical fiber sensor with the three different layers (glass fiber, primary polyamide coating and polyvinyl chloride coating) and their dimensions.

After acquiring the BFS, the strain profile between two consecutive measurements is given by

$$S(z) = (BFS_t - BFS_0) \cdot C_s \; [\mu\varepsilon] \tag{3}$$

where

- $S(z)$ is the strain at the z coordinate, commonly expressed in microstrain ($\mu\varepsilon = \varepsilon \cdot 10^{-6}$);
- BFS_0 is the Brillouin frequency shift profile acquired at time 0;
- BFS_t is the Brillouin frequency shift profile acquired at time t;
- C_s is the transduction coefficient of the optical fiber, equal to 20,000 µε/GHz;
- $\mu\varepsilon$ (microstrain) is the measurement unit of the strain.

The installed optical fiber sensor (Figure 6b) from the inside out is composed of i) a glass fiber with a total diameter of 125 μm, ii) a primary polyamide coating which brings the diameter to 250 μm and iii) a polyvinyl chloride (PVC) coating which brings the final diameter to 900 μm. The percentage of the rock mass strain transferred to the optical fiber sensor varies depending on the glue used, the protective coating and, finally, the bonded length. Parametric studies have shown that the higher the bonded length and the stiffer the coating and the adhesive, the more strain is transferred to the sensor [59,60]. In this work, the effects of such materials on the strain transfer from outside the cable to the sensing-fiber were not computed since the calibration of the strain coefficient (C_ϵ = 0.05 MHz/μɛ) performed in laboratory was used. However, several actions were executed in order to minimize the effects of the materials on the strain transfer: i) use of standard widely tested polymers, ii) coatings kept as thin as possible, iii) application of a glue characterized by a suitable value of the elastic modulus, and iv) implementation of a strong bonding between rock mass and sensors. To perform the temperature discrimination, two independent cables were installed parallel to each other. The cable for the strain measurement (500 linear meters long) was reinforced with strands of 316 stainless steel, with a diameter of 0.5 mm, and glued to the structure. The temperature compensation cable (additional 500 linear meters of length) consists of a tube containing the fiber and a gel to improve the heat exchange. This cable is arranged in parallel to the one for the strain measurement and provides only the temperature profile (the fiber inside this second cable is unconstrained).

The position of the optical fiber was designed by trying to intercept as many rock mass discontinuities as possible and following the location of topographic measurement prisms and geotechnical sensors on the quarry walls. Two different phases, pre-installation of the witness wire and installation of the effective sensor cable, were carried out by specialized climbing workers (Figures 7 and 8). In particular, the pre-installation phase took place through the installation of mechanical hooks and the engagement of the witness wire. Afterwards, the witness thread was removed, and a layer of "Sikaflex®-11 FC+", adhesive resin (Figure 7a), based on polyurethane and characterized by a Secant Modulus of Elasticity of ~0.60 N/mm^2 (after 28 days) (+23 °C) (ISO 8339), was applied through a special compressed air gun. Polyurethane adhesives are "structural adhesives" which can be used to join very different types of materials together with a long-lasting and strong bond. In particular, "Sikaflex®- 11 FC+" is an adhesive commonly and widely used in several works on optical fibers [61,62]. The sensor cable was placed on the adhesive and secured to metal plugs (Figure 7b). The cable was then covered with glue to improve adherence to the surface and to protect DOFS from both meteoric events and ultraviolet rays (Figure 8c). Since DOFS installation cannot have right angles or smaller, some different setting was necessary in zones characterized by artificial rock cuts and fracturing. Figure 8b, as an example, shows a typical setting adopted in these situations.

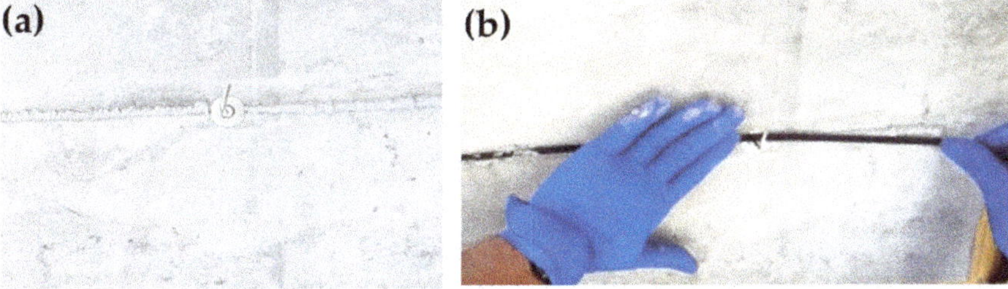

Figure 7. (a) Adhesive resin layer applied on the buttress along the path designed for optical fibers. **(b)** Positioning of the sensor cable on the glue layer.

Figure 8. (**a**) Installation of the optical fiber on the marble buttress carried out by specialized climbing workers. (**b**) Detail of a special optical fiber setting, reinforced by metal elements, in an area characterized by artificial rock cut and fracturing. (**c**) The black arrows indicate the sensors (light grey lines) glued to the buttress.

3. Results

3.1. Deliverables from UAV Photogrammetry

The photogrammetric processing led to the creation of a 3D dense point cloud, a DEM of the investigated area (GSD of 5.72 cm/pix), shown in Figure 9, and a textured 3D polygonal mesh type model (Figure 10). The polyline representing the spatial distribution of the optical fiber cable was discretized and analyzed in a GIS (Geographic Information System) environment through ESRI^{TM} ArcGIS Pro software. This operation was useful as a preparatory step for the analysis of displacements and strain. In fact, the vectorization and georeferencing of DOFS line allow to i) detect the exact location of deformation data and, ii) make it comparable to contemporaneous information coming from the topographic and geotechnical monitoring systems.

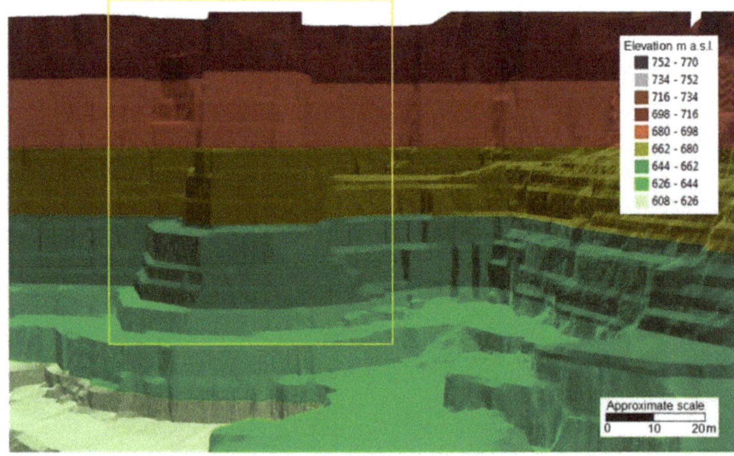

Figure 9. Three-dimensional representation of the elevation through a TIN (Triangulated Irregular Network) model. The yellow rectangle highlights the marble buttress area.

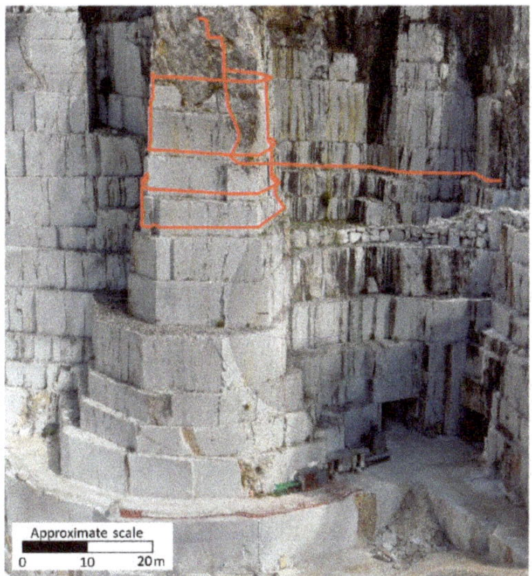

Figure 10. Textured 3D polygonal mesh of the marble buttress with the DOFS spatial distribution highlighted in red.

3.2. Data from RTS

The measurements recorded by the RTS are represented hereunder through graphs of time versus the slope distance (defined as the distance in a straight line in the space between two points).

In particular, the RTS displacement values are averaged on a mobile average line over a period of 24 h and filtered by choosing the same hour (at 00:00 h).

Figures 11–14 illustrate the results obtained from some measurement points located near the optical fiber installation. Figures 15 and 16 show the displacement time series of two reference prisms. The time span analyzed was from 2012 to 2018.

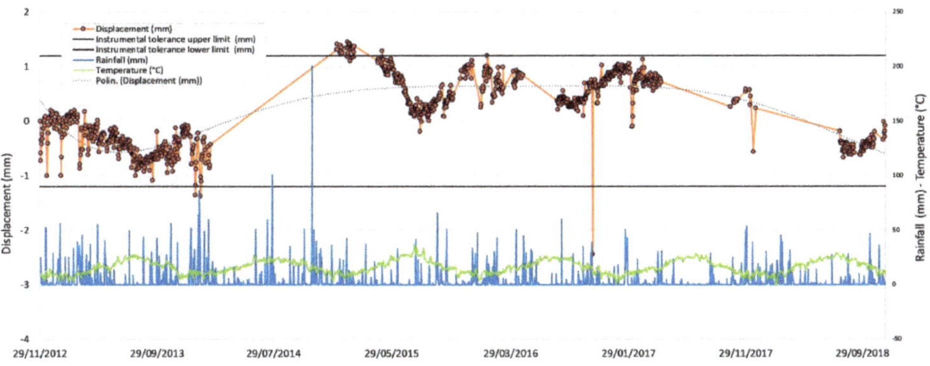

Figure 11. Displacement recorded by the RTS at the measurement point G02 compared to daily average rainfall and temperature. The black lines indicate the instrumental tolerance threshold values.

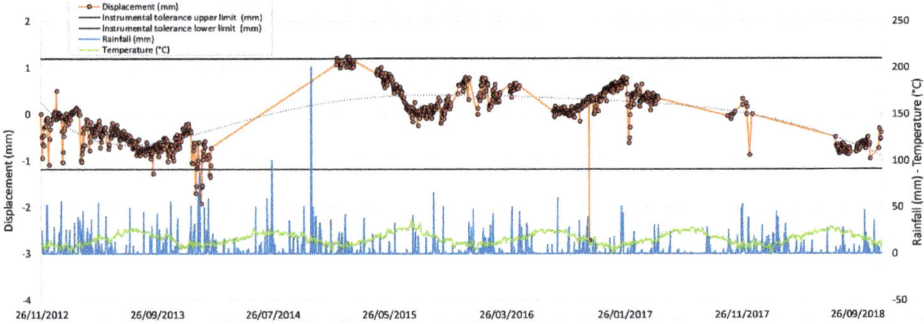

Figure 12. Displacement recorded by the RTS at the measurement point G15.

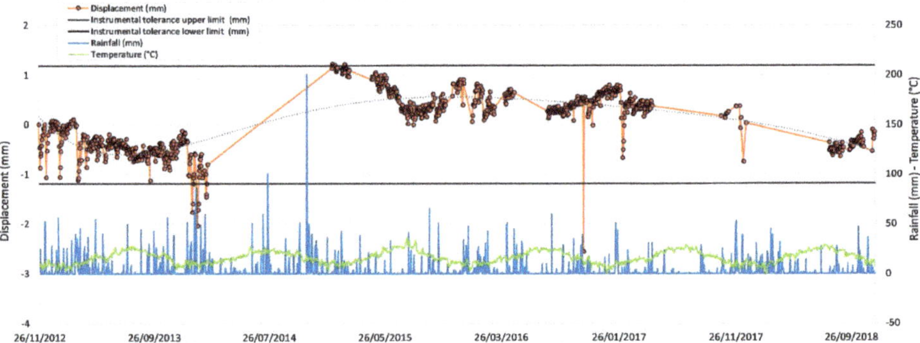

Figure 13. Displacement recorded by the RTS at the measurement point G17.

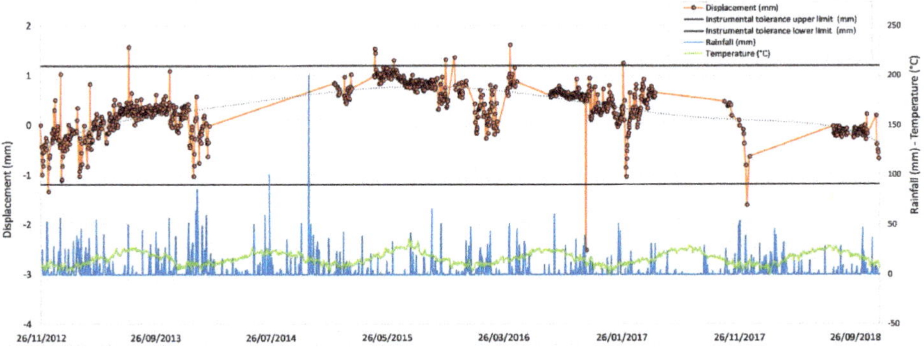

Figure 14. Displacement recorded by the RTS at the measurement point P07.

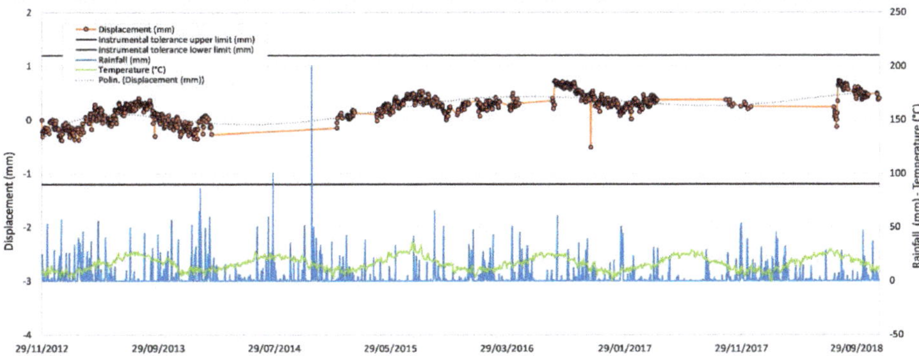

Figure 15. Displacement recorded by the RTS at the reference point R01.

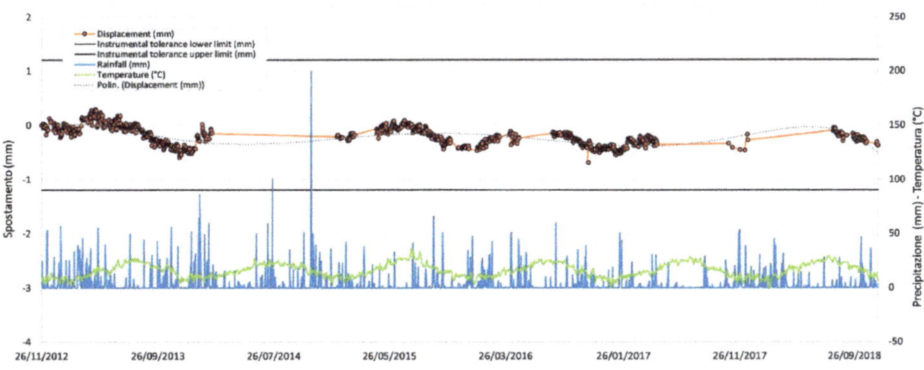

Figure 16. Displacement recorded by the RTS at the reference point R02.

The displacement recorded in every prism was correlated to the daily rainfall and the average daily temperature. The graphs also show the RTS instrumental tolerance threshold values as calculated for the individual prisms taking into consideration the instrumental angular accuracy and their distance from the RTS; these threshold values vary from ± 0.55 mm to ± 1.62 mm [15].

The meteorological data, useful to determine the environmental and climate conditions in the various measurement days, and from these to make the necessary deductions on observed displacements, were provided by the "Carrara Fossola" meteorological station [63].

As can be seen from the graphs in Figures 11–16, the RTS suffered periodic malfunctions and phases of inactivity during the whole 2012-2018 timespan. Moreover, this paper aimed at analyzing results that overlap with the available DOFS data. The availability of contemporaneous RTS and optical fiber measurements is limited to the summer and autumn seasons of 2018 and, more specifically to the time interval between 30/07/2018 and 28/11/2018 (about 4 months), with two interruption periods from 06/09/2018 to 22/09/2018 and from 30/10/2018 to 22/11/2018. The interruption periods were related to technical issues.

The analysis of the previous graphs (Figures 11–16) suggests a substantial difference in the displacement recorded by the RTS between reference (Figures 15 and 16) and measurement points (Figures 11–14) linked to the absence, in the first case, of anomalous values. The peaks of measurement points, which bring the associate prism closer to the observation point (i.e., the position of the RTS), occur in conjunction with thunderstorms. All the measurement points, installed in different positions with each other, show this very similar trend, suggesting an almost unanimous behavior of the entire marble structure [15]. Even if the anomalous values exceed the instrumental thresholds, these are values of slight entity (<3 mm) that do not represent a particularly critical situation, as demonstrated by the results from other sensors (as discussed later) and reality, since not rockfall occurred in the whole monitored time span.

Moreover, there is a generalized sinusoidal trend that can be related to the thermal response of marble slopes to the temperature variation. In fact, on an annual scale, it is possible to see that the displacement values are characterized by a decrease, at the beginning of the warm season, and by an increase with the arrival of the cold season. This behavior of the marble, known as thermoclasty, is most likely related to the anisotropy of calcite mineral, which tends to expand and contract in directions constrained by crystallographic axes [15,64–66]. The phenomenon is probably encouraged by different elements: i) the water infiltration within the rock discontinuities during the rain events; this can increase the inner pressure with a consequent growth in volume and expansion of the joints, thus creating a general stress that has repercussions on the whole buttress; ii) the site morphology and iii) the stress reduction due to both mining and natural causes (emersion condition linked to the geology of the area).

3.3. Data from the Geotechnical Monitoring System

The measurements recorded by the geotechnical sensors in the 2012-2019 time period (some examples are given in Figures 17 and 18 for crackmeter FS4 and extensometer ES3B1, respectively) confirm the general stability of the monitored site. The graphs show sinusoidal symmetrical trends that are yearly replicated with a good approximation. In particular, the values of the observed quantities reach the local minimums during the summer season and the local maximums in winter. This behavior may be attributed to the summer thermal expansion which tends to close fractures. On the contrary, in winter, joint aperture increases for the absence of rock thermal expansion and the presence of water and ice within fractures. Nothing particularly different from previous years was recorded during 2018. The deformation quantities confirm the absence of significant movements of the rock mass [43,44].

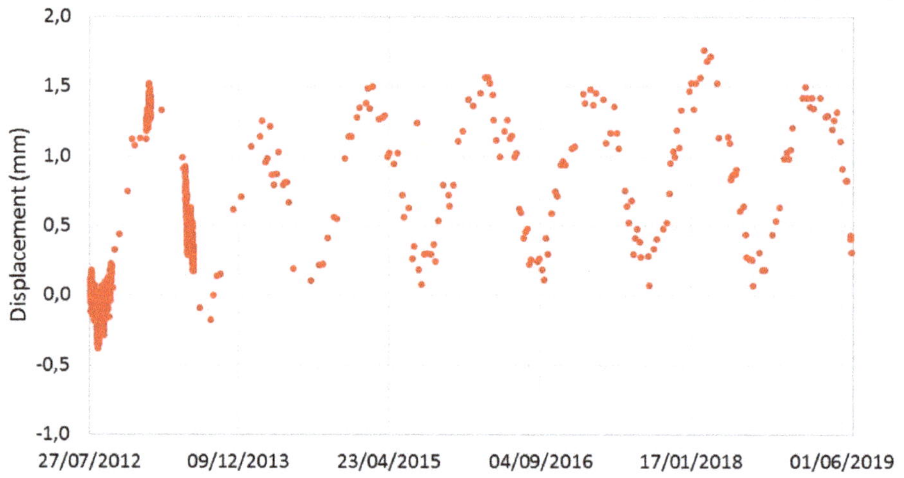

Figure 17. Displacement recorded by the crackmeter FS4 from 2012 to 2019.

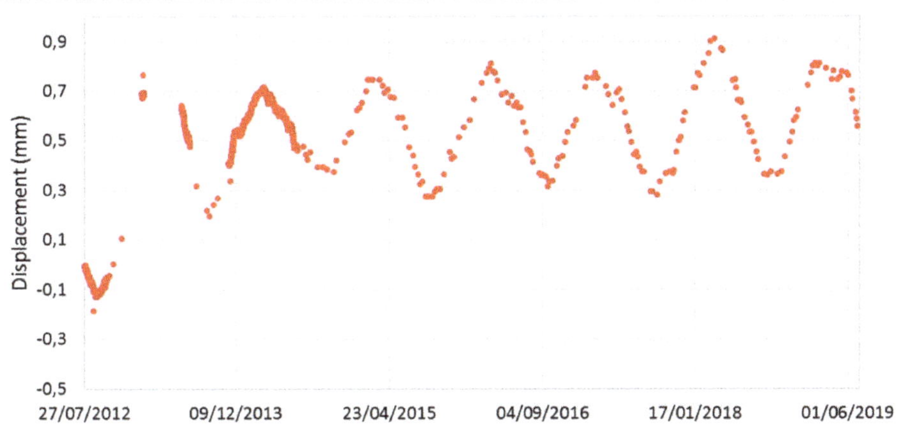

Figure 18. Displacement recorded by the extensometer ES3B1, 6 m long sensor, from 2012 to 2019.

3.4. Data from DOFS

In the Lorano marble quarry, BOTDA type measurements were performed at a spatial resolution of 20 cm. The first acquisition was executed on 4 January 2018 (Figure 19).

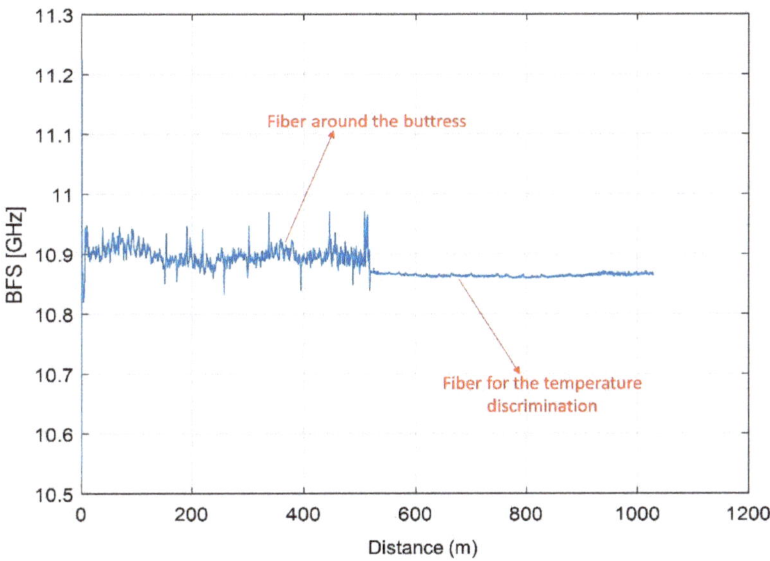

Figure 19. BFS profile acquired on 4 January 2018 (first measurement).

The first 500 m of the profile (Figure 19) correspond to the fiber sensor fixed around the buttress, and present the typical oscillations deriving from the deformations associated with the gluing. The profile continues for another 500 m, maintaining, in this second part, a much more stable trend which is typical of the fibers used for the temperature discrimination [67].

From January 2018 to February 2019, there are no significant changes in the BFS values that remain within ±0.1 GHz (Figure 20). The only evident deviation concerns the first acquisition (carried out on 4 January 2018) which, particularly in the points around 120 m and 260 m distant from the origin, shows slightly lower BFS values than the other acquisition dates. Since, in these points of the buttress, the other measurement sensors (RTS and geotechnical) were consistent with each other, it was assumed that this variation is due to the adjustment of the deformation state of the fiber following its recent installation. Therefore, for the calculation of the strain, it was decided to exclude the measurement of 4 January 2018, thus assuming as the first value that of 7 February 2018, after a certain adhesive resin hardening.

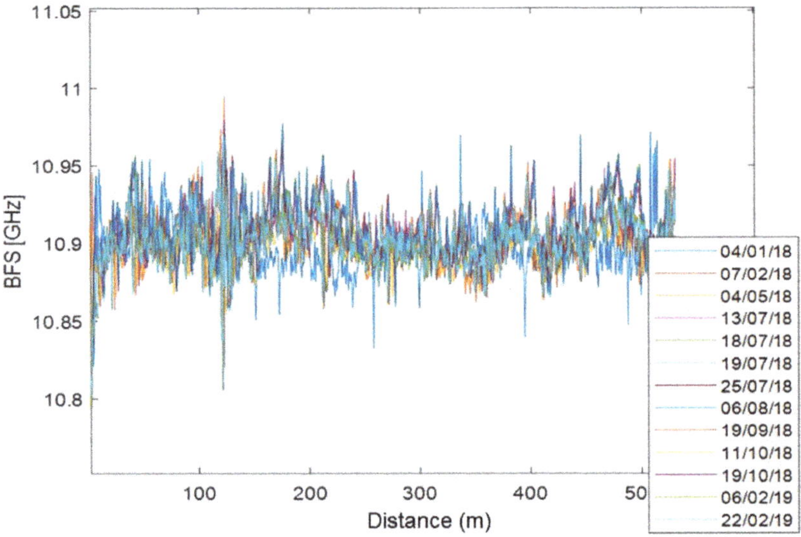

Figure 20. BFS profiles acquired from January 2018 to February 2019.

Strain was calculated by subtracting, for each measurement position (every 20 cm), the value of BFS measured on 7 February 2018 to that measured later and multiplying the variation of BFS thus obtained by the transduction coefficient equal to 20,000 $\mu\varepsilon$/GHz [51]. The obtained strain data was finally filtered using a moving average filter over 51 points (Figure 21), equivalent to a spatial resolution of 10 m.

As can be seen from Figure 21, the most significant deformations appear during the summer months with a reversible behavior. The strain values are relatively low if compared to the standard deviation of each profile (approximately 200 $\mu\varepsilon$). As suggested by the literature [68], it was decided to set the alarm threshold as a multiple of the standard deviation of the strain values. The obtained value of standard deviation suggests for the authors an alarm threshold for deformations five times greater (i.e., equal to 1000 $\mu\varepsilon$). At the Rocks Mechanics Laboratory of CGT, compressive strength tests were performed on marble samples from the site test of this work. The results of these analyses show an Elasticity Modulus equal to 60 GPa and a Compressive Strength (σ_u) value equal to 87 MPa. Similar values were confirmed by the literature data [69]. Considering the measured value of the Elasticity Modulus, the stress value corresponding to the strain of 1000 $\mu\varepsilon$ is given by

$$\sigma_{calc} = \text{strain} \ (\mu\varepsilon) \cdot 10^{-6} \cdot E \cong 59 \ \text{MPa} \tag{4}$$

Since $\sigma_{calc} < \sigma_u$, the value of 1000 $\mu\varepsilon$ was considered a suitable and precautionary alarm threshold value for the strain values in the Lorano marble quarry. Figure 21 shows how this threshold value was not reached in any position throughout the period of the DOFS monitoring survey.

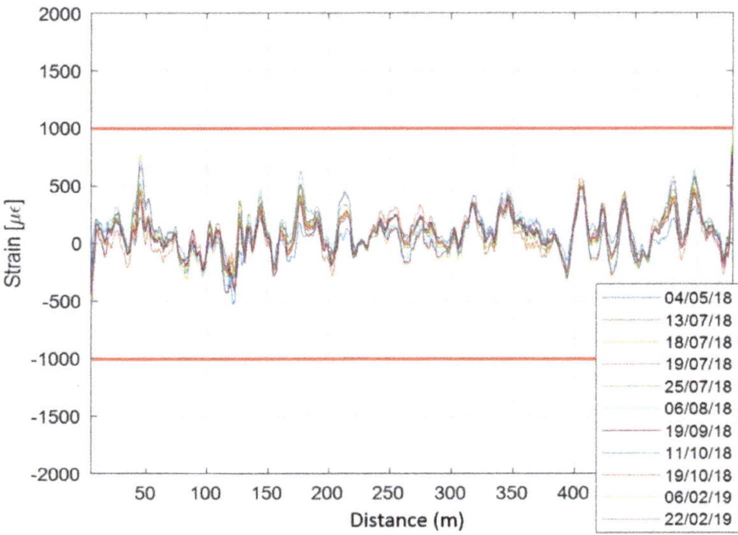

Figure 21. DOFS strain profiles along the marble buttress from January 2018 to February 2019 and alarm thresholds (red lines).

Profiles of temperature variation (Figure 22) were obtained starting from the BFS data and using 7 February 2018 as the reference measure. As in the case of strain, the temperature data were filtered using a moving average filter over 51 points (still achieving a spatial resolution of 10 m). The graph shows the natural difference between the temperature measurements taken in autumn and winter compared to those taken in spring and summer. The temperature data detected by optical fiber, in fact, vary between −10 °C and + 20 °C, as result of the natural seasonal variations in the case of an open pit.

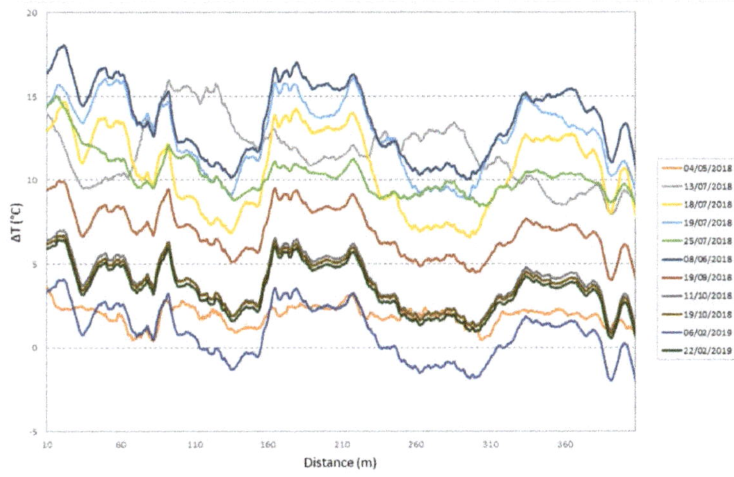

Figure 22. DOFS temperature variation profiles along the marble buttress from January 2018 to February 2019.

Figures 23 and 24 show two examples of the comparison between strain and temperature at two individual dates, one for the summer (Figure 23) and the other for the winter season (Figure 24). The graphs demonstrate how higher temperatures correspond to greater strain. It is important to underline that these strain values are always lower than the alarm threshold values of Figure 21.

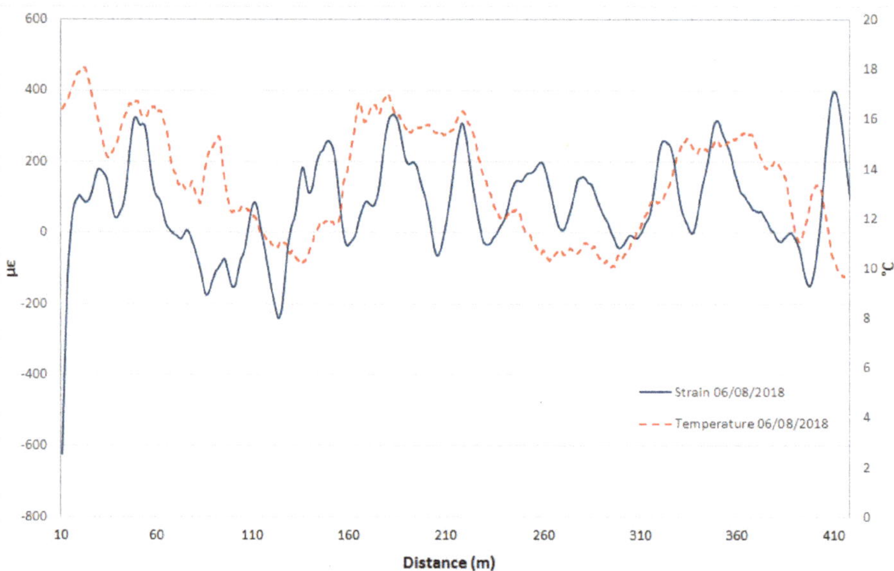

Figure 23. Example of comparison of strain and temperature profiles in summer.

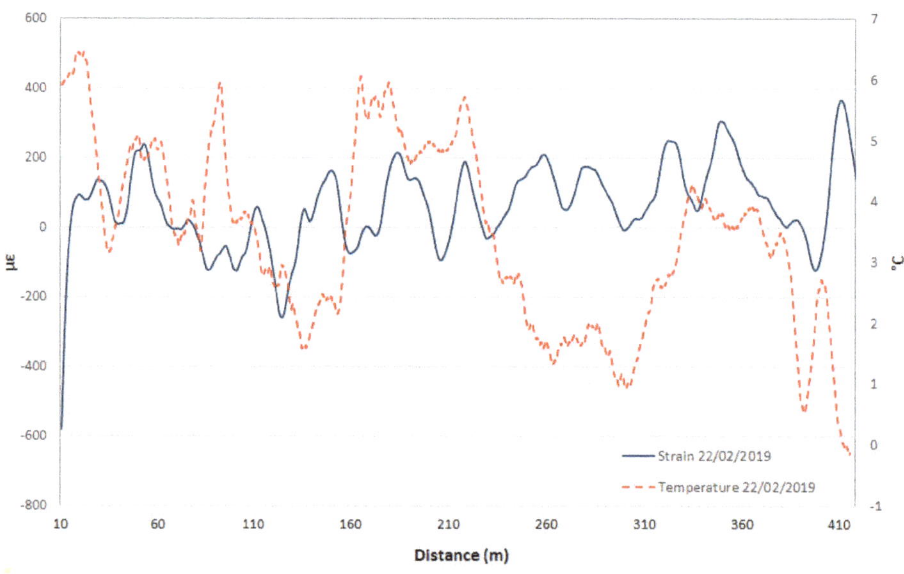

Figure 24. Example of comparison of strain and temperature profiles in winter.

4. Discussion

During the phase of DOFS data acquisition, some critical issues emerged due to both the complex morphology of the study area and the fragility of the instrumental components, both in terms of fiber cable and control unit. The intersection between fiber optics and rock mass asperities, in fact, led to several phenomena of failure of the sensor cable. However, these breaks were promptly repaired using a more robust protective sheath coupled with aluminum channels and steel cables (Figure 8b). Moreover, a damage to the OSD-1 system backup batteries and the surrounding components caused the interruption of data recording for over 3 months (i.e., from March to June 2018) waiting for the replacement parts.

Nevertheless, the data acquisition time span lasts in total for about one year, giving the possibility to analyze the obtained results, compare them with data from different monitoring techniques and highlight strengths and weaknesses of DOFS in such operative conditions.

Before giving some comments on the obtained results, some considerations about the possible errors due to the uncertainty of the strain transfer from the rock to the strain-sensing fiber must be made. Using the optical fiber as a sensor, it must be considered that the strain of the optical fiber is not that of the monitored structure due to both the glue and the coating. Moreover, because of geological stresses and gravity, the detection of the strain is a function of the fiber orientation with respect to the quarry walls; consequently, the deformation measured by the sensor is a percentage of that of the structure. Experimental studies [59,60] showed that the higher the adhesion length, the greater the strain transfer rate; the higher the shear modulus of the coating layer and the adhesive, the greater the strain transfer rate. In this work, it was attempted to minimize the effects of materials by choosing suitable standard polymers as polyurethane (adhesive layer), polyamide and PVC (coatings) and realizing suitable adhesion lengths. Specific analyses on the effects of the materials were not done since the strain coefficient as calculated in laboratory was used. Basing on this assumption, the uncertainty on the strain measurement is that of the OSD-1 measurement system (i.e., about 20 $\mu\varepsilon$). If, along the entire installed cable, the calibrated strain coefficient is constant (and therefore the strain transfer is also constant), there are no other errors to consider. Obviously, if the installation is faulty, there may be variations from point to point, but it is impossible to take them into account.

Below are some examples of comparison with displacements recorded by the other monitoring techniques in the overlapping time periods. Figures 25 and 26 show the DOFS strain values versus the RTS displacements, Figures 27–29 illustrate the DOFS strain values versus the geotechnical monitoring system displacements, and Figure 30 juxtaposes the temperature values measured by DOFS, the geotechnical monitoring system and the "Carrara Fossola" weather station.

At the observed scale, the trends of strain values from DOFS and that from displacements recorded by RTS and geotechnical sensors are very confusing even if quite similar if compared to each other. There is a close correspondence between the trend of the displacement values recorded by the crackmeters and the strain profiles (examples are reported in Figures 27 and 28), characterized by a certain cyclicity and higher values in the warmer months. Figure 29, instead, shows an example of a comparison between the extensometer displacements and the DOFS strain values. Both graphs show a decreasing trend, a local maximum reached in the hottest period (first recorded by the optical fiber) and a rising trend in the last part of the year.

The trends of the average daily temperature values measured by DOFS (average of all values recorded along the DOFS sensor cable), geotechnical monitoring system and weather station (Figure 30) are quite comparable, even if the DOFS values are lower than the others. This may be due to several reasons: first, to the different altitude between the station (55 m a.s.l.) and the marble buttress (about 600-700 m a.s.l.); secondly, to the coating and to the glue necessary to protect fiber optic from both meteoric events and ultraviolet rays. Moreover, the temperature measured by the fiber, resulting from the external temperature (air) and the temperature of the rock where the fiber is placed, vary according to the time of day, the position and the sunlight exposure: there may be shaded areas and zones exposed to the sun along the fiber path. These conditions have their effect in determining the average temperature values along the entire path of the optical fiber.

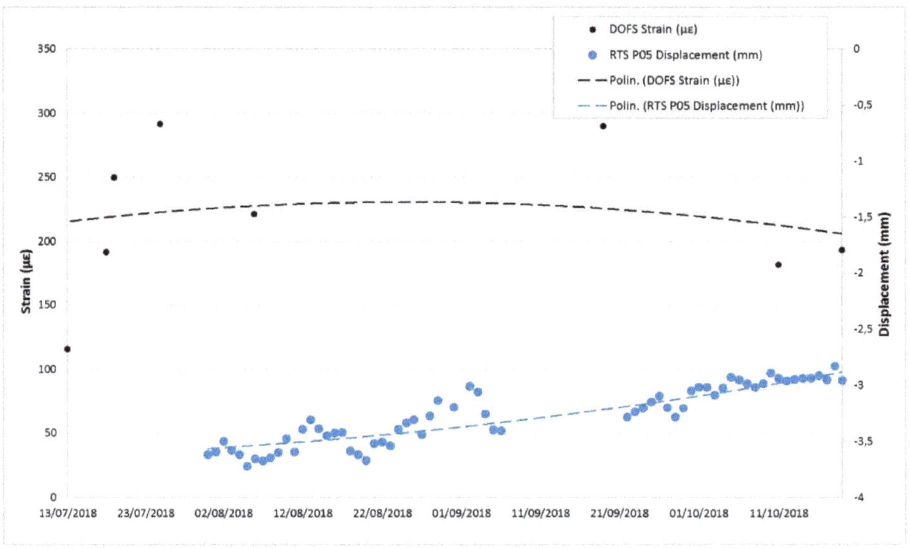

Figure 25. Comparison between the P05 RTS measuring point displacement (data interpreted according to a second order polynomial) and DOFS strain values in the same area.

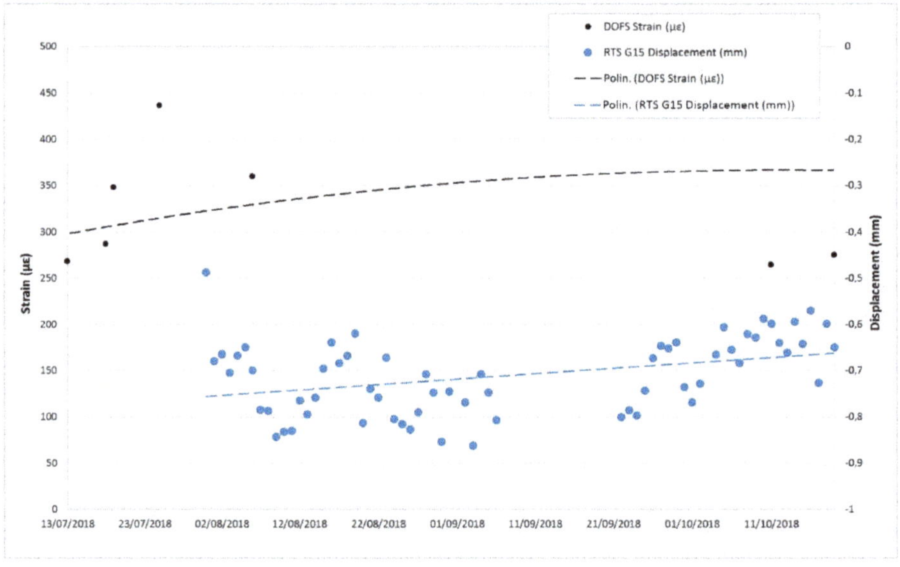

Figure 26. Comparison between the G15 RTS measuring point displacements (data interpreted according to a second order polynomial) and DOFS strain values in the same area.

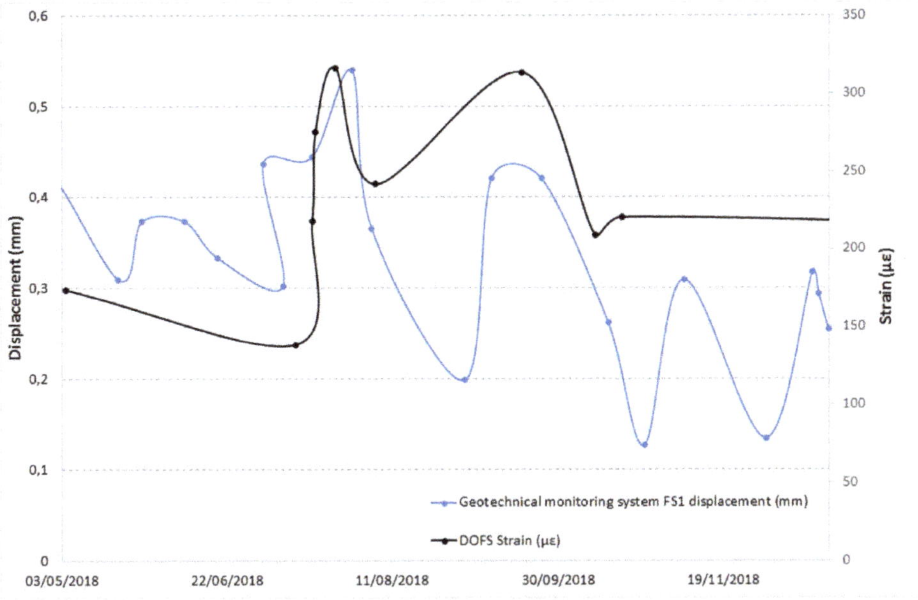

Figure 27. Comparison between the FS1 crackmeter displacements and DOFS strain values in the same area.

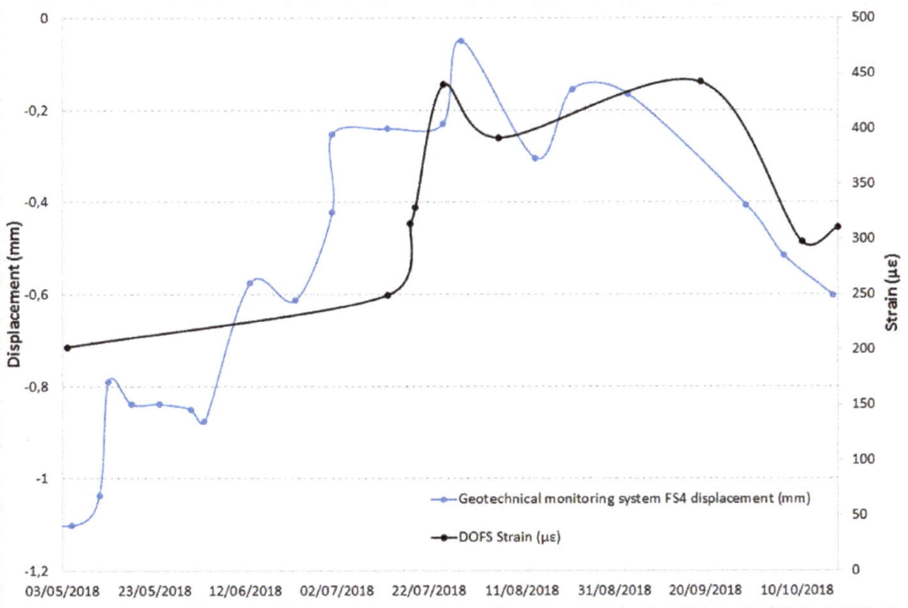

Figure 28. Comparison between the FS4 crackmeter displacements and DOFS strain values in the same area.

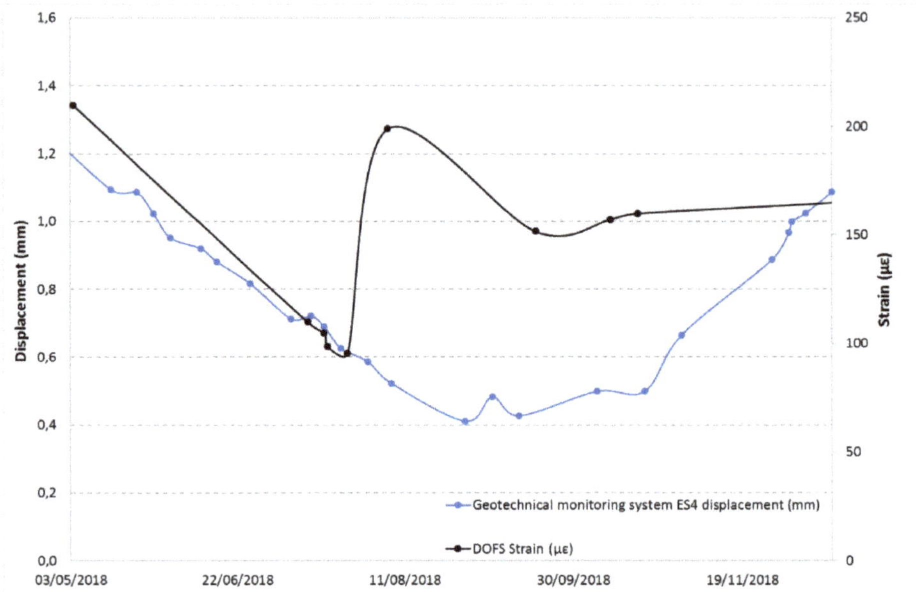

Figure 29. Comparison between the ES4 extensometer displacements and DOFS strain values in the same area.

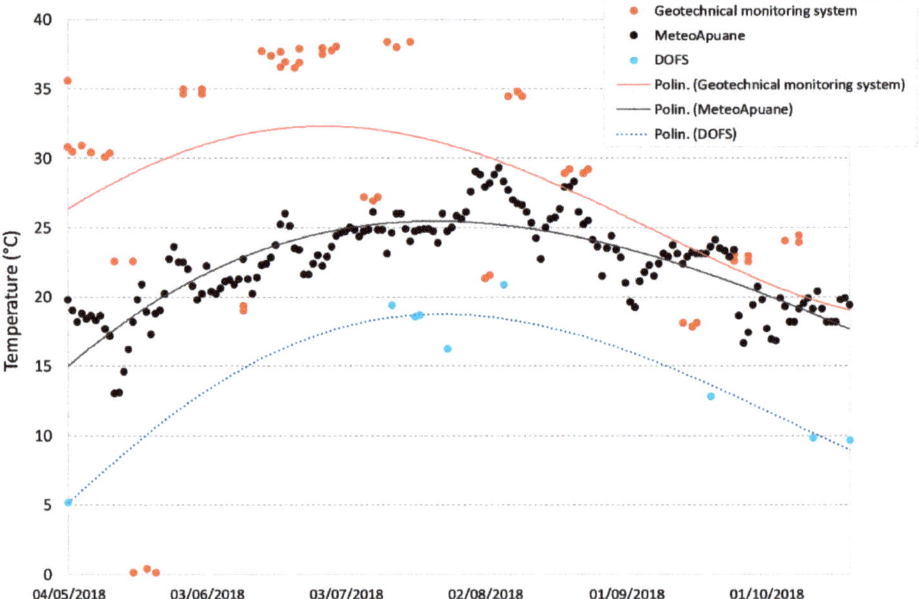

Figure 30. Comparison between the temperature values measured by DOFS, FS1 geotechnical sensor and "Carrara Fossola" weather station. The data were interpreted according to a fifth order polynomial.

Despite the short time of testing and small number of results, it can be affirmed that the proposed system has provided a reliable and accurate monitoring measure for active marble open pit deformation control over wide extension.

Conventional monitoring techniques, such as extensometers, crackmeters, and RTS, are capable of detecting the deformation in single points while DOFS can allow to monitor lines and, with suitable installation, surfaces, allowing to improve the accuracy, the completeness and the safety of the whole system.

Moreover, the high accuracy of the data shows that optical fibers can detect precursory signs of failure well before the collapse, paving the way for the development of more effective early warning systems. In this sense, the work presented here had the goal of improving the monitoring activity already implemented in the quarry in such a way as to allow planning of the mining activity development in the short and medium term in compliance with safety standards and in the interest of production.

The DOFS test for monitoring the Apuan Alps marble quarries is a new experiment. Moreover, due to their distinctive features, such as sub-vertical slopes and walls often jutting out, and tunnels, these extraction sites represent an ideal site for the experimental development of such a monitoring system.

These studies, together with additional data on risk assessment already available for the Lorano marble quarry, represent a great help for the excavation, pre-excavation and post-excavation periods, as well as for other displacement-dependent engineering-geological projects. For example, the DOFS monitoring system can be joined to distinct element numerical modelling to simulate the deformation features of an extraction front and to provide the scientific basis for detecting the early warning signs of a rockfall.

5. Conclusions

An innovative monitoring system was implemented by means of DOFS based on BFS on the walls of a marble buttress located in the Lorano quarry, in Italy. The system has been operative for about one year and the obtained results were compared with an integrated topographic-geotechnical monitoring system operative at the extraction site since 2012.

The evaluation of the effectiveness of the DOFS, with reference to their capability of data acquisition, is undoubtedly positive: the system, in fact, despite the encountered difficulties related to both the complex morphology of the area and the fragility of the components, proved to be able to carry out the detection of reliable and very precise BFS, strain and temperature data.

At the test site, BOTDA measurements were performed. The most significant deformations appeared during the summer months with a reversible behavior. The recorded strain values are relatively modest compared to the standard deviation of each profile (approximately equal to 200 $\mu\varepsilon$). This standard deviation value suggests that the choice of an alarm threshold for deformations at least five times greater, that is, equal to 1000 $\mu\varepsilon$, never reached in any position throughout the timespan of the monitoring. This evidence was confirmed by the displacement values found at specific measurement points by the RTS. Since these displacement values fall within or very near to the range defined by the instrumental tolerance threshold, no critical situations were detected, in accordance with results obtained with the optical fiber survey. Moreover, the comparative analysis of all the data acquired by the geotechnical monitoring system confirms the absence of significant displacements of the buttress in the studied time period.

The temperature data detected by optical fiber, variable in the range between $-10\ ^{\circ}C$ and $+20\ ^{\circ}C$, appear to be affected by natural seasonal variations, as it is natural to expect in the case of an open pit.

The obtained results confirm the potential of the DOFS monitoring system to detect and locate any deformation phenomenon that may occur along the fiber path, even in hostile and less hospitable environments like the marble extraction sites. However, some difficulties mainly related to the authors' inexperience were faced: considering the nature of a monitoring system (need for long-term analysis) and the encountered technical problems, it would be appropriate to continue the experiments in the future in order to increase the amount of the available data. In fact, a more substantial dataset could allow more in-depth assessments to be made regarding the effectiveness of the DOFS monitoring system and its durability.

The installed systems will be used continuing the measures to evaluate the progression of the deformation over time in order to provide useful indications for the realization of a pre-alarm system.

This latter can be a valid tool for guaranteeing safety conditions for quarry workers and planning the actions to be undertaken for the continuation of excavation work in adjacent areas.

Author Contributions: Both authors developed the research idea, contributed to the realization of the different project phases and to the writing of the manuscript. R.S. carried out the field activities including supervising of DOFS installation and UAV surveys. C.L. processed the photogrammetric data and set up the presentation of the topographic, geotechnical and DOFS monitoring data. Both the authors contributed to the analysis of the collected data. All authors have read and agreed to the published version of the manuscript.

Funding: This research was funded by Tuscany Region (Italy) though the POR FESR 2014-2020 plan within the project named "Real-time monitoring of quarry walls using fiber optic sensors".

Acknowledgments: The authors would like to thank Ferrari, M., Profeti, M. (Cooperativa Cavatori Lorano), Mirabile M. (Optosensing S.r.l.), Zeni L. (University of Campania), Vanneschi C., Mastrorocco G., Tufarolo E. and Carmignani L. (University of Siena) for their support of this research.

Conflicts of Interest: The authors declare no conflict of interest.

References

1. Istituto Nazionale di Statistica. Available online: https://www.istat.it/it/files//2017/04/Report_attivit%C3%A0-estrattive-19-04-2017.pdf (accessed on 30 December 2019).

2. Il Tirreno. Available online: http://iltirreno.gelocal.it/massa/cronaca/2018/02/12/news/ecco-le-produzioni-cava-per-cava-1.16467555 (accessed on 30 December 2019).

3. Centro di Geotecnologie. Available online: http://www.geotecnologie.unisi.it/index.php?id=1013 (accessed on 30 December 2019).

4. Geo Explorer, S.R.l. Available online: http://geoexplorer.cgtgroup.org/cav_ott/ (accessed on 30 December 2019).

5. Liu, T.; Wei, Y.; Song, G.; Hu, B.; Li, L.; Jin, G.; Wang, J.; Li, Y.; Song, C.; Shi, Z.; et al. Fibre optic sensors for coal mine hazard detection. *Measurement* **2018**, *124*, 211–223. [CrossRef]

6. Naruse, H.; Uehara, H.; Deguchi, T.; Fujihashi, K.; Onishi, M.; Espinoza, R.; Guzman, C.; Pardo, C.; Ortega, C.; Pinto, M. Application of a distributed fibre optic strain sensing system to monitoring changes in the state of an underground mine. *Meas. Sci. Technol.* **2007**, *18*, 3202–3210. [CrossRef]

7. Bin, T.; Hua, C. Application of Distributed Optical Fiber Sensing Technology in Surrounding Rock Deformation Control of TBM-Excavated Coal Mine Roadway. *J. Sens.* **2018**, *2018*, 8010746. [CrossRef]

8. Zhao, Z.; Zhang, Y.; Li, C.; Wan, Z.; Li, Y.; Wang, K.; Xu, J. Monitoring of coal mine roadway roof separation based on fiber Bragg grating displacement sensors. *Int. J. Rock Mech. Min. Sci.* **2015**, *74*, 128–132. [CrossRef]

9. Cheng, G.; Shi, B.; Zhu, H.H.; Zhang, C.C.; Wu, J.H. A field study on distributed fiber optic deformation monitoring of overlying strata during coal mining. *J. Civ. Struct. Health Monit.* **2015**, *5*, 553–562. [CrossRef]

10. Wang, S.; Luan, L. Analysis on the Security Monitoring and Detection of Mine Roof collapse based on BOTDR Technology. In Proceedings of the 2nd International Conference on Soft Computing in Information Communication Technology, Taipei, China, 31 May–1 June 2014; Chang, T., Hunter, A., Eds.; Atlantis Press: Beijing, China, 2014; pp. 83–86.

11. Zhigang, T.; Chun, Z.; Yong, W.; Jiamin, W.; Manchao, H.; Bo, Z. Research on Stability of an Open-Pit Mine Dump with Fiber Optic Monitoring. *Geofluids* **2018**, *2018*, 9631706. [CrossRef]

12. Arzu, A.K.; Mehmet, K.; Arif, E.; Haluk, A.; Mustafa, K. Optical Fiber Technology to Monitor Slope Movement. In *Engineering Geology for Society and Territory – Vol. 2 Landslide Processes*; Lollino, G., Giordan, D., Crosta, G.B., Corominas, J., Azzam, R., Wasowski, J., Sciarra, N., Eds.; Springer International Publishing Switzerland: Basel, Switzerland, 2015; Volume 2, pp. 1425–1429. ISBN 978-3-319-09057-3. [CrossRef]

13. Matano, F.; Caccavale, M.; Esposito, G.; Grimaldi, G.M.; Minardo, A.; Scepi, G.; Zeni, G.; Zeni, L.; Caputo, T.; Somma, R.; et al. An integrated approach for rock slope failure monitoring: The case study of Coroglio tuff cliff (Naples, Italy) – preliminary results. In Proceedings of the 1st IMEKO TC-4 International Workshop on Metrology for Geotechnics, Benevento, Italy, 17–18 March 2016; Curran Associates, Inc.: Red Hook, NY, USA, 2016; pp. 242–247.

14. Schenato, L.; Palmieri, L.; Camporese, M.; Bersan, S.; Cola, S.; Pasuto, A.; Galtarossa, A.; Salandin, P.; Simonini, P. Distributed optical fibre sensing for early detection of shallow landslides triggering. *Sci. Rep.* **2017**, *7*, 14686. [CrossRef]

15. Salvini, R.; Vanneschi, C.; Riccucci, S.; Francioni, M.; Gullì, D. Application of an integrated geotechnical and topographic monitoring system in the Lorano marble quarry (Apuan Alps, Italy). *Geomorphology* **2015**, *241*, 209–223. [CrossRef]

16. Carmignani, L.; Conti, P.; Cornamusini, G.; Meccheri, M. The internal Northern Apennines, the Northern Tyrrhenian Sea and the Sardinia-Corsica Block. In *Geology of Italy: Special Volume of the Italian Geological Society for the IGC 32 Florence-2004*; Crescenti, U., D'Offizi, S., Merlino, S., Sacchi, L., Eds.; Società Geologica Italiana: Roma, Italy, 2004; pp. 59–77.

17. Boccaletti, M.; Elter, P.; Guazzone, G. Plate tectonic models for the development of the Western Alps and Northern Apennines. *Nature* **1971**, *234*, 108–111. [CrossRef]

18. Scandone, P. Origin of the Tyrrhenian Sea and Calabrian Arc. *Boll. Soc. Geol. It.* **1979**, *98*, 27–34.

19. Mantovani, E.; Babbucci, D.; Farsi, F. Tertiary evolution of the Mediterranean region: Major outstanding problems. *Boll. Geof. Teor. Appl.* **1985**, *105*, 67–90.

20. Dercourt, J.; Zonenshain, L.P.; Ricou, L.E.; Kazmin, V.G.; Le Pichon, X.; Knipper, A.L.; Grandjacquet, C.; Sbortshikov, I.M.; Geyssant, J.; Lepvrier, C.; et al. Geological evolution of the Tethys belt from the Atlantic to the Pamirs since the Lias. *Tectonophysics* **1986**, *123*, 241–315. [CrossRef]

21. Stampfli, G.; Marcoux, J.; Baud, A. Tethyan margins in space and time. *Palaeogeogr. Palaeoclimatol. Palaeoecol.* **1991**, *87*, 373–409. [CrossRef]

22. Stampfli, G.M.; Borel, G.; Cavazza, W.; Mosar, J.; Ziegler, P.A. The Paleotectonic Atlas of the Peritethyan Domain. *Eur. Geophy. Soc.* **2001**.

23. Zaccagna, D. Descrizione Geologica delle Alpi Apuane. *Mem. Descr. Carta Geol. Ital.* **1932**, *25*, 1–440.

24. Gattiglio, M.; Meccheri, M.; Tongiorgi, M. Stratigraphic correlation forms of the Tuscan Palaeozoic basement. *Rend. Soc. Geol. Ital.* **1989**, *12*, 247–257.

25. Conti, P.; Di Pisa, A.; Gattiglio, M.; Meccheri, M. Prealpine basement in the Alpi Apuane (Northern Apennines, Italy). In *Pre Mesozoic Geology in the Alps*; Von Raumer, J.F., Neubauer, F., Eds.; Springer: Berlin, Germany, 1993; pp. 609–621. ISBN 978-3-642-84640-3.

26. Conti, P.; Carmignani, L.; Giglia, G.; Meccheri, M.; Fantozzi, P.L. Evolution of geological interpretations in the Alpi Apuane Metamorphic Complex, and their relevance for the geology of the Northern Apennines. In *The "Regione Toscana" Project of Geological Mapping*; Morini, D., Bruni, P., Eds.; Region of Tuscany: Florence, Italy, 2004; pp. 241–262.

27. Carmignani, L.; Giglia, G.; Kligfield, R. Nuovi dati sulla zona di taglio ensialica delle Alpi Apuane. *Mem. Soc. Geol. Ital.* **1980**, *21*, 93–100.

28. Carmignani, L.; Kligfield, R. Crustal extension in the northern Apennines: The transition from compression to extension in the Alpi Apuane Core Complex. *Tectonics* **1990**, *9*, 1275–1303. [CrossRef]

29. Pieruccioni, D.; Galanti, Y.; Biagioni, C.; Molli, G. Geology and tectonic setting of the Fornovolasco area, Alpi Apuane (Tuscany, Italy). *J. Maps* **2018**, *14*, 357–367. [CrossRef]

30. Molli, G.; Meccheri, M. Structural inheritance and style of reactivation at mid-crustal levels: A case study from the Alpi Apuane (Tuscany, Italy). *Tectonophysics* **2012**, *579*, 74–87. [CrossRef]

31. Toscana, E.R. *I Marmi Apuani. Schede Merceologiche*; Nuova Grafica Fiorentina: Firenze, Italy, 1980.

32. Meccheri, M. *Carta Geologico-Strutturale Delle Varietà Merceologiche dei Marmi del Carrarese, Scala 1:10.000*; Dipartimento di Scienze della Terra, Siena Univ.: Siena, Italy, 1996.

33. Carmignani, L.; Conti, P.; Mancini, S.; Massa, G.; Meccheri, M.; Simoncini, D.; Vaselli, L. *Carta Giacimentologica dei Marmi delle Alpi Apuane – Relazione Finale*; Technical Report; Centre of Geotechnologies, University of Siena: Siena, Italy, 2007.

34. D'Amato Avanzi, G.; Giannecchini, G.; Puccinelli, R. The influence of the geological and geomorphological settings on shallow landslides. An example in a temperate climate environment: The June 19, 1996 event in northwestern Tuscany (Italy). *Eng. Geol.* **2004**, *73*, 215–228. [CrossRef]
35. Salvini, R.; Riccucci, S.; Gullì, D.; Giovannini, R.; Vanneschi, C.; Francioni, M. Geological application of UAV photogrammetry and terrestrial laser scanning in marble quarrying (Apuan Alps, Italy). In *Engineering Geology for Society and Territory—Vol. 5 Urban Geology, Sustainable Planning and Landscape Exploitation*; Lollino, G., Manconi, A., Guzzetti, F., Culshaw, M., Bobrowsky, P., Luino, F., Eds.; Springer: Cham, Switzerland, 2014; Volume 5, pp. 979–983.
36. Profeti, M.; Cella, R. *Relazione Sulla Stabilità dei Fronti—Cava Lorano n. 22*; Technical Report; Cooperativa Cavatori Lorano: Carrara, Italy, 2010.
37. Bieniawski, Z.T. *Engineering Rock Mass Classifications*; John Wiley and Sons: New York, NY, USA, 1989.
38. Tufarolo, E.; Salvini, R.; Seddaiu, M.; Lanciano, C.; Bernardinetti, S.; Petrolo, F.; Carmignani, L.; Massa, G.; Pieruccioni, D. Structure from Motion technique of proximal sensing airborne data for 3D reconstruction of extraction sites. *IJEGE* **2017**, *2*, 97–107. [CrossRef]
39. Spetsakis, M.E.; Aloimonos, Y. A multi-frame approach to visual motion perception. *Int. J. Comput. Vis.* **1991**, *6*, 245–255. [CrossRef]
40. Boufama, B.; Mohr, R.; Veillon, F. Euclidean constraints on uncalibrated reconstruction. In Proceedings of the 4th International Conference on Computer Vision (ICCV '93), Berlin, Germany, 11–14 May 1993; IEEE Computer Society: Berlin, Germany, 1993; pp. 466–470.
41. Szeliski, R.; Kang, S.B. Recovering 3-D shape and motion from image streams using nonlinear least squares. *J. Vis. Commun. Image Represent.* **1994**, *5*, 10–28. [CrossRef]
42. Fonstad, M.A.; Dietrich, J.T.; Courville, B.C.; Jensen, J.L.; Carbonneau, P.E. Topographic structure from motion: A new development in photogrammetric measurement. *Earth Surf. Process. Landf.* **2012**, *38*, 421–430. [CrossRef]
43. Profeti, M. *Relazione Tecnica sul Monitoraggio Strumentale. Analisi dei dati dal 01/01/2018 al 05/06/2018*; Technical Report; Cooperativa Cavatori Lorano: Carrara, Italy, 2018.
44. Profeti, M. *Relazione Tecnica sul Monitoraggio Strumentale. Analisi dei dati dal 05/06/2018 al 31/12/2018*; Technical Report; Cooperativa Cavatori Lorano: Carrara, Italy, 2018.
45. Li, H.N.; Li, D.S.; Song, G.B. Recent applications of fiber optic sensors to health monitoring in civil engineering. *Eng. Struct.* **2004**, *26*, 1647–1657. [CrossRef]
46. Barrias, A.; Casas, J.R.; Villalba, S. A Review of Distributed Optical Fiber Sensors for Civil Engineering Applications. *Sensors* **2016**, *16*, 748. [CrossRef]
47. Kuzyk, M.G. *Polymer Fiber Optics: Materials, Physics, and Applications*, 1st ed.; CRC Press, Taylor & Francis Group LLC: Boca Raton, FL, USA, 2006; ISBN 9781574447064.
48. Aggarwal, I.D.; Lu, G. *Fluoride Glass Fiber Optics*, 1st ed.; Academic Press: Boston, MA, USA, 1991; ISBN 9781483259307.
49. Ballato, J.; Hawkins, T.; Foy, P.; Stolen, R.; Kokuoz, B.; Ellison, M.; McMillen, C.; Reppert, J.; Rao, A.M.; Daw, M.; et al. Silicon optical fiber. *Opt. Express* **2008**, *16*, 23–18675. [CrossRef]
50. Govind, P.A. *Fiber-Optic Communication Systems*, 3th ed.; Wiley-Interscience: New York, NY, USA, 2002; p. 546. ISBN 0-471-21571-6.
51. Minardo, A.; Coscetta, A.; Porcaro, G.; Giannetta, D.; Bernini, R.; Zeni, L. Distributed optical fiber sensors for integrated monitoring of railway infrastructures. *Struct. Monit. Maint.* **2014**, *1*, 173–182. [CrossRef]
52. Horiguchi, T.; Tateda, M. Development of a Distributed Sensing Technique Using Brillouin Scattering. *J. Lightwave Technol.* **1995**, *13*, 1296–1302. [CrossRef]
53. Lalam, N.; Ng, W.P.; Dai, X.; Wu, Q.; Fu, Q. Analysis of Brillouin Frequency Shift in Distributed Optical Fiber Sensor System for Strain and Temperature Monitoring. In Proceedings of the 4th International Conference on Photonics, Optics and Laser Technology (PHOTOPTICS), Rome, Italy, 27–29 February 2016. [CrossRef]
54. Minardo, A.; Catalano, E.; Coscetta, A.; Zeni, G.; Zhang, L.; Di Maio, C.; Vassallo, R.; Coviello, R.; Macchia, G.; Picarelli, L.; et al. Distributed Fiber Optic Sensors for the Monitoring of a Tunnel Crossing a Landslide. *Remote Sens.* **2018**, *10*, 1291. [CrossRef]
55. Zeni, L. Sensing Distribuito in Fibra Ottica. Scuola Dottorato GE, 2006, Benevento (Italy). Available online: http://ge2006.unisannio.it/download/GE_Zeni.pdf (accessed on 27 March 2020).

56. Bernini, R.; Minardo, A.; Zeni, L. Distributed fiber-optic frequency-domain Brillouin sensing. *Sens. Actuators A: Phys.* **2005**, *123–124*, 337–342. [CrossRef]

57. Galindez-Jamioy, C.A.; Lopez-Higuera, J.M. Brillouin distributed fiber sensors: An overview and applications. *J. Sens.* **2012**. [CrossRef]

58. Carlino, S.; Mirabile, M.; Troise, C.; Sacchi, M.; Zeni, L.; Minardo, A.; Caccavale, M.; Darányi, V.; De Natale, G.; Mcgonigle, A.; et al. Distributed-Temperature-Sensing Using Optical Methods: A First Application in the Offshore Area of Campi Flegrei Caldera (Southern Italy) for Volcano Monitoring. *Remote Sens.* **2016**, *8*, 674. [CrossRef]

59. Her, S.C.; Huang, C.Y. Effect of Coating on the Strain Transfer of Optical Fiber Sensors. *Sensors* **2011**, *11*, 6926–6941. [CrossRef] [PubMed]

60. Zhenglin, Z.; Gao, L.; Sun, Y.; Zhang, Q.; Zeng, P. Strain Transfer Law of Distributed Optical Fiber Sensor. *Chin. J. Lasers* **2019**, *46*. [CrossRef]

61. Kister, G.; Winter, D.; Gebremichael, Y.M.; Leighton, J.; Badcock, R.A.; Tester, P.D.; Krishnamurthy, S.; Boyle, W.J.O.; Grattan, K.T.V.; Fernando, G.F. Methodology and integrity monitoring of foundation concrete piles using Bragg grating optical fibre sensors. *Eng. Struct.* **2007**, *29*, 2048–2055. [CrossRef]

62. Hong, C.Y.; Zhang, Y.F.; Zhang, M.-X.; Leung, L.M.G.; Liu, L.Q. Application of FBG sensors for geotechnical health monitoring, a review of sensor design, implementation methods and packaging techniques. *Sens. Actuators A Phys.* **2016**, *244*, 184–197. [CrossRef]

63. Associazione "MeteoApuane". Available online: http://www.meteoapuane.it (accessed on 30 December 2019).

64. Bonazza, A.; Sabbioni, C.; Messina, P.; Guaraldi, C.; De Nuntiis, P. Climate change impact: Mapping thermal stress on Carrara marble in Europe. *Sci. Total Environ.* **2009**, *407*, 4506–4512. [CrossRef]

65. Malaga-Starzec, K.; Lindqvist, J.E.; Schouenburg, B. Experimental study on the variation in porosity of marble as a function of temperature. In *Natural Stone, Weathering Phenomena, Conservation Strategies and Case Studies*; Siegesmund, S., Weiss, T., Vollbrecht, A., Eds.; Geological Society: London, UK, 2002; Volume 205, pp. 81–88. ISBN 9781862394537.

66. Siegesmund, S.; Ullmeyer, K.; Weiss, T.; Tschegg, E.K. Physical weathering of marbles caused by anisotropic thermal expansion. *Int. J. Earth Sci.* **2000**, *89*, 170–182. [CrossRef]

67. Zaghloul, M.A.S.; Wang, M.; Milione, G.; Li, M.J.; Li, S.; Huang, Y.K.; Wang, T.; Chen, K.P. Discrimination of Temperature and Strain in Brillouin Optical Time Domain Analysis Using a Multicore Optical Fiber. *Sens. (Basel)* **2018**, *18*, 1176. [CrossRef] [PubMed]

68. Burr, T.; Hamada, M.S.; Howell, J.; Skurikhin, M.; Ticknor, L.; Weaver, B. Estimating Alarm Thresholds for Process Monitoring Data under Different Assumptions about the Data Generating Mechanism. *Sci. Technol. Nucl. Install. Hindawi Publ. Corp.* **2013**, 705878. [CrossRef]

69. Gullì, D.; Pellegri, M.; Marchetti, D. Mechanical behaviour of Carrara marble rock mass related to geo-structural conditions and in-situ stress. In *Integrating Innovations of Rock Mechanics, Proceedings of the 8th South American Congress on Rock Mechanics, Buenos Aires, Argentina, 15–18 November 2015*; Rocca, R.J., Flores, R.M., Sfriso, A.O., Eds.; IOS Press, 2015. [CrossRef]

Article

Load-Independent Characterization of Plate Foundation Support Using High-Resolution Distributed Fiber-Optic Sensing

Asmus Skar [1,*,†], Assaf Klar [1,2,†] and Eyal Levenberg [1,†]

[1] Department of Civil Engineering, Technichal University of Denmark, Nordvej, Building 119, 2800 Lyngby, Denmark
[2] Faculty of Civil and Environmental Engineering, Technion - Israel Institute of Technology, Haifa 32000, Israel
* Correspondence: asska@byg.dtu.dk; Tel.: +45-45-25-18-07
† These authors contributed equally to this work.

Received: 28 June 2019; Accepted: 8 August 2019; Published: 11 August 2019

Abstract: The evaluation of soil reaction in geotechnical foundation systems such as concrete pavements, mat- and raft foundations is a challenging task, as the process involves both the selection of a representative mechanical model (e.g., Winkler, Continuum, Pasternak, etc.) and identify its prevailing parameters. Moreover, the support characteristics may change with time and environmental situation. This paper presents a new method for the characterization of plate foundation support using high-resolution fiber-optic distributed strain sensing. The approach involves tracking the location of distinct points of zero and maximum strains, and relating the shift in their location to the changes in soil reaction. The approach may allow the determination of the most suited mechanical model of soil representation as well as model parameters. Routine monitoring using this approach may help to asses the degradation of the subsoil with time as part of structural health monitoring strategies. In this paper, fundamental expressions that relate between the location of distinct strain points and the variation of soil parameters were developed based on various analytical foundation support models. Finally, as an initial validation step and to underpin the idea basics, the proposed method was successfully demonstrated on a simple mechanical setup. It is shown that the approach allows for load-independent characterization of the soil response and, in that sense, it is superior to common identification methods.

Keywords: distributed fiber-optic strain sensing; soil-structure interaction; foundation support; structural health monitoring; geotechnical analysis; pavement analysis

---

1. Introduction

Common civil constructions, such as concrete pavements, mat- and raft foundations, involve precast or cast-in-place slabs resting on a prepared foundation support or so-called slab-on-grade construction. In engineering design, the mechanical behavior of slabs typically follows conventional elastic plate theory [1,2]. At the same time, the soil foundation behavior is most commonly represented by highly idealized response models, e.g., by employing the classical theories of elasticity and plasticity [3,4]. However, whereas the slab characteristics are engineered and well defined, the foundation support model is difficult to characterize.

In recent years, new sensing technologies have been developed for transforming conventional civil engineering structures into intelligent infrastructure. One leading technology in this connection is distributed fiber-optic strain sensing (see e.g., [5–7]). The development of this technology, and its capabilities to provide spatial profiles of strains along conventional telecommunication fibers, have led to a reevaluation of the manner in which strains can be used in civil engineering (see e.g., [8–10]).

Over the past decade, research in this area has been focusing on schemes of installation and interpretation of the spatially distributed data for various civil engineering problems, for example, pile foundations [11], evaluation of pipeline integrity to underneath excavation of a tunnel [12], stressing and deformation of secant pile walls [13], landslide localization [14], tunneling stressing [15], damage identification in concrete structures [16–18], strain measurement [19] and detection of cracks in asphalt pavements [20], and evaluation of tunneling induced ground deformations [21,22].

These interpretation methods are mostly based on analysis of static (slow-occurring) loading scenarios and on relatively low spatial resolutions of the order of one meter. However, new technological developments in this field now allows for a much higher spatial resolution at the order of a few millimeters [23–30]. This technological boost, not only enables new possibilities in structural health monitoring, but also makes the technology well suited for studying fundamental problems in a small-scale laboratory setting [31].

Foundation support models are relatively simple and easy-to-use approximations of the actual soil load-displacement response, and therefore play a significant role in geotechnical and pavement engineering research and practice (see e.g., [3,4,32–36]). It is generally assumed that, for serviceability design, the soil medium can be adequately represented by an elastic medium. For routine design purposes, Winkler's idealization [37], characterized by a single parameter called coefficient of subgrade reaction k, has been used almost exclusively [38]. Another idealization assumes continuum behavior of the soil, and the soil medium is thus represented by an elastic half-space [39]. These two foundation support models can be regarded as the two extreme cases of soil behavior, represented on the one hand by the completely discontinuous medium (i.e., composed of discrete springs), and on the other by the completely continuous elastic solid. Thus, several simplified soil foundation models have been developed to provide a transition between these two types of idealized soil behavior [40–44]. This class of mathematical models has an additional constant parameter and, hence, the models are called two-parameter foundation models.

The effectiveness of using foundation support models for analysis of soil–structure interaction problems depends on the accuracy with which model parameters can be determined. Over the years, significant research effort has been devoted to the development of empirical expressions, linking the coefficient of subgrade reaction k to the properties of an elastic continua (see e.g. [45–48]), as well as realistic field conditions (see e.g., [49,50]). The modulus of elasticity is often determined from the early stages of triaxial load tests. Plate loading tests, and other non-destructive methods, may also be used to determine the in-situ modulus of elasticity of the soil. The elastic material properties can then be related in a theoretically rigorous fashion to the two-parameter foundation model parameters [38,51–54]. Methods for determining the parameters from non-standard field tests have also been proposed [41,42,55].

However, since all foundation support models are essentially idealizations, their fundamental assumptions may not be completely satisfied in actual field conditions. Moreover, taking into account the sensitivity of the support characteristics to temperature and moisture changes, the governing properties continually evolve under usual service conditions. Consequently, for any given plate foundation system, it is not straightforward to identify the governing support model, and the prevailing support properties at a given time and environmental situation.

Common means for foundation characterization are based on large-scale experiments [56,57]. These involve application of a load having known dimensions and intensity, and measurement of the resulting mechanical responses. Such procedure is, by very nature, expensive and service-disruptive; it is essentially relevant for sparse time intervals. As a means of addressing these limitations, the current work offers a new method for characterizing support conditions that is non-destructive, non-disruptive, and load-independent. The development has two purposes: (*i*) identify the support model type that best applies in a given situation, and (*ii*) characterize the associated foundation (soil) properties (i.e., quantify the coefficient of subgrade reaction and the intensity of shear interaction between the Winkler springs).

The suggested approach involves a loaded plate instrumented with distributed strain-sensing gear. It is based on the ability of tracking the location (relative to the loading location) of distinct points of zero and maximal bending moment. The relative locations of these points are essentially associated with the support parameters. In the paper, fundamental expressions that relate between the shift of strain of the distinct points and the variation of support model parameters are developed based on various static analytical plate solutions. Furthermore, the proposed methodology is experimentally validated by instrumenting a small-scale plate foundation system with high-resolution distributed fiber-optic strain sensors, capable of detecting the effects of loading events.

2. Plate Foundation Support Models

2.1. Modeling Idealized Soil Response

The complex behavior of a real soil mass has led to the development of many idealized models for soil behavior, especially for the analysis of soil–structure interaction problems. The Winkler model is simple and practical to many engineering problems and has therefore been used extensively for routine design of foundations and pavements [38]. In the Winkler model, the soil foundation properties are idealized as independent springs on a rigid base neglecting the effect of shear deformation between the springs, as shown in Figure 1a.

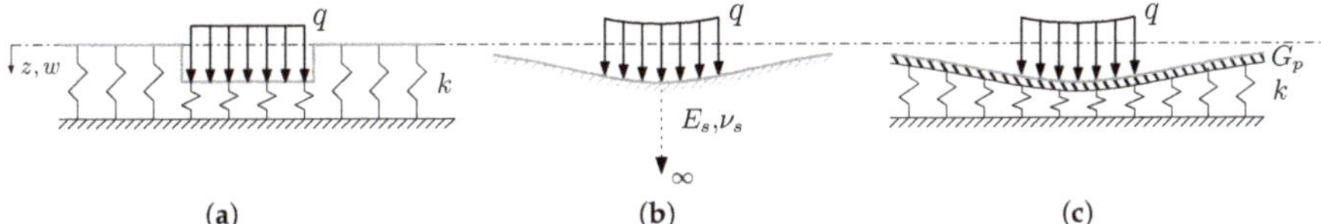

(a)   (b)   (c)

Figure 1. Response of foundation support models: (**a**) Winkler foundation model, consisting of independent springs characterized by a single parameter k, (**b**) the elastic half-space continuum model characterized by the Young's modulus E_s and Poisson's ratio ν_s, and (**c**) the Pasternak spring model with elastic layer capable of pure shear deformation, characterized by the two parameters k and G_p, respectively.

It is common experience that, in the case of soil media, surface deflections will occur not only immediately under the loaded region but also within certain limited zones outside the loaded region, as shown in Figure 1b. In attempts to account for this continuous behavior, soil media have often been idealized as three-dimensional continuous elastic isotropic solids. The first continuum representation of soil media stems from the work of Boussinesq [39].

However, both experimental and theoretical investigations have emphasized the need to provide a transition between these two types of idealized soil behavior since displacements outside the loaded region decrease more rapidly than that predicted by the elastic continuum model [3]. In this aspect, the mechanical two-parameter model proposed by Pasternak is attractive, shown in Figure 1c, offering an alternative to the elastic solid continuum by providing a degree of shear interaction between adjacent soil elements while remaining relatively simple to analyze [36].

2.2. Analytical Treatment of the Plate Foundation System

The mechanical behavior of slabs typically follows the classical Germain-Kirchhoff plate formulation [58]. Consider a plate of infinite size characterized by thickness h, Young's modulus E, Poisson's ratio ν, and therefore flexural rigidity D. The plate is loaded by a uniform vertical stress with intensity q distributed over a circular area with radius a. For this axisymmetric situation the vertical displacement field, u_z, depends only on the radial coordinate r with origin located directly under the load centroid. The plate moments (per unit length) in the radial (M_r) and tangential (M_θ) are obtained from the expressions

$$M_r = -D\left(\frac{d^2 u_z}{dr^2} + \frac{v}{r}\frac{du_z}{dr}\right)$$
$$M_\theta = -D\left(\frac{1}{r}\frac{du_z}{dr} + v\frac{d^2 u_z}{dr^2}\right)$$

$$(1)$$

The corresponding radial and tangential bending stresses are $\sigma_r = 12zM_r/h^3$ and $\sigma_\theta = 12zM_\theta/h^3$, respectively, where z is measured from the plate's mid-surface or neutral plane (positive = down). The extremal bending stresses are obtained at the bottom of the plate where $z = h/2$ and at the surface where $z = -h/2$. The notation is such that a positive moment is associated with tensile bending stress at the plate bottom under the loaded area, i.e., where $r = 0$ and $z = h/2$. Finally, radial strain (ε_r) and the tangential strain (ε_θ) at a given z are obtained from

$$\varepsilon_r = \frac{12z(M_r - vM_\theta)}{Eh^3}$$
$$\varepsilon_\theta = \frac{12z(M_\theta - vM_r)}{Eh^3}$$

$$(2)$$

In the case of a Pasternak foundation type, the vertical displacement field is [59]

$$u_z = \frac{qa}{lk}\int_0^\infty \frac{J_0(mr/l)J_1(ma/l)}{m^4 + 2bm^2 + 1}dm$$

$$(3)$$

where $l = \sqrt[4]{D/k}$ is the so-called radius of relative stiffness [60]. k is the coefficient of subgrade reaction (force/length^3), $J_n(\ )$ denotes a Bessel function of the first kind of order n, m is a unitless integration variable or wave number, and $b = G_p/2kl^2$ is positive and dimensionless wherein G_p (force/length) is the second parameter, and represents the intensity of the shear interaction between the Winkler springs. For the special case of $b = 0$, or equivalently $G_p = 0$, the expression provides the solution for a plate on Winkler foundation. The corresponding plate bending moments (per unit length) in the radial and tangential directions are

$$M_r = \frac{qaD}{l^2kr}\int_0^\infty \frac{mJ_1(ma/l)\left((mr/l)J_0(mr/l) - (1-v)J_1(mr/l)\right)}{m^4 + 2bm^2 + 1}dm$$
$$M_\theta = \frac{qaD}{l^2kr}\int_0^\infty \frac{mJ_1(ma/l)\left((vmr/l)J_0(mr/l) + (1-v)J_1(mr/l)\right)}{m^4 + 2bm^2 + 1}dm$$

$$(4)$$

Hogg assumed a plate with frictionless bottom bonded to a linear elastic isotropic half-space [61], herafter referred to as the half-space continuum model. The vertical displacement field is [3]

$$u_z = \frac{2qa(1 - v_0^2)}{l_e E_s r}\int_0^\infty \frac{J_0(mr/l)J_1(ma/l)}{m^4 + m}dm$$

$$(5)$$

where E_s and v_s are the elastic properties of the half-space support, and $l_e = \sqrt[3]{2D(1 - v_s^2)/E_s}$ is the characteristic length associated with the a plate on a half-space (analogous to the radius of relative stiffness). The corresponding plate bending moments (per unit length) in the radial and tangential directions are

$$M_r = \frac{2qa(1 - v_0^2)}{l_e E_s r} \int_0^\infty \frac{m J_1(ma/l)\left((mr/l)J_0(mr/l) - (1 - v)J_1(mr/l)\right)}{m^3 + 1} dm$$

$$M_\theta = \frac{2qa(1 - v_0^2)}{l_e E_s r} \int_0^\infty \frac{m J_1(ma/l)\left((vmr/l)J_0(mr/l) + (1 - v)J_1(mr/l)\right)}{m^3 + 1} dm$$

$$(6)$$

3. Proposed Interpretation Method

This section describes a method for characterising the foundation parameters based on distributed fiber-optic strain measurements of the radial strain (ε_r) at the top of plate ($z = -h/2$). It is assumed that the strains can be detected by fiber-optic cables attached to the slab. The methodology is composed of three elements: (i) a mechanical plate foundation model, (ii) distributed fiber-optic strain measurements, and (iii) an iterative interpretation scheme.

To exemplify the overall strain response of a standard support plate system, the radial strain is plotted versus the normalized radial distance (ρ) from the loaded point, shown in Figure 2a for three different load intensities (i.e., q = 0.5–1.5 MPa). In the Figure it was assumed that slabs are constructed on a 150 mm thick high-quality sub-base over subgrade soil. Moreover, the slab is composed of concrete having a Young's modulus E = 30,000 MPa and a Poisson's ratio v = 0.15. It's thickness is h = 300 mm, leading to a flexural rigidity of D = 6.91 × 10^{10} Nmm. Figure 2b shows the effect of the model on the the radial strain at the top when the slab is loaded by a single heavy wheel with radius, a = 150 mm. The foundation support parameters are given as: (i) Winkler-model: k = 0.055 MPa/mm, (ii) Pasternak-model: k = 0.055 MPa/mm and b = 0.5, and (iii) half-space continuum model: E_s = 102 MPa and v_s = 0.35. Some points of interest are also included with dotted blue lines, these are hereafter referred to as Distinct Points (DPs) of zero strain $DP_{0,i}^j$ (i.e., $\rho\left(\varepsilon_{0,i}\right)$), and maximum strain $DP_{m,i}^j$ (i.e., $\rho\left(\varepsilon_{m,i}\right)$), where i = 1, 2 is the number of zero/maximum strain location from the center of the load, and j = 'W', 'P', 'C', indicating the support type Winkler, Pasternak and Continuum, respectively. Zero strain is shown as a dashed dotted blue line along the abscissa-axis. The radial coordinate is normalized by the radius of relative stiffness (l) and characteristic length (l_e) both equal to 1059 mm.

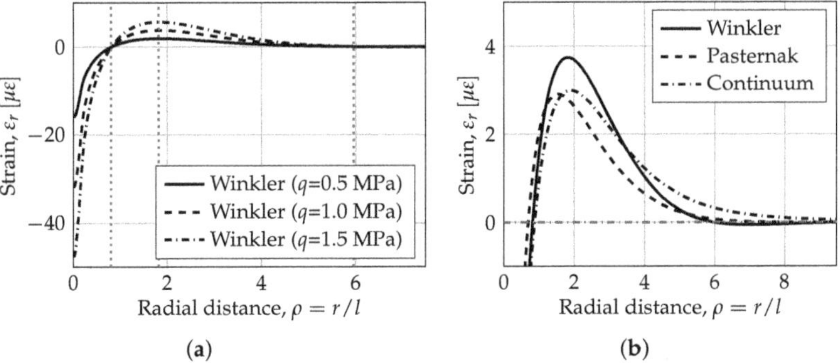

Figure 2. Influence of foundation model type on radial strain response for a standard concrete support plate system: (**a**) overview of strain response as a function of radial distance and (**b**) close-up of the region around first and second zero crossing.

From Figure 2a it is observed that the loading intensity q has no effect on the results. Moreover, separate analysis has shown that the results are essentially insensitive to the exact radius of loading as long as it is of the same order as the plate thickness or smaller (down to a point load). It was also found that the results are insensitive to the value of ν_s.

In Figure 2b a close-up of the strain response in the region around the first and second zero crossing is shown. A few characteristic features are revealed in the figure w.r.t. the distinct points for the different model types, i.e.: (i) the Winkler-model has a first ($\mathrm{DP}_{0,1}^W$) and second zero crossing ($\mathrm{DP}_{0,2}^W$) at a radial distance from the load of $\approx l$ and $6l$, respectively, (ii) the Pasternak-model has a first crossing at a radial distance lower than the Winkler-model (i.e., $\mathrm{DP}_{0,1}^P < \mathrm{DP}_{0,1}^W$), whereas the second crossing is larger (i.e., $\mathrm{DP}_{0,2}^P > \mathrm{DP}_{0,2}^W$), and ($iii$) the half-space continuum model as a first zero crossing larger than the Winkler-model (i.e., $\mathrm{DP}_{0,1}^C > \mathrm{DP}_{0,1}^W$) and no second zero crossing. It is also found that the magnitude of the second peak for the Winkler-model is higher than for the Pasternak-model and the half-space continuum model (i.e., $\varepsilon_{m,2}^W > \varepsilon_{m,2}^{P,C}$) due to the lack of shear transfer in the supporting medium.

The observations above show that the location of distinct points are closely related to the governing foundation model type and to the numerical values of the model parameters. To further investigate these features, the Pasternak-model is utilized as proposed in [62], considering two different cases, spanning two extreme yet realistic situations. The first case considered is a very thick concrete plate resting on a very 'soft' spring-bed with a large radius of relative stiffness of l = 2000 mm. The second case considered is a very thin concrete plate resting on a very 'hard' spring-bed support with a short radius of relative stiffness of l = 336 mm. First, the location of the distinct points are plotted versus the shear interaction parameter b, shown in Figure 3a. Presented next is the location ratio of the first to the second zero crossings (i.e., $\mathrm{DP}_{0,1}/\mathrm{DP}_{0,2}$) as a function of b for the two cases, shown in Figure 3b.

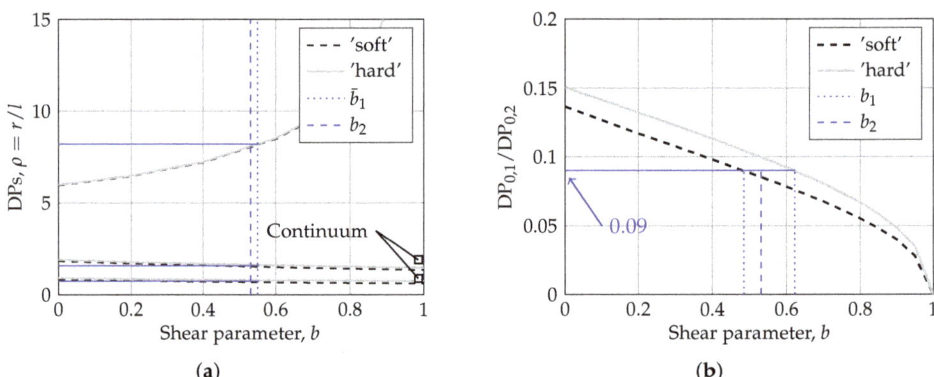

Figure 3. Parametric study of the Pasternak-model: (**a**) Influence of Pasternak's b parameter on the normalized locations of the DPs defined in Figure 2 for two cases (**b**) influence of Pasternak's b parameter on the location ratio of first to second zero crossings defined in in Figure 2 for the two cases.

From Figure 3a it can be seen that the results for the two cases essentially overlap, i.e., that the location of distinct points are relatively insensitive to the value of l. The first zero crossing location is influenced by b, dropping from about $0.86l$ at $b = 0$ to about $0.67l$ at $b = 1$. The second peak location also drops with increasing b, $1.86l$ at $b = 0$ to about $1.40l$ at $b = 1$. A pronounced dependence on b is exhibited by the location of the second zero crossing, increasing from about $5.89l$ at $b = 0$ towards infinity as b approaches unity. It is also found that as b increase, the discrepancy between the Pasternak-model and the half-space continuum model increases. This shows that although the Pasternak-model possess some of the characteristic features of continuous elastic solids, it is a simplification which cannot capture all complexities. Finally, from Figure 3b it is observed that the two curves do not collapse onto one unique line, the gap between them is considered small, establishing an almost unique relationship between the ratio and parameter b.

The relations shown in Figure 3 form the basis of the non-disruptive fiber-optic based test method for characterizing the plate support conditions suggested in herein. Specifically, the problem is to determine k and b such that a best match is achieved between measured and modeled location of distinct points/strain response.

The starting point for the method is identifying the locations of the distinct points (see Figure 2). This step is non-destructive, non-disruptive, load-independent (given that the load magnitude and radius are not needed), and can be done repeatedly. Next, Figure 3b is entered for the first iteration, with the location ratio of the first to second zero crossings to provide a range of estimated values for the shear parameter, i.e., b_1, marked with blue dotted lines. The average of this range (i.e., $\bar{b}_1$) is then utilized in Figure 3a to resolve the value of the radius of relative stiffness l. Once l is identified, a second iteration is to be carried out to refine the estimation of the shear parameter (and subsequently l), i.e., b_2, marked with a blue dashed line in the Figure. Finally, given that the plate properties are known, it becomes possible to calculate the Winkler coefficient of subgrade reaction $k = Dl^{-4}$, and the intensity of shear interaction between the Winkler springs $G_p = 2kbl^2$.

In the Figure the procedure proposed are shown for the plate foundation system in Figure 2 with $k = 0.055$ MPa/mm and $b = 0.5$ (i.e., the 'Pasternak-model'). The distinct points are calculated as $\mathrm{DP}_{0,1} = 736.5$ mm, $\mathrm{DP}_{m,2} = 1637.2$ mm and $\mathrm{DP}_{0,1} = 8204.6$ mm which yields a location ratio of the first to the second zero crossings of 0.09, and resulting foundation model parameters $k = 0.064$ MPa/mm, $b = 0.54$ and $l = 1017$. Thus, k, b and l was estimated with 17%, 8% and 4% accuracy, respectively.

4. Experimental Investigation

4.1. Experimental Setup

In order to validate the proposed interpretation method a small-scale experiment was designed applying high-resolution distributed fiber-optic strain sensing. The test was designed and carried out to provide a first-order demonstration of the proposed characterization concept with off-the-shelf equipment; it was not designed to mimic real-life situation. Therefore, practical aspects such as embedding fiber optic cables in concrete and dealing with multiple load situations were not considered.

The experimental setup consisted of an aluminum plate over a finite thickness support material on a concrete floor, shown in Figure 4a. The aluminum plate was instrumented with a fiber-optic cable, glued to the top of the plate in both directions (at right angles), and connected to the measurement device on one end. Finally, a dead-load was applied, using 36×0.5 kg weights placed in a grid, to ensure contact between the plate and the support material. Since the loading and theory are axisymmetric, the fibre lines in the experiment were positioned to capture radial strains and not in a grid arrangement. A grid arrangement is envisioned for field applications, where the load position cannot be a priori known. The availability of a strain grid can be utilized to identify the loading location based on a criterion of maximal bending strain. Afterwards, strain analysis is to be performed according to the proposed theory for fibres that measure radial strains.

The plate was loaded at the center (i.e., far from the edges) with hand-held weights on a small rubber pad with area, $A_{\mathrm{load}} = 36$ mm^2. The load was applied at the center of the test plate to minimize edge effects and better comply with the theoretical derivation. Strain measurements were then recorded with a commercial Optical Backscatter Reflectometer (OBR) device [63] depicted in Figure 4b. Strain values were recorded in intervals of 1mm with a gauge length of 10 mm. In the specific case, the load is of short time duration compared to changes in the support, and thus, no temperature compensation is needed. In cases where loads are stationary for a long period of time, e.g., in the case of foundations, temperature compensation will be needed.

(a)

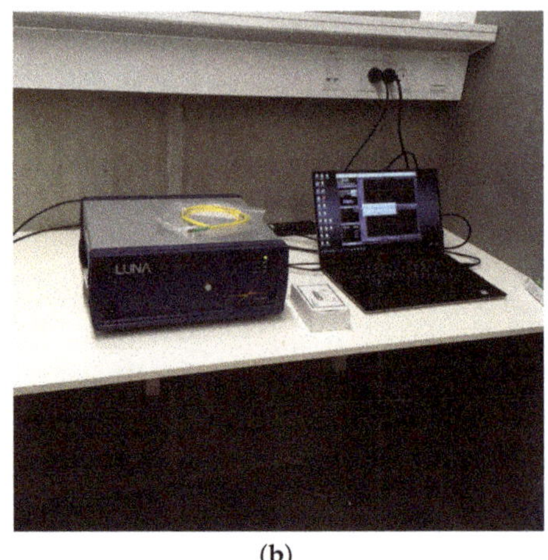

(b)

Figure 4. Small-scale slab on grade system: (**a**) aluminum plate instrumented with fiber-optic cables and supported by a thick polystyrene mat and (**b**) OBR device and laptop.

A thin rubber mat and a thick polystyrene mat were selected as foundation material in an attempt to resemble the two outer extreme cases, i.e., a Winkler-type and a Continuum-type foundation, hereafter referred to as support type 'thin' and 'thick', respectively. Moreover, a thin flexible plate, was selected to limit the size of the test setup, and at the same time, comply with the plate formulation presented in Section 2 (i.e., avoid edge effects). The geometrical and material properties for the different structural elements of the system are given in Table 1.

Table 1. Geometrical and material properties used in experimental study.

Structural Element	Material	Young's Modulus [MPa]	Poisson's Ratio [-]	Thickness [mm]	Length [mm]	Width [mm]
Plate	Aluminium	68,300	0.33	1.5	1100	1100
Support ('thin')	Rubber [64]	2–4	0.45	10.0	1250	1250
Support ('thick')	Polystyrene [65]	1–2	0.05	100.0	1200	1200

4.2. Experimental Results

The raw fiber-optic strain measurements are presented in the 1-D plots in Figure 5. In the Figure the strain data for support type 'thin' (gray lines) and 'thick' (black lines) are plotted at a load level of 2–4 kg along one of the lengths of the aluminum plate.

Comparing the raw strain signal for foundation support type 'thin' and 'thick' in Figure 5a,b, respectively, it is observed that the overall shape is similar and symmetric. However, the noise level is higher for foundation type 'thin', and also higher than the expected/specified level of app. ± 1 $\mu\varepsilon$. Thus, subsequent analysis of data were performed on strain measurements averaged over 4 load and unloading tests (i.e., for noise reduction). Moreover, for further analysis a load of 4 kg (i.e., $q \approx 1.09$ MPa) was selected in order to maximize the signal to noise ratio.

In Figure 6, the detailed experimental results are presented. Figure 6a shows the peak strains measured for the two support types at four different load levels. Figure 6b shows the effect of the support type on the developed strain. Finally, Figure 6c,d present a close-up the raw data signal and the moving average of strain measurements is shown for support type 'thin' and 'thick', respectively. The moving average data is calculated using a base length of L_b = 50 mm.

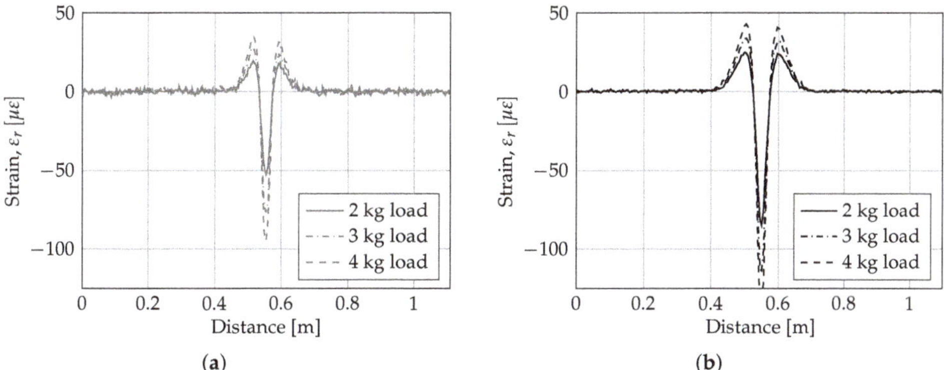

Figure 5. 1-D representation of raw fiber-optic strain measurements at different load levels (2–4 kg) for (**a**) support type 'thin' (gray) and (**b**) support type 'thick' (black).

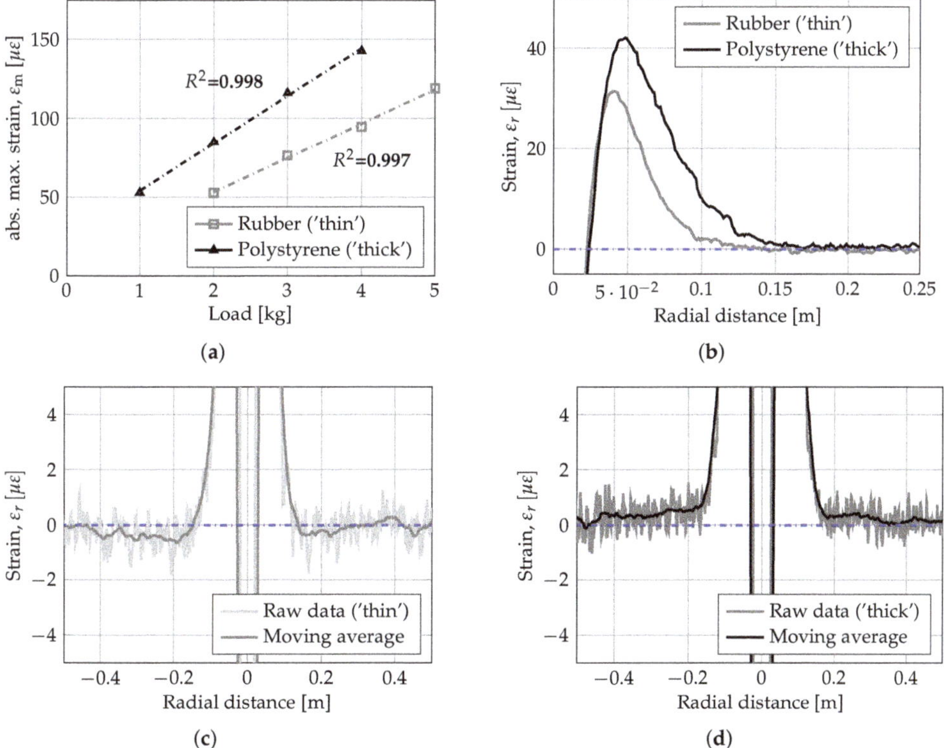

Figure 6. Comparison of distributed fiber-optic strain measurements for different support materials: (**a**) linearity of materials, (**b**) close-up of region around first and second zero crossing, (**c,d**) close-up of second zero crossing, showing point and moving average strain measurements for foundation type 'thin' and 'thick', respectively.

From Figure 5a, it is observed that both plate foundation systems behave linearly for the applied loading magnitudes. This is a basic prerequisite for further analysis of the strain data using the proposed framework, avoiding influence of shift in distinct points due to nonlinear behavior.

In Figure 2b a close-up of the strain response in the area around the first and second zero crossing is shown. From the Figure it is observed that the support type 'thick' has a first zero crossing slightly larger than support type 'thin', i.e., $DP_{0,1}^{'thick'} = 24.4$ mm $> DP_{0,1}^{'thin'} = 19.9$ mm. Moreover, support type 'thin' has a clear zero crossing at $DP_{0,2}^{'thin'} = 130.9$ mm, whereas support type 'thick' has a no clear zero crossing, although the abscissa is crossed at a radial offset of app. ± 400 mm. Thus, the two support types 'thin' and 'thick' show some of the characteristic features of the Winkler-model and the half-space continuum model, respectively (i.e., equivalent to a Pasternak-model with high and low shear interaction). The location of the second peak is $DP_{m,2}^{'thin'} = 37.9$ mm and $DP_{m,2}^{'thick'} = 47.0$ mm.

4.3. Interpretation of Results

The results obtained from the small-scale experiment are next interpreted using the iterative scheme proposed in Section 3. First, the results visualised in Figure 3 are reproduced for the aluminium plate with a flexural rigidity of $D = 2.16 \times 10^4$ Nmm and load radius $a = 3.38$ mm. The support conditions are taken as $k = 0.001$ MPa/mm and $k = 10.0$ MPa/mm to ensure a sufficient upper and lower limit of l, i.e., 68.10 mm and 6.81 mm, respectively. The analytical results for the experimental plate support system, i.e., the location of the distinct points, as well as the ratio of first zero crossing to second zero crossing, are shown in Table 2.

Table 2. Location of the distinct points for the aluminium plate support system.

	$l = 68.10$ mm					$l = 6.81$ mm				
b	$DP_{0,1}$	$DP_{m,2}$	$DP_{0,2}$	$DP_{0,1}/DP_{0,2}$	b	$DP_{0,1}$	$DP_{m,2}$	$DP_{0,2}$	$DP_{0,1}/DP_{0,2}$	
0				0.152						
0	0.811	1.821	5.962	0.136	0	0.924	1.910	6.006	0.154	
0.01	0.808	1.815	5.986		0.01	0.921	1.904	6.030	$\rightarrow$	$l_1^{'thin'} = 21.72$ mm
					0.02				0.152	
0.20	0.753	1.694	6.456	0.117	0.20	0.879	1.798	6.504	0.135	
0.40	0.705	1.585	7.202	0.098	0.40	0.843	1.703	7.254	0.116	
0.60	0.664	1.490	8.477	0.078	0.60	0.813	1.622	8.532	0.095	
0.70	0.645	1.447	9.561	0.067	0.70	0.800	1.585	9.617	0.083	
0.75				0.063						
0.80	0.628	1.408	11.350		0.80	0.788	1.552	11.408	$\rightarrow$	$l_1^{'thick'} = 33.82$ mm
0.80	0.628	1.407	11.384	0.055	0.80	0.788	1.551	11.442	0.090	
					0.85				0.063	
0.85	0.620	1.387	12.941	0.048	0.85	0.783	1.535	12.982	0.060	
0.90	0.613	1.369	15.242	0.040	0.90	0.777	1.520	15.591	0.050	
0.95	0.605	1.351	21.533	0.029	0.95	0.772	1.505	21.730	0.037	
1.00	0.598	1.333	∞	0.000	1.00	0.767	1.491	∞	0.000	

The ratio of first zero crossing to second zero crossing is 0.152 and 0.063, for support type 'thin' and 'thick', respectively. These are entered in Table 2, and the corresponding parameters $b_1^{'thin'} = 0–0.019$ and $b_1^{'thick'} = 0.751–0.845$ are found by interpolation (marked with gray cells in the Table). These values are then used to provide a first estimate of the shear parameter of $\bar{b}_1^{'thin'} = 0.010$ and $\bar{b}_1^{'thick'} = 0.798$ after averaging, providing three different estimates of the distinct points (marked with gray cells in the Table 2). Consequently, the radius of relative stiffness is estimated, via averaging of the three different possible values, to be 27.2 mm and 33.82 mm, for support type 'thin' and 'thick', respectively.

Next, another iteration is performed with 0.152 and 0.063, and a value of $b_2^{'thin'} = 0.005$ and $b_2^{'thick'} = 0.843$ is obtained via interpolation (considering that $l_1^{'thin'} = 21.72$ mm and $l_1^{'thick'} = 33.82$ mm). The location of the distinct points are reentered, this time with $b_2^{'thin'} = 0.005$ and $b_2^{'thick'} = 0.843$, and the above described calculations are repeated. The final result is $l_2^{'thin'} = 21.71$ mm and $l_2^{'thick'} = 32.77$ mm, which leads to a modulus of subgrade reaction of $k^{'thin'} = 0.097$ MPa/mm and $k^{'thick'} = 0.019$ MPa/mm.

The results of the analysis are summarized in Table 3. Specified material properties (see Table 1) and measured values are shown in brackets. The expression for the characteristic length (see Section 2) is utilized for calculating the 'equivalent' Young's modulus for each support type.

From Table 3, it is observed that predicted Young's modulus for support type 'thin' and 'thick' is 3.36 MPa and 1.22 MPa, respectively, matching well the known material properties (see Table 1). Moreover, the estimated shear parameter b is 0.005 and 0.843, respectively, indicating that the 'thin' support system is dominated by compression (i.e., the support material acts similar to the Winkler-model), whereas the 'thick' support system is highly affected by shear. Thus, the methodology enables identification of suitable model type. It is also found that the estimated location of distinct points match well with the experimental values. This outcome provides basic confidence in the proposed method and experimental results obtained.

Table 3. Summary of analysis results.

Support Type	$DP_{0,1}$ [mm]	$DP_{m,2}$ [mm]	$DP_{0,2}$ [mm]	k [MPa/mm]	b [-]	E_s [MPa]	v_s [-]
'thin'	18.8 (19.9)	40.4 (37.9)	130.2 (130.9)	0.097	0.005	3.38 (2–4)	(0.45)
'thick'	23.0 (24.4)	48.0 (47.0)	417.8 (400.0)	0.019	0.843	1.22 (1–2)	(0.05)

To visualize the results the analytical (using the estimated model parameters from Table 3) and experimental strain curves are plotted in the region around the first and second zero crossings, shown in Figure 7.

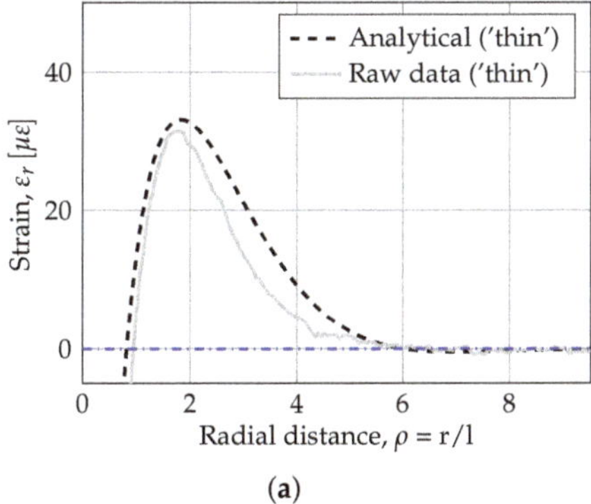

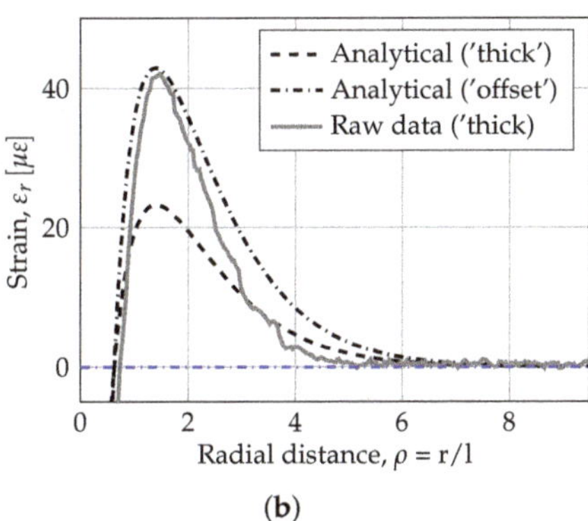

Figure 7. Comparison of analytical results with estimated support conditions and experimental fiber-optic strain measurements: (**a**) Pasternak-model with $k = 0.097$ MPa/mm and $b = 0.005$ vs. rubber foundation ('thin) and (**b**) Pasternak-model with $k = 0.019$ MPa/mm and $b = 0.843$ vs. polystyrene foundation ('thick). In generating these plots the loading intensity of $q = 1.09$ MPa was used.

From Figure 7a, it is observed that the analytical strain response resembles the experimental curve. The discrepancy between curves is especially pronounced after the second peak. This could indicate that other effects (not only compression and shear) also influence the system. One potential effect is friction, as the coefficient of friction in the experiment (i.e., between rubber and aluminium) is much higher than actual field conditions (i.e., between sand subbase/polyethylene sheet and concrete). The potential effect of friction was investigated in a separate analysis utilizing a detailed Finite Element (FE) model of the problem. The FE computations showed that friction between the aluminium plate and foundation material have little influence in the location of the first zero crossing. However, increasing friction results in an increasing negative slope after the second peak. The shift was 0–10%, decreasing rapidly as $\rho \to \infty$. Thus, increasing friction has a positive effect on the overall fit between analytical and experimental results.

From Figure 7b, it is observed that the magnitude of the second peak (i.e., $\varepsilon_{m,2}$) is significantly lower than the experimental peak strain. This discrepancy can be explained by the sensitivity of the experimental setup to offsets in the position of the fiber-optic cable relative to the plate's mid-surface (i.e., position of strain measurement), e.g., created by small variations in thickness of the film of gluing between fiber-optic cable and plate. The difference in peak strain is equivalent to an offset error of app. 0.4 mm, exemplified in the Figure with a dashed dotted curve named 'offset'. Thus, this error is negligible for real world applications. Furthermore, it is observed that the difference in the position of strain measured do not influence the position of distinct points, showing the robustness of the methodology selected.

5. Conclusions

This study focused on the development of a non-destructive interpretation method for characterizing the plate foundation support conditions using static analytical foundation support models and high-resolution distributed fiber-optic strain sensing.

The proposed methodology was based on tracking a few distinct points of zero and maximum strain. This is the first time such a tracking idea has been utilized for parameter identification in geotechnical infrastructure. This approach has the advantage that it allows for load-independent characterization of the soil response, and in that sense, it is superior to other system identification methods that rely on response magnitudes.

As a first step towards routine engineering application, the method was tested and validated in a small scale experiment of aluminium plate resting on a thin rubber and thick polystyrene support systems. The experimental results showed that a second zero crossing was identified for the thin rubber foundation. Whereas, the strain approaches zero after the second peak (as the radial distance increases), and that no clear second zero crossing can be identified for the polystyrene foundation. These findings comply well with the theory that the thin rubber should resemble a Winkler support system while the thick polystyrene a continuum domain.

In the experiments presented, the third peak for foundation type 'thin' was ≈ 1 με, whereas the theoretical third peak in realistic concrete plate on foundation systems is ≈ 0.1 με. In this aspect identification of the second zero crossing may be difficult considering a real-scale plate foundation health monitoring system. On the other hand, true sized problems do not require such high-resolution, and a resolution of a few centimeters should be sufficient to clearly identify the characteristic points. It can also be shown that the highly idealized modeling result in discrepancy in the overall strain response in the region around the second peak. This is a result of the materials selected for the conceptual small-scale experiment presented in this paper (i.e., high friction coefficient between materials). In realistic slab-on-grade construction friction contact will likely have a minor effect (i.e., considering the relatively low friction between soil and structure).

The current research demonstrated, both theoretically and experimentally, that shape features of the spatial strain profile (under load) contain relevant information for foundation characterization. Thus, a conceptually novel monitoring technique can be envisioned that is non-destructive, non-disruptive, and load-independent.

There are many challenges in line before the idea can be applied in real life situations, e.g., sensing placement, sensing resolution, and sensing range. The present work serves as an initial first step towards a full-scale health monitoring, underpinning the idea basics, and therefore identifying those practical aspects that require further development. The next development phase should involve application of the proposed interpretation method to real soils and validation against an independent measurement system. Future efforts should also be expended on improving the technology, enabling analysis of moving and dynamic loads, as well as fiber-optic optimization for finding the required trade-off between resolution and accuracy. The methodology provided within this paper can be the basis for such future research efforts.

Author Contributions: Conceptualization, A.K.; methodology, A.S., A.K. and E.L.; writing final paper (original draft), figures, data analysis and post-processing, planning and preparation of experiments, A.S.; conducting experiments, A.S. and A.K.; writing first draft on basic concept, E.L.; Support and review of paper, A.K. and E.L.

Funding: This research received no external funding

Conflicts of Interest: The authors declare no conflict of interest.

References

1. Horvath, J. *Soil-Structure Interaction Research Project: Basic SSI Concepts and Applications Overview*; Manhattan College School of Engineering: Bronx, NY, USA, 2002.
2. NCHRP. *Guide for Mechanistic-Empirical Design of New and Rehabilitated Pavement Structures: Structural Response Modeling of Rigid Pavements—Appendix QQ*; National Cooperative Highway Research Program, Transportation Research Board, National Research Council: Washington, DC, USA, 2003.
3. Selvadurai, A.P. *Elastic Analysis of Soil-Foundation Interaction*; Elsevier: Amsterdam, The Netherlands, 1979.
4. Horvath, J.S. Subgrade models for soil-structure interaction analysis. In *Foundation Engineering: Current Principles and Practices*; ASCE: Reston, VA, USA, 1989; pp. 599–612.
5. Güemes, A.; Fernández-López, A.; Soller, B. Optical fiber distributed sensing—Physical principles and applications. *Struct. Health Monit.* **2010**, *9*, 233–245. [CrossRef]
6. Motil, A.; Bergman, A.; Tur, M. [INVITED] State of the art of Brillouin fiber-optic distributed sensing. *Opt. Laser Technol.* **2016**, *78*, 81–103. [CrossRef]
7. Meng, L.; Wang, L.; Hou, Y.; Yan, G. A Research on Low Modulus Distributed Fiber Optical Sensor for Pavement Material Strain Monitoring. *Sensors* **2017**, *17*, 2386. [CrossRef] [PubMed]
8. Klar, A.; Levenberg, E.; Tur, M.; Zadok, A. Sensing for smart infrastructure: Prospective engineering applications. In *Transforming the Future of Infrastructure Through Smarter Information, Proceedings of the International Conference on Smart Infrastructure and Construction (ICSIC 2016), Cambridge, UK, 27–29 June 2016*; Institution of Civil Engineers: London, UK, 2016; pp. 289–295. [CrossRef]
9. Barrias, A.; Casas, J.R.; Villalba, S. A review of distributed optical fiber sensors for civil engineering applications. *Sensors* **2016**, *16*, 748. [CrossRef] [PubMed]
10. Soga, K.; Luo, L. Distributed fiber optics sensors for civil engineering infrastructure sensing. *J. Struct. Integr. Maint.* **2019**, *3*. [CrossRef]
11. Klar, A.; Bennett, P.J.; Soga, K.; Mair, R.J.; Tester, P.; Fernie, R.; St John, H.D.; Torp-Petersors, G. Distributed strain measurement for pile foundations. *Proc. Inst. Civ. Eng. Geotech. Eng.* **2006**, *159*, 135–144. [CrossRef]
12. Vorster, T.E.; Soga, K.; Mair, R.J.; Bennett, P.J.; Klar, A.; Choy, C.K. The use of fibre optic sensors to monitor pipeline response to tunnelling. *Geocongress 2006 Geotech. Eng. Inf. Technol. Age* **2006**, *2006*, 33. [CrossRef]
13. Mohamad, H.; Bennett, P.J.; Soga, K.; Klar, A.; Pellow, A. Distributed optical fiber strain sensing in a secant piled wall. *Geotech. Spec. Publ.* **2007**, 81. [CrossRef]
14. Iten, M.; Puzrin, A.M. BOTDA road-embedded strain sensing system for landslide boundary localization. *Proc. SPIE Int. Soc. Opt. Eng.* **2009**, *7293*, 729316. [CrossRef]
15. Mohamad, H.; Bennett, P.J.; Soga, K.; Mair, R.J.; Bowers, K. Behaviour of an old masonry tunnel due to tunnelling-induced ground settlement. *Geotechnique* **2010**, *60*, 927–938. [CrossRef]
16. Goldfeld, Y.; Klar, A. Damage identification in reinforced concrete beams using spatially distributed strain measurements. *J. Struct. Eng.* **2013**, *139*, 04013013. [CrossRef]
17. Glisic, B.; Inaudi, D. Development of method for in-service crack detection based on distributed fiber optic sensors. *Struct. Health Monit.* **2012**, *11*, 161–171. [CrossRef]
18. Regier, R.; Hoult, N.A. Distributed strain behavior of a reinforced concrete bridge: Case study. *J. Bridge Eng.* **2014**, *19*, 05014007. [CrossRef]
19. Xiang, P.; Wang, H.P. Optical fibre-based sensors for distributed strain monitoring of asphalt pavements. *Int. J. Pavement Eng.* **2016**, *19*, 842–850. [CrossRef]

20. Chapeleau, X.; Blanc, J.; Hornych, P.; Gautier, J.L.; Carroget, J. Assessment of cracks detection in pavement by a distributed fiber optic sensing technology. *J. Civ. Struct. Health Monit.* **2017**, *7*, 459–470. [CrossRef]

21. Klar, A.; Dromy, I.; Linker, R. Monitoring tunneling induced ground displacements using distributed fiber-optic sensing. *Tunn. Undergr. Space Technol.* **2014**, *40*, 141–150. [CrossRef]

22. Hauswirth, D.; Puzrin, A.M.; Carrera, A.; Standing, J.R.; Wan, M.S. Use of fibre-optic sensors for simple assessment of ground surface displacements during tunnelling. *Geotechnique* **2014**, *64*, 837–842. [CrossRef]

23. Lu, Y.; Zhu, T.; Chen, L.; Bao, X. Distributed Vibration Sensor Based on Coherent Detection of Phase-OTDR. *J. Light. Technol.* **2010**, *28*, 3243–3249. [CrossRef]

24. Zadok, A.; Antman, Y.; Primerov, N.; Denisov, A.; Sancho, J.; Thevenaz, L. Random-access distributed fiber sensing. *Laser Photonics Rev.* **2012**, *6*, L1–L5. [CrossRef]

25. Cohen, R.; London, Y.; Antman, Y.; Zadok, A. Brillouin optical correlation domain analysis with 4 millimeter resolution based on amplified spontaneous emission. *Opt. Express* **2014**, *22*, 12070–12078. [CrossRef]

26. Monsberger, C.; Woschitz, H.; Hayden, M. Deformation Measurement of a Driven Pile Using Distributed Fibre-optic Sensing. *J. Appl. Geod.* **2016**, *10*, 61–69. [CrossRef]

27. Stern, Y.; London, Y.; Preter, E.; Antman, Y.; Diamandi, H.H.; Silbiger, M.; Adler, G.; Levenberg, E.; Shalev, D.; Zadok, A. Brillouin Optical Correlation Domain Analysis in Composite Material Beams. *Sensors* **2017**, *17*, 2266. [CrossRef] [PubMed]

28. Schenato, L.; Palmieri, L.; Camporese, M.; Bersan, S.; Cola, S.; Pasuto, A.; Galtarossa, A.; Salandin, P.; Simonini, P. Distributed optical fibre sensing for early detection of shallow landslides triggering. *Sci. Rep.* **2017**, *7*, 14686. [CrossRef] [PubMed]

29. Zadok, A.; Preter, E.; London, Y. Phase-Coded and Noise-Based Brillouin Optical Correlation-Domain Analysis. *Appl. Sci.* **2018**, *8*, 1482. [CrossRef]

30. Monsberger, C.; Lienhart, W.; Hirschmüller, S.; Marte, R. Monitoring of soil nailed slope stabilizations using distributed fiber optic sensing. *Proc. SPIE Int. Soc. Opt. Eng.* **2018**, *10598*, 1059835. [CrossRef]

31. Uchida, S.; Levenberg, E.; Klar, A. On-specimen strain measurement with fiber optic distributed sensing. *Meas. J. Int. Meas. Confed.* **2015**, *60*, 104–113. [CrossRef]

32. Westergaard, H.M. New Formulas for Stress in Concrete Pavements of Airfields. *Am. Soc. Civ. Eng. Trans.* **1948**, *113*, 425–439.

33. Ioannides, A.; Thompson, M.R.; Barenberg, E.J. Westergaard solutions reconsidered. *Transp. Res. Rec.* **1985**, *1043*, 13–23.

34. Khazanovich, L.; Ioannides, A.M. Finite element analysis of slabs-on-grade using higher order subgrade soil models. In Proceedings of the Conference on Airport Pavement Innovations, Vicksburg, MS, USA, 8–10 September 1993.

35. Khazanovich, L. Finite element analysis of curling of slabs on Pasternak foundation. In Proceedings of the 16th ASCE Engineering Mechanics Conference, Seattle, WA USA, 16–18 July 2003; pp. 16–18.

36. Ioannides, A.M. Concrete pavement analysis: The first eighty years. *Int. J. Pavement Eng.* **2006**, *7*, 233–249. [CrossRef]

37. Winkler, E. *Die Lehre von der Elasticitaet und Festigkeit*; Verlag Dominicus: Prag, Czech Republic, 1868.

38. Colasanti, R.J.; Horvath, J.S. Practical subgrade model for improved soil-structure interaction analysis: Software implementation. *Pract. Period. Struct. Des. Constr.* **2010**, *15*, 278–286. [CrossRef]

39. Boussinesq, J. *Application des Potentiels a L'équilibre et du Mouvement des Solides Élastiques...*; Gauthier-Villars: Paris, France, 1885; Volume 4.

40. Filonenko-Borodich, M. Some approximate theories of the elastic foundation. *Uchenyie Zap. Mosk. Gos. Univ. Mekhanica* **1940**, *46*, 3–18.

41. Pasternak, P. *On a New Method of Analysis of an Elastic Foundation by Means of Two Constants [Gosudarstvennoe Izdatelstvo Literaturi po Stroitelstvu I Arkhitekture]*; USSR: Moscow, Russia, 1954.

42. Kerr, A.D. A study of a new foundation model. *Acta Mech.* **1965**, *1*, 135–147. [CrossRef]

43. Reissner, E. A note on deflections of plates on a viscoelastic foundation. *J. Appl. Mech.* **1958**, *25*, 144–145.

44. Vlasov, V.Z.; Leont'ev, N. *Beams, Plates, and Shells on an Elastic Base*; Fizmatgiz: Moscow, Russia, 1960.

45. Biot, M.A. Bending of an infinite beam on an elastic foundation. *J. Appl. Math. Mech.* **1922**, *2*, 165–184.

46. Klar, A.; Vorster, T.; Soga, K.; Mair, R. Soil-pipe interaction due to tunnelling: Comparison between Winkler and elastic continuum solutions. *Geotechnique* **2005**, *55*, 461–466. [CrossRef]

47. Yu, J.; Zhang, C.; Huang, M. Soil-pipe interaction due to tunnelling: Assessment of Winkler modulus for underground pipelines. *Comput. Geotech.* **2013**, *50*, 17–28. [CrossRef]
48. Daloglu, A.T.; Vallabhan, C.V.G. Values of k for slab on Winkler foundation. *J. Geotech. Geoenviron. Eng.* **2000**, *126*, 463–471.:5(463). [CrossRef]
49. Terzaghi, K. Evaluation of coefficients of subgrade reaction. *Geotechnique* **1955**, *5*, 297–326. [CrossRef]
50. Vesic, A. Bending of beams resting on isotropic elastic solid. *J. Eng. Mech. Div.* **1961**, *87*, 35–54.
51. Vallabhan, C.G.; Das, Y. Parametric study of beams on elastic foundations. *J. Eng. Mech.* **1987**, *114*, 2072–2082. [CrossRef]
52. Vallabhan, C.V.G.; Das, Y.C. Modified Vlasov model for beams on elastic foundations. *J. Geotech. Eng.* **1991**, *117*, 956–966. [CrossRef]
53. Fwa, T.; Shi, X.; Tan, S. Use of Pasternak foundation model in concrete pavement analysis. *J. Transp. Eng. ASCE* **1996**, *122*, 323–328. [CrossRef]
54. Skar, A.; Poulsen, P.; Olesen, J. Cohesive cracked-hinge model for simulation of fracture in one-way slabs on grade. *Int. J. Pavement Eng.* **2017**. [CrossRef]
55. Loof, H. The theory of the coupled spring foundation as applied to the investigation of structures supported on soil. *Heron* 1965, *3*, 29–49.
56. Setiadji, B.H.; Fwa, T.F. Examining k-E relationship of pavement subgrade based on load-deflection consideration. *J. Transp. Eng.* **2009**, *135*, 140–148.:3(140). [CrossRef]
57. *ASTM D1196/D1196M-12(2016). Standard Test Method for Nonrepetitive Static Plate Load Tests of Soils and Flexible Pavement Components, for Use in Evaluation and Design of Airport and Highway Pavements*; ASTM International: West Conshohocken, PA, USA, 2016. [CrossRef]
58. Timoshenko, S.P.; Woinowsky-Krieger, S. *Theory of Plates and Shells*; McGraw-Hill Book Company, Inc.: New York, NY, USA, 1959.
59. Van Cauwelaert, F.; Stet, M.; Jasienski, A. The General Solution for a Slab Subjected to Centre and Edge Loads and Resting on a Kerr Foundation. *Int. J. Pavement Eng.* **2002**, *3*, 1–18. [CrossRef]
60. Westergaard, H. Stresses in concrete pavements computed by theoretical analysis. *Public Roads* **1926**, *7*, 25–35.
61. Hogg, A.H.A. Equilibrium of a thin plate, symmetrically loaded, resting on an elastic foundation of infinite depth. *Lond. Edinb. Dublin Philos. Mag. J. Sci.* **1938**, *25*, 576–582. [CrossRef]
62. Levenberg, E.; Klar, A.; Skar, A. Foundation Characterization for Slab-On-Grade Constructions with Fiber-Optic Distributed Strain-Sensing. In Proceedings of the ASCE Geo-Congress 2020. Submitted.
63. Luna OBR 4600 Fiber Optic Sensing Device. Available online: https://lunainc.com/product/sensing-solutions/obr-4600/ (accessed on 27 May 2019).
64. Regupol® Rubber Material Specification. Available online: http://vibratec.se/products/regupol-vibration-1000/ (accessed on 27 May 2019).
65. Jackopor® Polystyrene Material Specification. Available online: https://www.jackon.no/assets/FileUploads/Teknisk-tabell-Jackon-EPS-05-2017.pdf (accessed on 27 May 2019).

Article

Evaluation of Horizontal Stresses in Soil during Direct Simple Shear by High-Resolution Distributed Fiber Optic Sensing

Assaf Klar [1,2,*,†], Michael Roed [1], Irene Rocchi [1] and Ieva Paegle [1]

[1] Department of Civil Engineering, Technichal University of Denmark, 2800 Kgs. Lyngby, Denmark
[2] Faculty of Civil and Environmental Engineering, Technion-Israel Institute of Technology, Haifa 32000, Israel
*   Correspondence: askla@byg.dtu.dk or klar@technion.ac.il
†   On Leave from the Technion-Israel Institute of Technology.

Received: 20 June 2019; Accepted: 16 August 2019; Published: 24 August 2019

Abstract: This paper presents an approach for evaluating the horizontal stresses that develop in geotechnical Direct Simple Shear (DSS) tests through the use of high-resolution distributed fiber optic sensing. For this aim, fiber optics were embedded in 3D printed rings used for confining the soil in the test procedure. An analytical approach linking the measured spatially-distributed strain profile and the internal soil-ring contact stresses is developed in the paper. The method is based on representation of the contact stresses by a Fourier series expansion, and determining the coefficients of the series by minimizing the difference between the measured strain and the analytical strain within the linear elastic ring. The minimization problem results in a linear set of equations that can easily be solved for a given measurement. The approach is demonstrated on a set of drained DSS tests on clean sand specimens. Stress paths using the evaluated horizontal stresses are plotted together with Mohr circles at failure. These illustrate how, in these specific tests, the horizontal stress increases and principal stress direction rotates, until failure occurs along horizontal planes.

Keywords: distributed diber-optic sensing; direct simple shear; geotechnical engineering; contact problems; soil properties

---

1. Introduction

Distributed fiber optic sensing has been recognized as one of the potential technologies that could revolutionize structural and geotechnical monitoring and testing (e.g., [1–3]). Most of the research in the area has focused on large-scale problems, such as pile foundations [4–7], pipeline stressing [8–11], stressing and deformation of secant pile walls [12,13], landslides and sinkholes [14–17], tunneling stressing [18–20], strain measurement and detection of cracks in asphalt pavements [21–24], and evaluation of tunneling-induced ground deformations [25–27]. These applications are based on static, and low (~1 m) sensing resolution limiting themselves to large-scale problems.

New sensing developments, however, allow for higher spatial resolution of a few millimeters to centimeters [28–32]. These developments open the window to new possibilities of smaller-scale applications, such as on-sample measurements [33], high-resolution profiling [34], and key-point characterization methods [35].

This paper suggests an additional new application for high-resolution fiber optic sensing, aiming to enrich the information obtained from a Direct Simple Shear (DSS) test. The DSS test is a common soil laboratory test used to evaluate the geotechnical properties of geomaterials. In the test, a soil sample is sheared in a simple shear mode, such that parallel planes remain parallel and maintain a constant distance, while translating relative to each other (as demonstrated in Figure 1). Guidelines for performing the tests can be found in ASTM D6528-18 [36]. While it is traditionally assumed that failure and the associated deformation occurs due to sliding of horizontal planes, other modes of shearing can lead to kinematically compatible deformations, as illustrated in Figure 1.

A DSS test can be performed either under fixed vertical stress (allowing the soil to compact or dilate), or under a fixed vertical strain (in which case the vertical stress is measured and may decrease or increase). Overall, four parameters are controlled or monitored in the test procedure, namely: the shear strain γ, the shear stress τ, the vertical stress σ_v, and the vertical strain ϵ_v (which supposedly constitutes the volumetric strain in this specific mode of shearing). Two types of DSS devices exist—the first being the complex, non-commercial, Cambridge cuboidal simple shear apparatus [37], and the other the Geonor DSS device [38], which uses a cylindrical specimen confined by a wire-reinforced membrane. As the complete stress condition is not known due to the rotation of principal stresses and strains (not even sufficient to construct a Mohr circle of stresses), measurements need to be performed to evaluate the value of the horizontal stresses at the time of the shearing.

This is achieved by load cells [39] in the Cambridge device, while a resistance wire has been sometimes applied [40] to the Geonor device to evaluate the average radial stress using a Wheatstone bridge. However, typical DSS devices that are available commercially do not have these capabilities. This hinders the interpretation of the test results, as certain assumptions need to be made regarding the stress state at the time of failure.

Over the years, various models have been suggested to analyze the test results based on a mix of analytical and empirical relations. For example, semi-empirical expressions for calculating the horizontal stress, based on the direct proportionality between the inclination of the principal stress axes and the shear stress ratio, were developed by [39,41]. The assumption of collinearity between stresses and strains was used by [42] to develop an analytical incremental expression that allows evaluation of the horizontal stress in DSS devices. Coupled 3D computer simulations of DSS with the measured test response were used by [43] to calibrate the constitutive behavior.

In general, shear tests are used to establish soil stiffness, strength, and shear-volumetric relations (e.g., dilation). These may be a function of the stress level, stress ratios, and strains, which generally constitute the variables of a constitutive model for soil representation. Out of six stress components, only four components are known in typical DSS devices (two measured and two being zero, due to the symmetry of the device and loading). Evaluation of the horizontal stresses can enrich the knowledge to six stress components, and thus allow better interpretation of the results. This may include the evaluation of the confining stress and its effect on the response, the specific effect of the intermediate stress, as well as stress path analysis and its relation to the mechanism of failure, all of which are currently missing from standard DSS devices. Therefore, it is clear that measurements of the horizontal stresses within the sample could contribute to the interpretation of the test results, regardless of the exact test conditions (i.e., drained, undrained, monotonic, or cyclic) or type of soil (i.e., sands or clays).

This paper suggests a new approach for evaluating the horizontal stresses in the sample, and presents the first steps of its development. The approach utilizes recent advances in distributed fiber optic sensing technologies that allow for high accuracy and resolution sensing, together with an approach for determination of the contact stresses. More specifically, the approach links between spatially continuous profiles of the developed strain in the apparatus confining rings with an analytical solution of the rings' response to contact stresses. The contact stress is resolved by an optimization problem for which the solution can be obtained by a set of linear equations. The paper presents both the analytical development, as well as a demonstration of the approach on the DSS apparatus.

The paper is composed of five main sections. Firstly, the approach of evaluating the contact stresses by an optimization problem of the measured high resolution distributed strain profile linked to the analytical solution of the elastic response of the rings is presented. Secondly, the upgraded apparatus, including fiber-embedded 3D printed rings, is described. Thirdly, test results of the direct simple shear on clean sand are presented. Fourthly, analysis of the measured strain profile and interpretation of the results are presented. Finally, discussion and conclusions are provided.

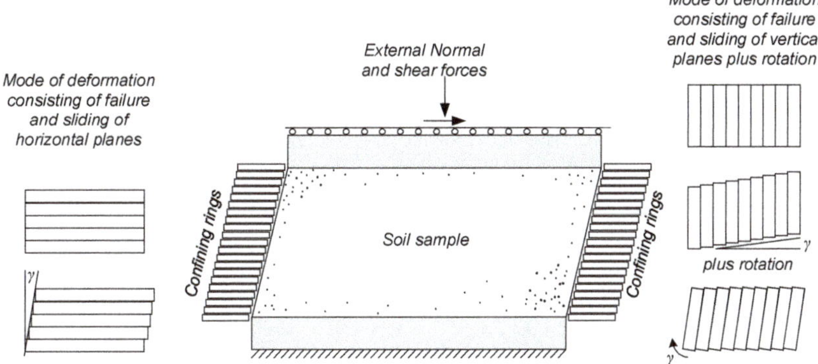

Figure 1. Illustration of the direct simple shear test and possible kinematically compatible modes of deformation (following [44,45]).

2. Fourier-Based Optimization for Evaluating Contact Stresses

It is assumed that the contact between the membrane (enclosing the soil) and the confining rings in the DSS apparatus is smooth, and that the membrane is sufficiently flexible not to confine the soil by itself, allowing for the contact stress on the ring to be equal to the stresses in the soil. Under this condition, the horizontal traction between the soil and the confining rings is made of normal stresses.

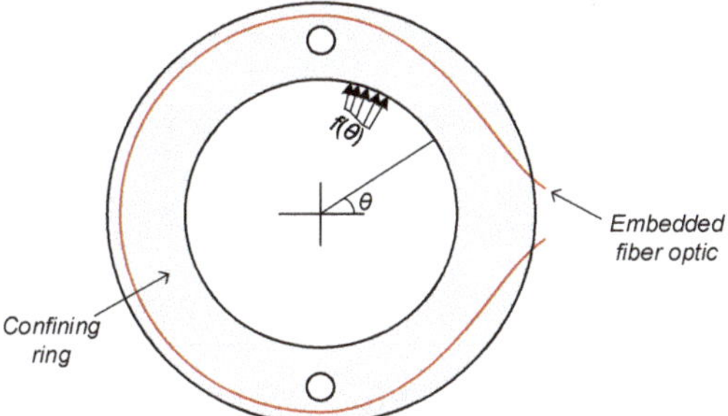

Figure 2. Illustration of a confining ring involved in a direct simple shear device.

Let us focus on the behavior of a given ring within the direct simple shear apparatus, and represent the normal stress acting on the ring by a Fourier series:

$$f(\theta) = a_0 + \sum_{i=2}^{N} a_i \cos(i\theta) \tag{1}$$

where $f(\theta)$ is the normal stress acting at an angle θ, as illustrated in Figure 2 which represents the plain view of a single ring. a_i are the amplitudes of the sinusoids composing the Fourier series and N is the number of terms in the Fourier series. Note that the mode of $i = 1$ is dropped from the series on purpose. This is because it is the only mode that does not satisfy a horizontal equilibrium. All other modes satisfy a horizontal equilibrium, and because there is no external support to the rings (assuming the friction between the rings is negligible), then the first mode must be, by definition, zero (i.e., $\int_0^{2\pi} a_1 \cos(\theta) \cos(\theta) = 0$ and therefore, $a_1 = 0$). In other words, the above Fourier series satisfies global horizontal equilibrium, irrespective of the coefficients' value. It is also assumed that a symmetry exists relative to $\theta = 0$, considering that the horizontal translation and shearing occur in this direction (otherwise, the Fourier series expansion should be enriched by sine functions).

Considering a linear elastic response of the ring, a function can be established to represent the longitudinal strain response along the fiber optic for a given unit mode (i.e., for a given frequency when the amplitude is unit). Such a representation is feasible due to the superposition principle applicable to linear elastic materials. Let us refer to this function as $\Gamma_i(x)$, such that the strain along the monitored fiber, stretched from $x = 0$ to $x = L_f$, is:

$$\epsilon(x) = a_0\Gamma_0(x) + \sum_{i=2}^{N} a_i\Gamma_i(x) \tag{2}$$

The Γ functions can be derived analytically if a closed form solution exists, or numerically for a general structure. In the current work, $\Gamma_i(x)$ was established by repeated linear elastic finite element simulations, each of which for a different unit mode of the internal normal stress distribution.

For each case, the $\Gamma_i(x)$ was established by extracting the longitudinal strain value along the fiber arc. This can be preformed either by investigating the local stretching of the arc using the displacement field, or by strain transformation (i.e., $\mathbf{n}^T\mathbf{En}$, where $\mathbf{E}$ is the strain tensor presented in a matrix form, and $\mathbf{n}$ is a unit vector along the arc). The linear elastic finite element simulations were preformed using the mechanical model in COMSOL Multyphysics [46], taking advantage of its parametric sweep option to automate the process of creating $\Gamma_i(x)$ for the various spatial frequencies of the Fourier series. The ability of COMSOL to incorporate functional input to define loads for the boundary conditions facilitated the process of creating $\Gamma_i(x)$. The simulations were performed under small displacements and small strain definitions, ignoring any geometrical nonlinearity aspect.

Figure 3 shows the first five modes of $\Gamma_i(x)$ for the location of the fiber plotted in Figure 2, as established by finite element simulations of the 3D printed rings used in the experiments. The embedded fiber length in the ring, L_f, is equal to 281 mm. The $\Gamma_i(x)$ values in the figure are normalized by Young's modulus of the ring E, assuming a Poisson's ratio of 0.36, associated with the PLA material from which the rings were fabricated using a 3D printer. The Young's modulus of the PLA is $E = 3150$ MPa. It can be seen from the figure that the strain response is highly affected by the mode (or spatial frequency), where Γ_0 roughly results in a constant value, and Γ_5 oscillates five times. The effect of the small holes in the ring is also apparent in all modes, at about $x = 0.23L_f$ and $x = 0.77L_f$.

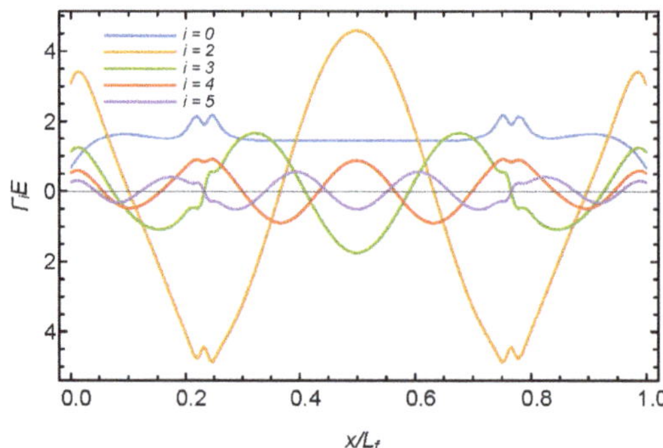

Figure 3. First five modes of Γ functions presented in a normalized manner.

In order to evaluate the profile of the contact stresses in an experiment, one may ask what are the best Fourier series coefficients that will lead to an agreement between the measured strain and the analytical strain. This can be formulated mathematically as:

$$a_i^* = \underset{a_i}{\mathrm{argmin}} \int_0^{L_f} (\epsilon(x) - \epsilon_m(x))^2 \, dx \tag{3}$$

where $\epsilon(x)$ is the analytical strain profile and $\epsilon_m(x)$ is the measured strain profile. Replacing $\epsilon(x)$ with the Fourier series (i.e., Equation (2)) and replacing the summation with a dot product of relevant vectors results in:

$$a_i^* = \underset{a_i}{\mathrm{argmin}} \int_0^{L_f} \left(\mathbf{a}^T\mathbf{b} - \epsilon_m(x) \right)^2 dx = \int_0^{L_f} \mathbf{a}^T\mathbf{b}\mathbf{b}^T\mathbf{a} - 2\mathbf{a}^T\mathbf{b}\epsilon_m(x) + \epsilon_m^2(x)dx \tag{4}$$

where $\mathbf{a} = \{a_0, a_2, ..., a_N\}^T$ and $\mathbf{b} = \{\Gamma_0(x), \Gamma_2(x), ..., \Gamma_N(x)\}^T$. The expression can be further developed to be written in a matrix form:

$$a_i^* = \underset{a_i}{\mathrm{argmin}} \left(\mathbf{a}^T\mathbf{B}\mathbf{a} - 2\mathbf{a}^T\mathbf{b_1} + C \right) \tag{5}$$

where

$$\mathbf{B} = \begin{bmatrix} \int_0^{L_f} \Gamma_0(x)\Gamma_0(x)dx & \int_0^{L_f} \Gamma_0(x)\Gamma_2(x)dx & \int_0^{L_f} \Gamma_0(x)\Gamma_3(x)dx & \cdots & \int_0^{L_f} \Gamma_0(x)\Gamma_N(x)dx \\ \int_0^{L_f} \Gamma_2(x)\Gamma_0(x)dx & \int_0^{L_f} \Gamma_2(x)\Gamma_2(x)dx & \int_0^{L_f} \Gamma_2(x)\Gamma_3(x)dx & \cdots & \int_0^{L_f} \Gamma_2(x)\Gamma_N(x)dx \\ \vdots & \vdots & \vdots & \ddots & \vdots \\ \int_0^{L_f} \Gamma_N(x)\Gamma_0(x)dx & \int_0^{L_f} \Gamma_N(x)\Gamma_2(x)dx & \int_0^{L_f} \Gamma_N(x)\Gamma_3(x)dx & \cdots & \int_0^{L_f} \Gamma_N(x)\Gamma_N(x)dx \end{bmatrix} \tag{6}$$

and $\mathbf{b_1} = \{\int_0^{L_f} \Gamma_0(x)\epsilon_m(x)dx, \int_0^{L_f} \Gamma_2(x)\epsilon_m(x)dx, ..., \int_0^{L_f} \Gamma_N(x)\epsilon_m(x)dx\}^T$. C is a constant independent of a_i (and equal to $\int_0^{L_f} \epsilon_m^2(x)dx$). Equation (5) is a classical quadratic objective function for which the optimal a_i (gathered in vector $\mathbf{a}$) are established through a linear solution, as follows:

$$\mathbf{a} = \mathbf{B}^{-1}\mathbf{b_1} \tag{7}$$

Note that matrix $\mathbf{B}$ characterizes the fiber-embedded ring (and does not change from one experiment to another), while vector $\mathbf{b_1}$ includes information from the test itself, embodied in a $\epsilon_m(x)$ profile which is integrated with the various $\Gamma_i(x)$ functions. The above formulation was used in the demonstration experiments, as detailed in the following sections.

3. Experimental Setup

To demonstrate the approach, the DSS apparatus by the Geocomp Corporation [47] was modified to include fiber-embedded rings. This DSS device uses confining rings to support the soil while sheared, as opposed to an NGI DSS device [38] that uses a reinforced membrane to confine the soil. The inner diameter of the rings is 64.5 mm, and the external diameter is 101.5 mm. Figure 4 shows the original Teflon-coated aluminium rings next to the 3D PLA printed rings. The rings were printed with a thickness of 3 mm, except for the central one which had double the thickness, to provide higher bending resistance. Figure 5 shows a fiber optic embedded ring. A 0.9 mm single mode Corning G652D LSZH tight-buffer fiber optic cable was placed in the 3D printed groove, pre-strained, and glued with a cyanoacrylate adhesive.

The fibers were installed on the top and bottom of the ring. Earlier experiments involved a one-sided, fiber-embedded ring, but it was found that the measurements were sensitive to potential bending of the rings. Consequently, the fibers were installed on both sides of the rings, allowing for compensation of the bending response by summing the strain from both profiles. Other installation configurations are possible, but their investigation is beyond the scope of this paper, which merely presents and demonstrates the new concept.

Note that the solid model used for the 3D printing was also used for generating the $\Gamma_i(x)$ response functions with the finite element code COMSOL Multyphysics [46], as described in the previous section. For the interpretation of the results, the PLA properties were taken as $E = 3150$ MPa following the mechanical properties provided by the technical data sheet of the manufacturer (Ultimaker). The rings were printed under 100% filling configuration to create a true continuum structure.

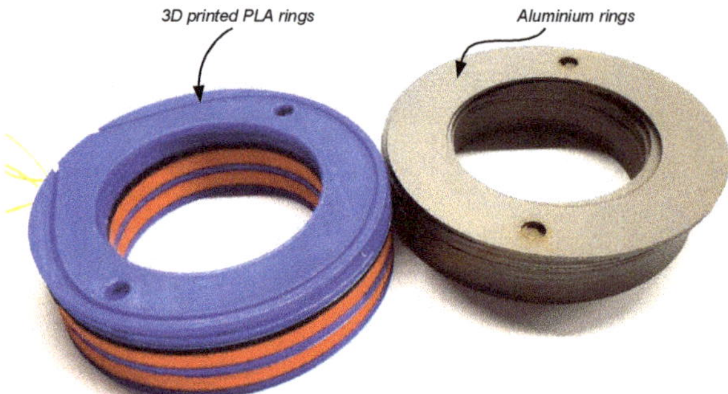

Figure 4. Printed PLA rings next to original Aluminium rings.

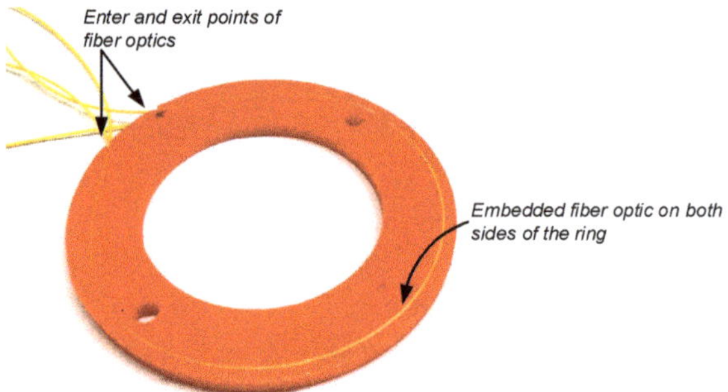

Figure 5. Fiber optic-embedded confining ring.

Figure 6 shows the integrated test setup. A single fiber-embedded ring was positioned in the middle of the shear sample. The high-resolution strain profiles were recorded with a commercial Optical Backscatter Reflectometer (OBR) device—LUNA 4600 [48]—utilizing the principle of Optical Frequency Domain Reflectometry (OFDR). In this technique, an interferometer is formed by a reference arm internal to the analyzer and the measured fiber optic. By sweeping through the optical frequency range, using a tunable laser, interference fringe data is collected and analyzed both in time and frequency domain to establish changes in length/strain in the fiber. In the current work, strain values were recorded at intervals of 1 mm with a gauge length of 10 mm (i.e., averaged strain over a centered spatial window of 10 mm). In addition to the fiber optic measurements, conventional readings, of horizontal translation, vertical deformation, vertical load, and horizontal forces, were recorded throughout the tests.

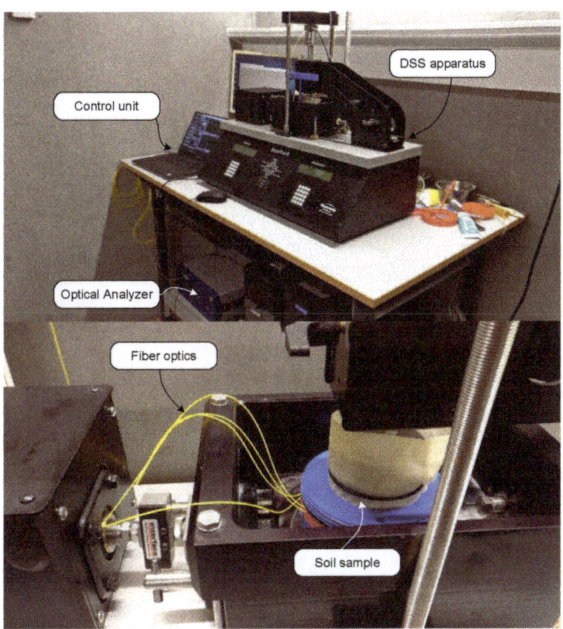

Figure 6. Integrated test system.

4. Direct Simple Shear Tests and Results

DSS tests were performed on dry, clean Fontainebleau sand, characterized by quartz particles in the range from 0.063 mm to 0.25 mm, and a uniformity index of $U < 2$ (minimum void ratio of $e = 0.549$ and maximum of $e = 0.853$). Six tests (named T1 to T6 in the paper) were performed on loose- to medium-dense specimens with a relative density (D_r) roughly ranging over 30% to 50%. The shear tests were performed on normally consolidated samples, loaded to a vertical stress of 485 kPa. Additional tests were performed with the original rings. No significant difference was observed in the overall recorded response of the sample, suggesting that the modifications made to the system did not significantly alter the soil response. Therefore, focus was placed only on the modified system.

Figure 7 shows conventional test results in terms of stress-strain curves and dilation response. As can be seen, the stress-strain response is rather similar among the tests, while the volumetric response varies, although all sample showed the same trends of initial contraction followed by dilation.

Figure 8 shows a typical reading of the developed fiber optic strain in the shearing process (specifically demonstrated for experiment T1). The fiber strain response is also affected by the bending behavior of the ring, and therefore, a post-processing stage was performed in which the fibers were superimposed to nullify (or at least reduce) the bending contribution before they were introduced to the stress evaluation algorithm (which does not include such modes of deformations). Strain changes also appeared outside the ring, although to a much smaller extent. These relate to the fact that the base of the carriage translates and forces movement upon the "free" fiber inside the box, as well as outside (leading to minute displacements and small strains outside the ring area). Note that no temperature compensation was considered, as the experiments were performed in a temperature controlled room.

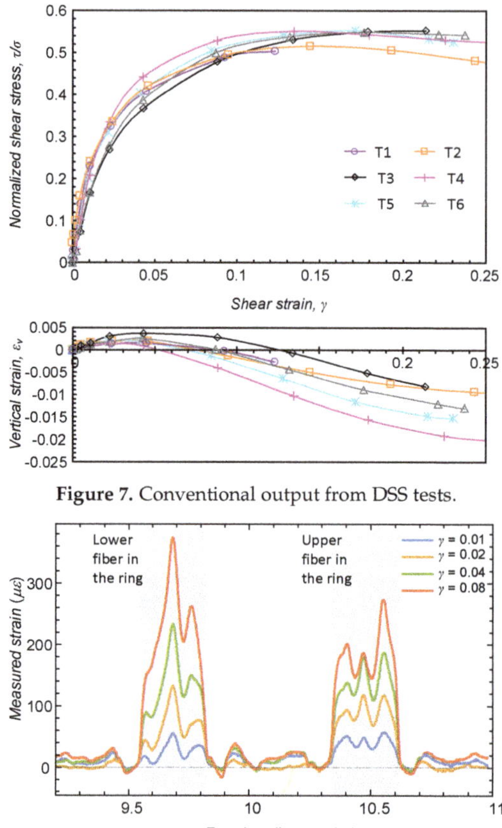

Figure 7. Conventional output from DSS tests.

Figure 8. Example of raw measurements along the optical fiber.

The strain data was accumulated for the six tests, with measurements being taken at translation intervals of about 0.1 mm until overall translation of 3 mm, and 0.5 mm intervals until translation of 8 mm. The analysis and interpretation of the results are provided in the following section.

5. Interpretation of the Test Results and Discussion

The interpretation here focuses on the stress condition at the time of shearing, assuming that a K_0 condition prevailed prior to shearing. A K_0 value of 0.54 was considered using Jaky's formula ($K_0 = 1 - \sin \phi'$) [49], based on the measured friction angle of the sand, leading to an initial horizontal stress of 264 kPa.

Figure 9 shows a graphical representation of the horizontal stress distribution applied to the soil sample, as analyzed by the Fourier-based optimization at different stages of testing, expressed by the shear strain, γ.

Figure 9. Contact stress distributions at different stages of the shear tests.

Specifically, values at γ = 0.01, 0.02, 0.04, and 0.08 were plotted based on analyses with 3 and 5 modes. The radial distance from the center circle represents the amplitude of the horizontal normal stress to the soil sample (equal to the stress applied to the confining ring). The K_0 (initial) condition is marked by a blue line, where the length of the light blue arrows represents its magnitude. A scale is given for reference, and the light red arrows extend to the largest calculated stress. Let us refer to a Cartesian coordinate system in which z represents the vertical direction, and x the direction of translation (y being the other horizontal direction). These are plotted in the figure.

As can be seen, with increasing shear strain, the horizontal stress increases in all directions. Overall, the magnitude of the developed stress profiles and their distribution are similar among the different tests. The magnitude of variation is similar to that observed in the applied shear stress in Figure 7. The analysis with five modes shows greater discrepancy between the tests, specifically in the higher spatial frequencies. This is because higher frequencies are more sensitive to measurement errors. While the five modes analysis indicates a "square" type of distribution, the stress values in the translation direction (σ_x) and the perpendicular one (σ_y) are rather similar to those based on the three modes analysis. Hereafter, all analyses are performed with the five modes data, although similar results and trends are observed from the three modes analyses.

Figure 10 shows the development of the mean σ_x, taken as $0.5(f(\theta = 0) + f(\theta = \pi))$. As can be seen, as the shearing process progresses, the horizontal stress increases, eventually becoming larger than the vertical stress (equal to 485 kPa) by a factor of almost two.

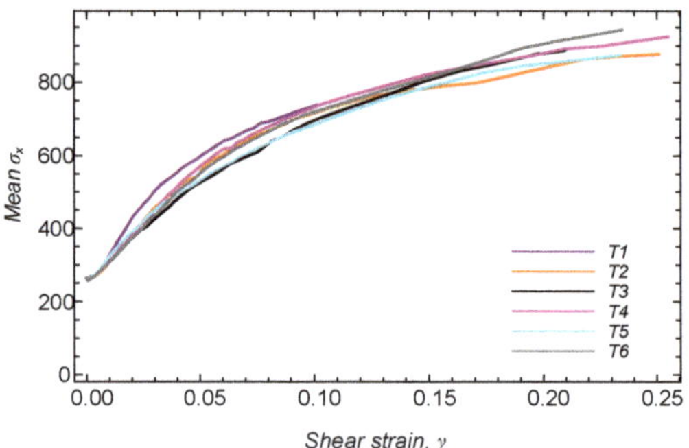

Figure 10. Mean σ_x throughout the tests, taken as $0.5(f(\theta = 0) + f(\theta = \pi))$.

Assuming that this mean σ_x represents the average normal stress applied to sheared vertical planes, a stress path can be plotted in terms of (σ_x, τ_{xz}). This stress path is unknown from conventional DSS testing, and is a result of the new suggested approach. It can supplement the conventional (σ_z, τ_{zx}) stress path to allow constructions of Mohr stress circles throughout the shearing process.

Figure 11a shows the various stress paths established from the analysis of all tests. The Mohr circles associated with the peak shear stress, τ_{zx}, are presented, as well as that of the initial K_0 condition (identical for all samples). The (σ_z, τ_{zx}) stress path is essentially identical for all samples, with a slight variation in the peak τ_{zx} value (as also demonstrated in stress-strain curves in Figure 7). The interpretation of the different samples is rather similar, and as an example, Figure 11b shows the interpretation for test *T3*, entailing the stress paths, initial and final Mohr circles, their poles, and a tangent Mohr-Coulomb cohesionless failure criterion.

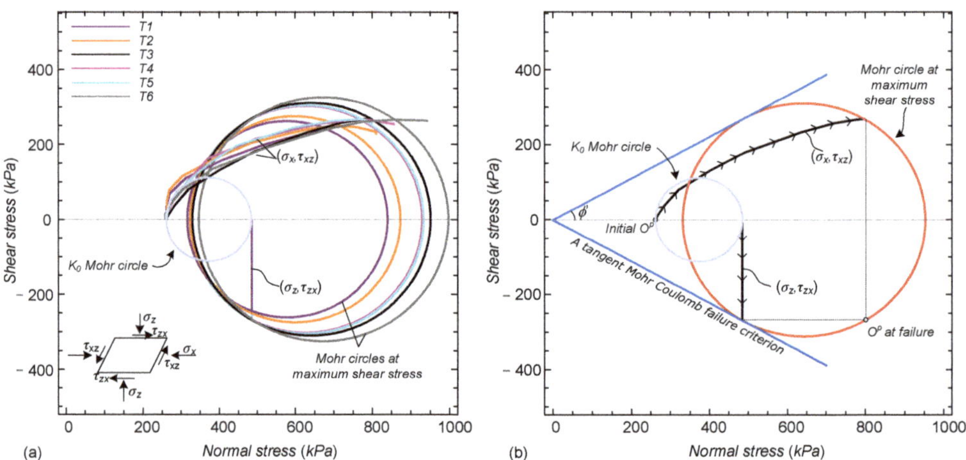

Figure 11. (**a**) (σ_z, τ_{zx}) and (σ_x, τ_{xz}) stress paths for all tests; (**b**) interpretation illustrated on T3.

It can clearly be seen that the (σ_z, τ_{zx}) stress path reaches the Mohr-Coulomb failure line at tangent point (although this condition was not forced). This indicates that failure occurred along horizontal planes. The principal stress directions rotated significantly throughout the shearing process of the sample, by about 60 degrees, leading the pole to change its position (marked in the figure as "initial O^p" and "O^p at failure").

Repeating the same process on the other samples resulted in the same trends. The obtained friction angles, by the tangent line to the Mohr circle of each test, ranged from 27 to 29 degrees among the test. These values are similar to those obtained with the conventional setup and slightly less than those measured on a range of sands in direct shear [50].

The fact that the (σ_z, τ_{zx}) stress path reaches the failure envelope at the same location as the tangent to the Mohr circle infers that a direct evaluation of the strength properties from a conventional analysis assuming $\tau_{zx}/\sigma_z = \tan\phi'$ is appropriate for these specific tests. However, this may not always be the case, especially in undrained (or zero volumetric strain) conditions, where failure occurs along vertical planes and the greatest shear stress experienced does not correspond to the maximum shear stress measured [44]. The ability to evaluate such a stress path development, as well as rotation in principal directions, may be important for analysis of undrained shearing and cyclic loading, where there might be interest in the change of confining mean stress due to compaction and dilation at shear direction reversal [51].

6. Conclusions

This paper presented a new approach for evaluating the horizontal stress condition in direct simple shear tests using high-resolution fiber optic sensing coupled to a Fourier based elastic solution of the confining rings. An optimization problem was formulated aiming to evaluate the best stress profile (in terms of Fourier coefficients) generating an analytical stress profile that would match the measured strain profile.

The approach was demonstrated with a commercial DSS apparatus, for which a new set of fiber optic-embedded confining rings was produced using 3D printing. Six drained direct simple shear tests were performed on clean sand to demonstrate the applicability of the approach. It was demonstrated how a stress path and Mohr circles can be evaluated with the new information derived from fiber optic sensing. Specifically, it was shown that in these particular (drained) tests, the soil truly fails along horizontal plane when considering a Mohr Coulomb yield condition.

The main focus of the paper was to conceptually develop a suitable approach for utilizing high-resolution fiber optic sensing for laboratory geotechnical testing, and to demonstrate its feasibility. Clearly, further research could be performed to improve the suggested approach. For example, studies could be performed to seek the best position of the optical fiber within the rings. Investigation of different ring materials or shape could be considered, as well as utilization of other high-resolution sensing techniques. These investigations, as well as investigations into more complex soil behavior, are beyond the scope of this paper, which simply attempts to take the first steps of advancing the use of high-resolution fiber optic sensing for laboratory geotechnical testing.

Author Contributions: Conceptualization, A.K.; Investigation, A.K., M.R., I.R. and I.P.; Methodology, A.K. and M.R.; Supervision, A.K. and I.R.; Writing—Original draft, A.K., M.R. and I.R.; Writing—Review & editing, A.K., M.R., I.R. and I.P.

Funding: This research received no external funding.

Conflicts of Interest: The authors declare no conflict of interest.

References

1.   Klar, A.; Levenberg, E.; Tur, M.; Zadok, A. Sensing for smart infrastructure: Prospective engineering applications. In Proceedings of the International Conference on Smart Infrastructure and Construction, Cambridge, UK, 27–29 June 2016; pp. 289–295, doi:10.1680/tfitsi.61279.289. [CrossRef]

2. Barrias, A.; Casas, J.R.; Villalba, S. A review of distributed optical fiber sensors for civil engineering applications. *Sensors* **2016**, *16*, 748. [CrossRef] [PubMed]

3. Soga, K.; Luo, L. Distributed fiber optics sensors for civil engineering infrastructure sensing. *J. Struct. Integr. Maint.* **2019**, *3*, 1–21, doi:10.1080/24705314.2018.1426138. [CrossRef]

4. Ohno, H.; Naruse, H.; Kurashima, T.; Nobiki, A.; Uchiyama, Y.; Kusakabe, Y. Application of Brillouin Scattering Based Distributed Optical Fiber Strain Sensor to Actual Con- crete Piles. *IEICE Trans. Elecrtron. E* **2002**, *85*, 945–951.

5. Klar, A.; Bennett, P.J.; Soga, K.; Mair, R.J.; Tester, P.; Fernie, R.; St John, H.D.; Torp-Petersors, G. Distributed strain measurement for pile foundations. *Proc. Inst. Civ. Eng. Geotech. Eng.* **2006**, *159*, 135–144, doi:10.1680/geng.2006.159.3.135. [CrossRef]

6. Mohamad, H.; Soga, K.; Amatya, B. Thermal Strain Sensing of Concrete Piles Using Brillouin Optical Time Domain Reflectometry. *Geotech. Test. J.* **2014**, *37*, 333–346, doi:10.1520/GTJ20120176. [CrossRef]

7. Gao, L.; Han, C.; Xu, Z.; Jin, Y.; Yan, J. Experimental Study on Deformation Monitoring of Bored Pile Based on BOTDR. *Appl. Sci.* **2019**, *9*, 2435, doi:10.3390/app9122435. [CrossRef]

8. Vorster, T.E.; Soga, K.; Mair, R.J.; Bennett, P.J.; Klar, A.; Choy, C.K. The use of fibre optic sensors to monitor pipeline response to tunnelling. In Proceedings of the GeoCongress 2006: Geotechnical Engineering in the Information Technology Age, Atlanta, GA, USA, 26 February–1 March 2006; p. 33. doi:10.1061/40803(187)33. [CrossRef]

9. Glisic, B.; Yao, Y. Fiber optic method for health assessment of pipelines subjected to earthquake-induced ground movement. *Struct. Health Monit.* **2012**, *11*, 696–711, doi:10.1177/1475921712455683. [CrossRef]

10. Simpson, B.; Hoult, N.A.; Moore, I.D. Distributed Sensing of Circumferential Strain Using Fiber Optics during Full-Scale Buried Pipe Experiments. *J. Pipeline Syst. Eng. Pract.* **2015**, *6*, 04015002. doi:10.1061/(ASCE)PS.1949-1204.0000197. [CrossRef]

11. Klar, A.A.; Linker, R.; Herrmann, S. Leakage-induced pipeline stressing and its potential detection by distributed fiber optic sensing. *Geotech. Eng.* **2019**, *50*, 85–90.

12. Mohamad, H.; Bennett, P.J.; Soga, K.; Klar, A.; Pellow, A. Distributed optical fiber strain sensing in a secant piled wall. In Proceedings of the Seventh International Symposium on Filed Measurements in Geomechanics, Boston, MA, USA, 24–27 September 2007; p. 81, doi:10.1061/40940(307)81. [CrossRef]

13. Mohamad, H.; Soga, K.; Pellew, A.; Bennett, P.J. Performance Monitoring of a Secant-Piled Wall Using Distributed Fiber Optic Strain Sensing. *J. Geotech. Geoenviron. Eng.* **2011**, *137*, 1236–1243. [CrossRef]

14. Iten, M.; Puzrin, A.M. BOTDA road-embedded strain sensing system for landslide boundary localization. In Proceedings of the Smart Sensor Phenomena, Technology, Networks, and Systems 2009, San Diego, CA, USA, 7 April 2009; Volume 7293, p. 729316, doi:10.1117/12.815266. [CrossRef]

15. Linker, R.; Klar, A. Detection of Sinkhole Formation by Strain Profile Measurements Using BOTDR: Simulation Study. *J. Eng. Mech.* **2017**, *143*, B4015002, doi:10.1061/(ASCE)EM.1943-7889.0000963. [CrossRef]

16. Xu, J.; He, J.; Zhang, L. Collapse prediction of karst sinkhole via distributed Brillouin optical fiber sensor. *Meas. J. Int. Meas. Confed.* **2017**, *100*, 68–71, doi:10.1016/j.measurement.2016.12.046. [CrossRef]

17. Schenato, L.; Palmieri, L.; Camporese, M.; Bersan, S.; Cola, S.; Pasuto, A.; Galtarossa, A.; Salandin, P.; Simonini, P. Distributed optical fibre sensing for early detection of shallow landslides triggering. *Sci. Rep.* **2017**, *7*, 1–7, doi:10.1038/s41598-017-12610-1. [CrossRef] [PubMed]

18. Mohamad, H.; Bennett, P.J.; Soga, K.; Mair, R.J.; Bowers, K. Behaviour of an old masonry tunnel due to tunnelling-induced ground settlement. *Géotechnique* **2010**, *60*, 927–938, doi:10.1680/geot.8.R074. [CrossRef]

19. Gue, C.; Wilcock, M.; Alhaddad, M.; Elshafie, M.; Soga, K.; Mair, R. Monitoring the effects of tunnelling under an existing tunnel-fibre optics. In *Geotechnical Aspects of Underground Construction in Soft Ground*; CRC Press: Boca Raton, FL, USA, 2014; pp. 357–361, doi:10.1201/b17240-65.

20. Minardo, A.; Catalano, E.; Coscetta, A.; Zeni, G.; Zhang, L.; Di Maio, C.; Vassallo, R.; Coviello, R.; Macchia, G.; Picarelli, L.; et al. Distributed Fiber Optic Sensors for the Monitoring of a Tunnel Crossing a Landslide. *Remote Sens.* **2018**, *10*, 1291, doi:10.3390/rs10081291. [CrossRef]

21. Artières, O.; Bacchi, M.; Bianchini, P.; Hornych, P.; Dortland, G. Strain measurement in pavements with a fibre optics sensor enabled geotextile. *Rilem Bookseries* **2012**, *4*, 201–210.

22. Xiang, P.; Wang, H.P. Optical fibre-based sensors for distributed strain monitoring of asphalt pavements. *Int. J. Pavement Eng.* **2016**, *19*, 1–9. [CrossRef]

23. Meng, L.; Wang, L.; Hou, Y.; Yan, G. A Research on Low Modulus Distributed Fiber Optical Sensor for Pavement Material Strain Monitoring. *Sensors* **2017**, 17, 2386. [CrossRef]

24. Chapeleau, X.; Blanc, J.; Hornych, P.; Gautier, J.L.; Carroget, J. Assessment of cracks detection in pavement by a distributed fiber optic sensing technology. *J. Civ. Struct. Health Monit.* **2017**, *7*, 459–470, doi:10.1007/s13349-017-0236-5. [CrossRef]

25. Klar, A.; Linker, R. Feasibility study of automated detection of tunnel excavation by Brillouin optical time domain reflectometry. *Tunn. Undergr. Space Technol.* **2010**, *25*, 575–586, doi:10.1016/j.tust.2010.04.003. [CrossRef]

26. Klar, A.; Dromy, I.; Linker, R. Monitoring tunneling induced ground displacements using distributed fiber-optic sensing. *Tunn. Undergr. Space Technol.* **2014**, *40*, 141–150, doi:10.1016/j.tust.2013.09.011. [CrossRef]

27. Hauswirth, D.; Puzrin, A.M.; Carrera, A.; Standing, J.R.; Wan, M.S. Use of fibre-optic sensors for simple assessment of ground surface displacements during tunnelling. *Geotechnique* **2014**, *64*, 837–842, doi:10.1680/geot.14.T.009. [CrossRef]

28. Lu, Y.; Zhu, T.; Chen, L.; Bao, X. Distributed Vibration Sensor Based on Coherent Detection of Phase-OTDR. *J. Lightwave Technol.* **2010**, *28*, 3243–3249, doi:10.1109/JLT.2010.2078798. [CrossRef]

29. Zadok, A.; Antman, Y.; Primerov, N.; Denisov, A.; Sancho, J.; Thevenaz, L. Random-access distributed fiber sensing. *Laser Photonics Rev.* **2012**, *6*, L1–L5. doi:10.1002/lpor.201200013. [CrossRef]

30. Cohen, R.; London, Y.; Antman, Y.; Zadok, A. Brillouin optical correlation domain analysis with 4 millimeter resolution based on amplified spontaneous emission. *Opt. Express* **2014**, *22*, 12070–12078, doi:10.1364/OE.22.012070. [CrossRef]

31. Stern, Y.; London, Y.; Preter, E.; Antman, Y.; Diamandi, H.H.; Silbiger, M.; Adler, G.; Levenberg, E.; Shalev, D.; Zadok, A. Brillouin Optical Correlation Domain Analysis in Composite Material Beams. *Sensors* **2017**, *17*, 2266, doi:10.3390/s17102266. [CrossRef]

32. Zadok, A.; Preter, E.; London, Y. Phase-Coded and Noise-Based Brillouin Optical Correlation-Domain Analysis. *Appl. Sci.* **2018**, *8*, 1482, doi:10.3390/app8091482. [CrossRef]

33. Uchida, S.; Levenberg, E.; Klar, A. On-specimen strain measurement with fiber optic distributed sensing. *Measurement* **2015**, *60*, 104–113, doi:10.1016/j.measurement.2014.09.054. [CrossRef]

34. Klar, A.; Uchida, S.; Levenberg, E. In Situ Profiling of Soil Stiffness Parameters Using High-Resolution Fiber-Optic Distributed Sensing. *J. Geotech. Geoenviron. Eng.* **2016**, *142*, 04016032.10.1061/(ASCE)GT.1943-5606.0001493. [CrossRef]

35. Skar, A.; Klar, A.; Levenberg, E. Load-Independent Characterization of Plate Foundation Support Using High-Resolution Distributed Fiber-Optic Sensing. *Sensors* **2019**, *19*, 3518, doi.org/10.3390/s19163518 [CrossRef]

36. ASTM. *Standard Test Method for Consolidated Undrained Direct Simple Shear Testing of Cohesive Soils*; ASTM: West Conshohocken, PA, USA, 2007; pp. 1–9, doi:10.1520/D6528-17. [CrossRef]

37. Roscoe, K. An Apparatus for the Application of Simple Shear to Soil Samples. In Proceedings of the 3rd International Conference on Soil Mechanics and Foundation Engineering, Switzerland, 1–27 August 1953; Volume 1, pp. 181–191.

38. Bjerrum, L.; Landva, A. Direct Simple-Shear Tests on a Norwegian Quick Clay. *Geotechnique* **1966**, *16*, 1–20, doi:10.1680/geot.1966.16.1.1. [CrossRef]

39. Wood, D.; Drescher, A.; Budhu, M. On the determination of the stress state in the simple shear apparatus. *Geotech. Test. J.* **1979**, *2*, 211–222.

40. Airey, D.W.; Wood, D.M. An evaluation of direct simple shear tests on clay. *Géotechnique* **1987**, *37*, 25–35, doi:10.1680/geot.1987.37.1.25. [CrossRef]

41. Roscoe, K.H.; Bassett, R.H.; Cole, E.R.L. Principal Axes Observed During Simple Shear of a Sand. In Proceedings of the Geotechnical Conference on Shear Strength Properties of Natural Soils and Rocks, Oslo, Norway, 19–22 September 1967; pp. 213–237.

42. Frydman, S.; Talesnick, M. Simple shear of isotropic elasto–plastic soil. *Int. J. Numer. Anal. Methods Geomech.* **1991**, *15*, 251–270, doi:10.1002/nag.1610150404. [CrossRef]

43. Woo Moon, S.; Hashash, Y.M.A. From Direct Simple Shear Test to Soil Model Development and Supported Excavation Simulation: Integrated Computational-Experimental Soil Behavior Characterization Framework. *J. Geotech. Geoenviron. Eng.* **2015**, *141*, 04015050. [CrossRef]

44. Wroth, C.P. The interpretation of in situ soil tests. *Géotechnique* **1984**, *34*, 449–489.10.1680/geot.1986.36.4.605. [CrossRef]

45. De Josselin de Jong, G. Discussion to session II. In Proceedings of the Roscoe Memorial Symposium Stress-Strain Behaviour of Soils, Cambridge, UK, 29–31 March 1971; Parry, R., Ed.; pp. 258–261.

46. *COMSOL Multiphysics Reference Manual, Version 5.3*; 2015.

47. Geocomp. Corporation. *Direct Simple Shear Software, Sheartrac-ii DSS*; Geocomp. Corporation: New York, NY, USA, 2015.

48. Luna OBR 4600 Fiber Optic Sensing Device. Available online: https://lunainc.com/product/sensing-solutions/obr-4600/ (accessed on 27 May 2019).

49. Jaky, J. Pressure in silos. In Proceedings of the 2nd International Conference on Soil Mechanics and Foundation Engineering, Rotterdam, The Netherlands, 21–30 June 1948.

50. Takahashi, A.; Jardine, R. Assessment of standard research sand for laboratory testing. *Q. J. Eng. Geol. Hydrogeol.* **2007**, *40*, 93–103, doi:10.1144/1470-9236/06-017. [CrossRef]

51. Budhu, M. Nonuniformities imposed by simple shear apparatus. *Can. Geotech. J.* **1984**, *21*, 125–137, doi:10.1139/t84-010. [CrossRef]

 *sensors*

Article

Localization and Discrimination of the Perturbation Signals in Fiber Distributed Acoustic Sensing Systems Using Spatial Average Kurtosis

Fei Jiang [1,2], Honglang Li [2,*], Zhenhai Zhang [1,*], Yixin Zhang [3] and Xuping Zhang [3]

[1] School of Mechatronics Engineering, Beijing Institute of Technology, Beijing 100081, China; flyjiang92@gmail.com

[2] Institute of Acoustics, Chinese Academy of Sciences, Beijing 100190, China

[3] The Key Laboratory of Intelligent Optical Sensing and Manipulation, Nanjing University, Nanjing 210008, China; zyixin@nju.edu.cn (Y.Z.); xpzhang@nju.edu.cn (X.Z.)

[*] Correspondence: lhl@mail.ioa.ac.cn (H.L.); zhzhang@bit.edu.cn (Z.Z.)

Received: 13 July 2018; Accepted: 21 August 2018; Published: 28 August 2018

Abstract: Location error and false alarm are noticeable problems in fiber distributed acoustic sensing systems based on phase-sensitive optical time-domain reflectometry (Φ-OTDR). A novel method based on signal kurtosis is proposed to locate and discriminate perturbations in Φ-OTDR systems. The spatial kurtosis (SK) along the fiber is firstly obtained by calculating the kurtosis of acoustic signals at each position of the fiber in a short time period. After the moving average on the spatial dimension, the spatial average kurtosis (SAK) is then obtained, whose peak can accurately locate the center of the vibration segment. By comparing the SAK value with a certain threshold, we may to some degree discriminate the instantaneous destructive perturbations from the system noise and certain ambient environmental interferences. The experimental results show that, comparing with the average of the previous localization methods, the SAK method improves the pencil-break and digging locating signal-to-noise ratio (SNR) by 16.6 dB and 17.3 dB, respectively; and decreases the location standard deviation by 7.3 m and 9.1 m, respectively. For the instantaneous destructive perturbation (pencil-break and digging) detection, the false alarm rate can be as low as 1.02%, while the detection probability is maintained as high as 95.57%. In addition, the time consumption of the SAK method is adequate for a real-time Φ-OTDR system.

Keywords: fiber distributed acoustic sensing; Φ-OTDR; localization; discrimination; kurtosis

---

1. Introduction

Fiber distributed acoustic sensing systems based on phase-sensitive optical time domain reflectometry (Φ-OTDR) have been widely used in many fields such as oil and gas pipeline monitoring, structure health monitoring, and perimeter security [1–4] due to their high sensitivity, large dynamic range, fully distributed manner, and simple configuration. The principal of Φ-OTDR is based on the interference of Rayleigh scattering lights returned from the fiber. The phase of the Rayleigh scattering lights can be affected by external vibration events, which will result in the amplitude variation of the Rayleigh scattering traces. In Φ-OTDR systems, locating the vibration events is an essential problem. Ideally, the vibration location can be obtained by subtracting a Rayleigh trace from an earlier stored trace [1]. However, this method may cause false locations due to the amplitude fluctuation in Rayleigh traces and the system noise of Φ-OTDR.

Various signal processing methods have been introduced to solve this problem. Some focus on the time-domain signal denoising. For example, Lu et al. proposed the moving averaging and moving differential method to reduce the noise power and also increase the frequency response range of the system [5]. Qin et al. introduced a wavelet denoising method to reject the random noises induced by varied polarization states in different position and detectors [6]. Shi et al. [7] and Qin et al. [8] proposed using empirical mode decomposition (EMD) to improve the signal-to-noise ratio (SNR) of location information for the vibration events. Qin et al. also proposed a curvelet denoising method to reduce the time domain noise and improve the detection performance of Φ-OTDR systems [9]. Ölçer et al. presented an adaptive temporal matched filtering method to reduce the effect of fading noise without any impact on the frequency response of the detection system [10]. Some locate the perturbations using the 2-D image characters of Rayleigh traces. For instance, two-dimensional edge detection can improve both the spatial resolution and the SNR of location information [11,12]. Adaptive 2-D image processing method of bilateral filtering algorithm can also enhance the SNR of vibration/intrusion location in Φ-OTDR [13]. Some pay attention to the spatial frequency information of the Rayleigh signals while locating, such as [14–16], which can simultaneously extract the vibration location and frequency information with an improved SNR. In addition, correlation dimension method has been proposed to solve the coherent fading problem and improve the multiple location results' standard deviation [17].

However, the methods above are sensitive to all kinds of perturbations. Perturbations caused by environmental interference may be located as well. For instance, the fiber cables buried underground may be influenced by pedestrians, vehicles, and trains; the fiber cables hanging in the air are usually perturbed by the wind and rain; and fibers that are embedded in concrete and stuck to pipelines could be affected by natural vibrations of the structures. All of these harmless perturbations may also cause the same disturbance results, which may obscure the real harmful perturbations and cause high false alarm rates (FARs). A new method based on multi-scale wavelet decomposition and backpropagation neural network [18] has been proposed to solve this problem. It can locate the perturbations by multi-scale wavelet decomposition and determine them by a 3-layer BP neural network. The results show that the FAR is decreased efficiently. In addition to this method, various event recognition methods [19–22] have been proposed to discriminate the signals in Φ-OTDR systems.

In this paper, a new localization and discrimination method based on the spatial kurtosis of Rayleigh signals is introduced for Φ-OTDR systems. The spatial kurtosis (SK) along the fiber is first obtained by calculating the kurtosis at each position of the fiber in a short time period, and then spatial average kurtosis (SAK) can be obtained by computing the moving average of SK on the spatial dimension. As a result, both the SK peaks and the SAK peaks may show the perturbation locations. Comparing with SK method, the SAK method can reduce the location error caused by coherent fading problem. We can also simply evaluate whether the perturbation is instantaneous destructive by comparing the SAK value with a certain threshold to some degree. This is because instantaneous destructive perturbations generally have high positive kurtosis, while stationary environmental noises usually have low negative kurtosis. Experimental results show that, the SAK method has higher SNR and lower location standard deviation compared with the previous localization methods. The time consumption of SAK method is low enough to achieve perturbation localization and discrimination in real-time, which is promising for a practical Φ-OTDR system.

2. Methods

2.1. Kurtosis of Acoustic Signals in Phase-Sensitive Optical Time-Domain Reflectometry

The kurtosis is a fourth-order statistic that can be used to describe the shape of a probability distribution of a random variable. It is defined as:

$$Kurt[X] = \frac{E\left[(X - \mu)^4\right]}{\left(E\left[(X - \mu)^2\right]\right)^2} - 3,$$

$$(1)$$

where μ is the mean of random variable X, and $E\ [t]$ represents the expected value of quantity t. The kurtosis of any univariate normal distribution is 0. Thus we can evaluate a distribution by comparing its kurtosis with 0. Distributions with a positive kurtosis (called leptokurtic distributions) have infrequent extreme deviations (or outliers), while distributions with a negative kurtosis (called platykurtic distribution) produce no outliers. In order to analyze the kurtosis of Φ-OTDR signals, we simulated these three main types of signals in Φ-OTDR systems:

(1) Random noise. In Φ-OTDR systems, noises are mainly caused by phase noise of the laser, partial interferometric problem, and electrical noises such as thermal noise and shot noise [5]. These noises nearly follow a normal distribution, thus their kurtosis should be around 0. A Gaussian white noise with the amplitude of 2 mV is used to simulate the random noise in Φ-OTDR systems, shown in Figure 1a. Its kurtosis is 0.09.

(2) Instantaneous destructive perturbations. These signals (structural cracks, gas explodes, excavation damage, destructive intrusions, etc.) are virtually short-time and non-stationary, whose frequency distribution change rapidly over time. They generally have certain extreme values, thus their kurtosis should be positive. Transient signal in Figure 1b is used to simulate the instantaneous destructive perturbations, which is generated by a 5–400 Hz broadband damped sinusoidal signal plus a Gaussian white noise with the amplitude of 1 mV, and the kurtosis of it is 6.25.

(3) Environmental interferences. These signals (wind, the vehicle engine, structural free vibration, etc.) are mostly continuous and stationary, whose frequency distribution is fixed in a short time period. They usually have no extreme values, thus their kurtosis should be negative. Signals in Figure 1c–f are used to simulate environmental interferences. Signals in Figure 1c,d are simulated by a sinusoidal signal with the amplitude of 1 mV plus a Gaussian white noise with the amplitude of 1 mV, which are used to simulate the natural vibrations of the structures. Their sinusoidal signal frequencies are 5 Hz and 200 Hz and their kurtosis are -1.14 and -1.01 respectively. Signals in Figure 1e,f are band-limited white noises, which are used to simulate the environmental interferences such as wind. Their frequency band are 5–15 Hz and 45–65 Hz and their kurtosis are -1.10 and -1.01, respectively.

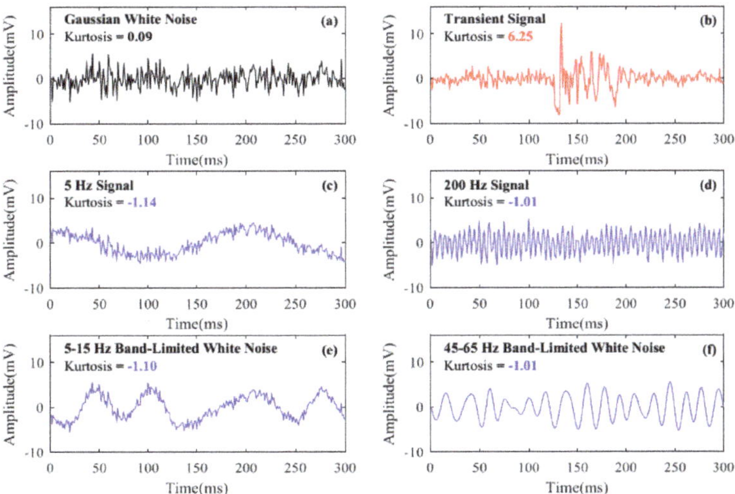

Figure 1. The simulated signals in phase-sensitive optical time-domain reflectometry (Φ-OTDR) systems. (**a**) Gaussian noise; (**b**) Transient signal; (**c**) Signal with the frequency of 5 Hz; (**d**) Signal with the frequency of 200 Hz; (**e**) 5–15 Hz band-limited white noise; (**f**) 45–65 Hz band-limited white noise.

We can see that the kurtosis of random noise is around 0, and the reason why it is not totally equal to 0 is that the length of it is too short (only 300). We can also see that signals caused by certain stationary environmental interferences have negative kurtosis, while transient signal (corresponding to instantaneous destructive perturbations) has high positive kurtosis. Thus we may discriminate such types of signals in Φ-OTDR systems by comparing their kurtosis with zero.

2.2. Spatial Kurtosis of Rayleigh Optical Time-Domain Reflectometry Traces

In Φ-OTDR systems, the Rayleigh backscattering signals can be composed into a matrix. The data picked up from each Rayleigh backscattering trace with the same time delay will form the time series of the corresponding sensing point. Suppose that each Rayleigh backscattering trace has L sampling points, which are determined by the sampling rate of the data acquisition card and the fiber length, the matrix with N received Rayleigh backscattering traces can be defined as:

$$\mathbf{R_{N \times L}} = \begin{bmatrix} x_1^1 & x_2^1 & \cdots & x_L^1 \\ x_1^2 & x_2^2 & \cdots & x_L^2 \\ \vdots & \vdots & \ddots & \vdots \\ x_1^N & x_2^N & \cdots & x_L^N \end{bmatrix}, \tag{2}$$

where x_l^n represents the amplitude of Rayleigh signal at position l at time n.

We accumulated w adjacent Rayleigh traces as one processing unit to calculate the kurtosis of the amplitude signal. The kurtosis of the signal at position l at time n can be defined as:

$$k_l^n = \frac{\frac{1}{w} \sum_{i=n-w+1}^{n} \left(x_l^i - \frac{1}{M} \sum_{i=n-w+1}^{n} x_l^i \right)^4}{\left(\frac{1}{w} \sum_{i=n-w+1}^{n} \left(x_l^i - \frac{1}{w} \sum_{i=n-w+1}^{n} x_l^i \right)^2 \right)^2} - 3, \tag{3}$$

where w can be also regarded as the window size of kurtosis computing, which determines the temporal resolution of the localization and discrimination. The spatial kurtosis of the Rayleigh traces at time n is defined as:

$$\mathbf{SK}(n) = [k_1^n, k_2^n, \cdots, k_L^n]. \tag{4}$$

By identifying the local maxima and minimum of SK at time n, we could estimate whether there are perturbations and locate them accurately on the sensing fiber. Moreover, we introduced a time step s to reduce the computation in real-time Φ-OTDR systems. It means that the time interval of calculating SK is $s \times fs$ (where fs represents the pulse repeat frequency). Composing the SK vectors at different times into a new matrix

$$\mathbf{SK} = \begin{bmatrix} k_1^1 & k_2^1 & \cdots & k_L^1 \\ k_1^{1+s} & k_2^{1+s} & \cdots & k_L^{1+s} \\ k_1^{1+2s} & k_2^{1+2s} & \cdots & k_L^{1+2s} \\ \vdots & \vdots & \cdots & \vdots \end{bmatrix}, \tag{5}$$

a two-dimensional image of SK over the monitoring duration can be obtained. Then the perturbation locations could be identified clearly from this image.

2.3. Moving Average on the Spatial Dimension

In Φ-OTDR systems, the accurate location of perturbations can be figured out by [7]:

$$z = z_{cent} - \frac{T_p v}{4}, \tag{6}$$

where z_{cent} is the center of the vibration response segment, T_p is the optical pulse width, and v is the light velocity in fiber. In order to obtain the accurate location of the perturbation in Φ-OTDR systems, we should figure out the center of the vibration segment. However, some subsections of the fiber vibration segment may be in insensitive states due to the coherent fading problem [23]. Figure 2a shows 100 consecutive averaged Rayleigh traces, which were acquired by our Φ-OTDR system. The optical pulse width used here is 100 ns, which means that the spatial resolution is 10 m. The vibration segment is from 5213 m to 5223 m. We can see that two subsections of the vibration segment are in sensitive state while the middle subsection is in insensitive state. This may make it difficult to locate the accurate vibration segment center and result in a location error and a bad locating repeatability.

For the SK localization method this problem exists as well. As shown in Figure 2b, the SK curve of the original Rayleigh traces splits into two peaks in the vibration segment. The two SK peaks correspond to the two sensitive subsections in Figure 2a, while the SK valley corresponds to the insensitive subsection in Figure 2a. It is hard to figure out the accurate vibration location by the SK curve. Virtually, we can solve this problem by calculating the moving average of SK on the spatial dimension. The result can be called SAK, which is defined as:

$$\mathbf{SAK}(n) = \frac{1}{M}\left[\sum_{l=1}^{M} k_l^n, \sum_{l=2}^{M+1} k_l^n, \cdots, \sum_{l=L-M+1}^{L} k_l^n\right] \tag{7}$$

where M represents the moving average number, which depends on T_p and the data acquisition card (DAQ) sampling rate (SR) of the Φ-OTDR system. It should be equal to the number of sampling points in an optical pulse width ($M = T_p \times SR$). When the vibration occurs, the location of SAK peak should be the center of the vibration segment. We can see from Figure 2b that the SAK peak virtually locates at the vibration segment center. Then we can obtain the accurate location of the vibration according to Equation (6).

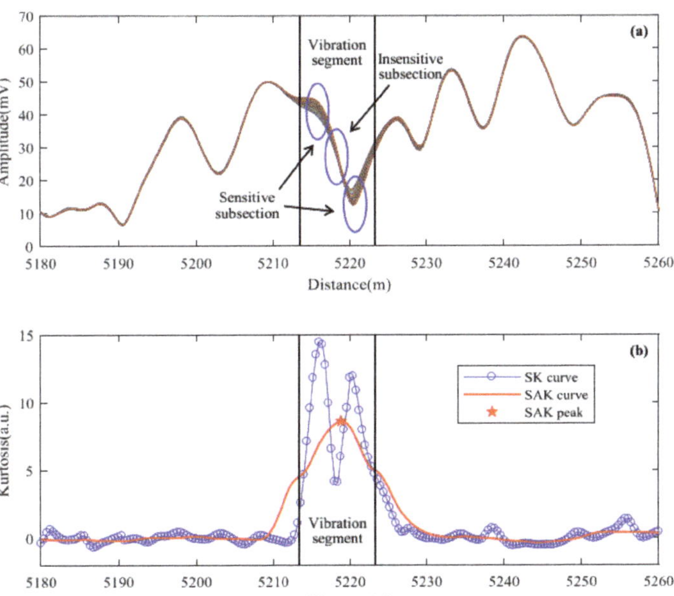

Figure 2. The vibration locating results when coherent fading problem exists. (**a**) 100 averaged Rayleigh traces; (**b**) The spatial kurtosis (SK) and special average kurtosis (SAK) curves of the Rayleigh traces.

3. Experimental Results

3.1. Experimental Setup

The experimental setup of our Φ-OTDR system is shown in Figure 3. A narrow line-width laser (NLL) is used as the light source. The laser light is divided into two parts by a 10:90 optical coupler (OC1). Then 10% of the light is used as the local light to perform a coherent detection, and 90% of the light is modulated into optical pulses with a 200 MHz frequency shift by an acoustic optical modulator (AOM). After being amplified by an erbium-doped fiber amplifier (EDFA), the optical pulses are launched into an optical circulator. The Rayleigh backscattering light from the sensing fiber then interferes with the local light at a 50:50 optical coupler (OC2) and result in a beat light. The beat light is then detected by a balanced photodetector (BPD) through a differential way, which could improve the SNR by 3 dB [4]. The output electric signal from the BPD is then sampled by a DAQ with a 1 GS/s sampling rate. The Rayleigh trace amplitude is finally demodulated by Hilbert transform and downsampled to 250 MS/s to reduce the computation load. In our experiments, the pulse repeat frequency is 1 kHz and the pulse width is 100 ns, which means that the maximum frequency response and the spatial resolution of our system are 500 Hz and 10 m respectively.

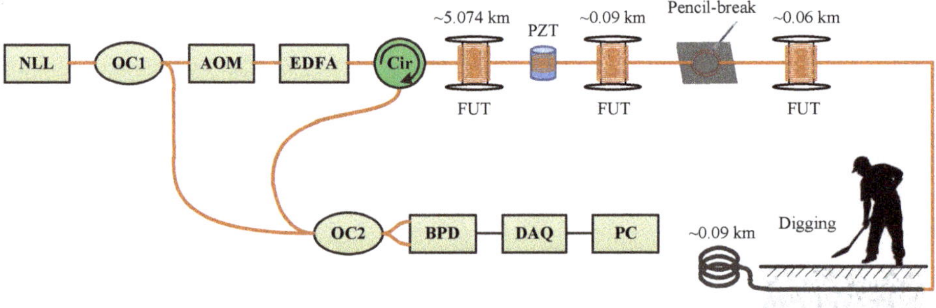

Figure 3. The experimental setup. NLL: narrow line-width laser; OC: optical coupler; AOM: acoustic-optic modulator; EDFA: Erbium-doped fiber amplifier; Cir: circulator; FUT: fiber under test; BPD: balanced photo detector; DAQ: data acquisition; PC: personal computer.

Both indoor experiment and outdoor experiment were conducted to evaluate the proposed method. In the indoor experiment, a piezoelectric transducer (PZT) cylinder glued with 10 m long fiber was added after the ~5.074 km long fiber under test (FUT) in our system. The PZT was used to simulate the environmental interferences such as structural free vibration and vehicle engine. Pencil-break vibration was used to simulate the civil structure crack, which was felt by a 2 m long fiber loops glued to a flame retardant (FR-4) plate. The pencil-break vibration was conducted by an HB pencil lead with 0.5 mm diameter. The fiber loops was put at ~5.164 km. In the outdoor experiment, a ~100 m long fiber cable was placed outdoor. The first ~10 m of it was buried beneath a flower bed with 0.3 m deep, while the rest ~90 m of it was placed in a corner. Digging the soil near the buried fiber cable was inducted as the instantaneous destructive perturbations. The flower bed is near to the road, so the fiber cable buried in the soil might be affected by the pedestrians and vehicles, which could be regard as the harmless perturbations.

3.2. Localization Results of Instantaneous Destructive Perturbations

A total of 200 tests, including 100 pencil-break tests (50 of which were under 100 ns pulse width and 50 of which were under 500 ns pulse width) and 100 digging tests (50 of which were under 100 ns pulse width and 50 of which were under 500 ns pulse width) were carried out to evaluate the localization performance of proposed method. The pencil-break tests and digging tests were performed at two fixed positions (~5164 m for pencil-break and ~5220 m for digging).

In this part, we compared the proposed method with four previous methods, including moving average and differential (MAD) [5], two-dimensional edge detection (TED) [11], wavelet denoising (WD) [6], and multi-scale wavelet decomposition (MWD) [18]. The comparisons include locating SNR, locating repeatability, and time consumption. The parameters of these methods are tuned to get the relatively high locating SNR. The moving average number of MAD was 30; the Sobel operator size of TED was 3 × 3 for pencil-break and 5 × 5 for digging; the wavelet basis function of WD and MWD was db3, and the decomposition level was 6. The window length used to compute kurtosis was 300, the time step was 150, and the moving average number of SAK was 25.

3.2.1. Signal-to-Noise Ratio Improvement of the Locating Information

Locating results of pencil-break vibration and digging are shown in Figures 4–7. Here, SNR is defined as the ratio between the signal peak intensity (S) and the maximum background noise intensity (N) with the equation of 20 lg(S/N) [18]. Figures 4 and 5 respectively show the locating results of pencil-break vibration and digging under 100 ns pulse width. We can see that the SAK method obtains the highest SNR (23.68 dB for pencil-break and 24.32 dB for digging) among all the methods. Figures 6 and 7 respectively show the locating results of pencil-break vibration and digging under 500 ns pulse width. Likewise, the SAK method has the highest SNR (22.28 dB for pencil-break and 23.36 dB for digging) among all the methods. We can see that when the pulse width is increased from 100 ns to 500 ns, the SNR of MAD, TED, and MWD significantly reduce, whereas the SNR of SK and SAK maintain high. The SNR of SAK method is higher than SK, which means that moving average on the spatial dimension may improve the locating SNR.

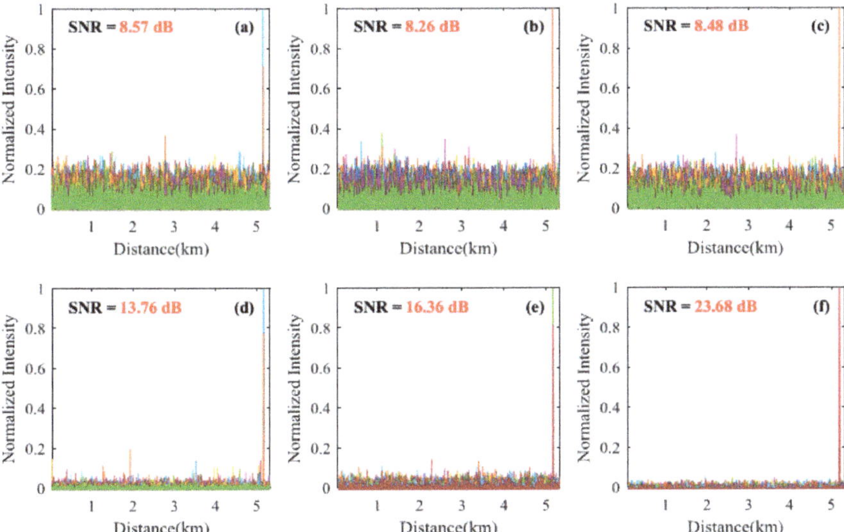

Figure 4. The locating results of pencil-break vibration under 100 ns pulse width. (**a**) The result of moving average and differential (MAD) method; (**b**) The result of two-dimensional edge detection (TED) method; (**c**) The result of multi-scale wavelet decomposition (MWD) method; (**d**) The result of wavelet denoising (WD) method; (**e**) The result of SK method; (**f**) The result of SAK method. SNR: signal-to-noise ratio.

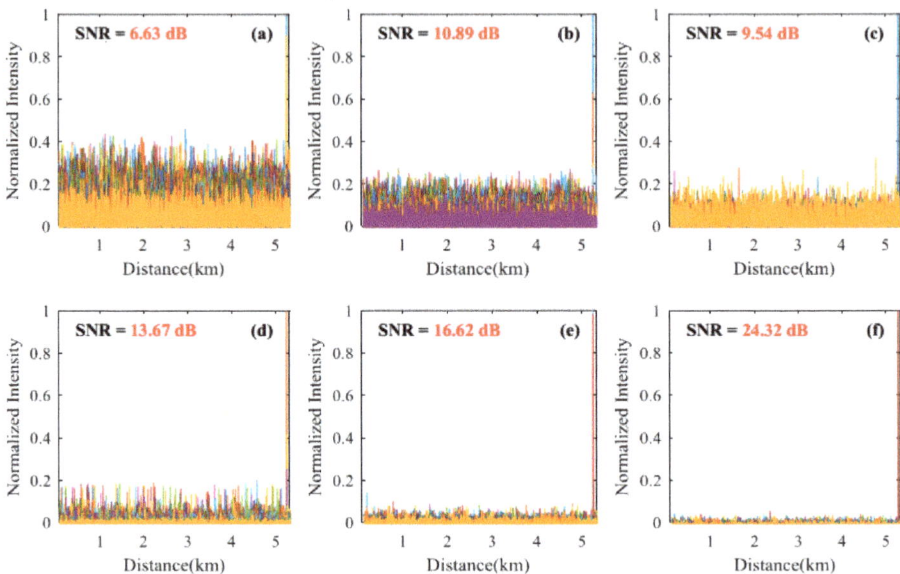

Figure 5. The locating results of digging event under 100 ns pulse width. (**a**) The result of MAD method; (**b**) The result of TED method; (**c**) The result of MWD method; (**d**) The result of WD method; (**e**) The result of SK method; (**f**) The result of SAK method.

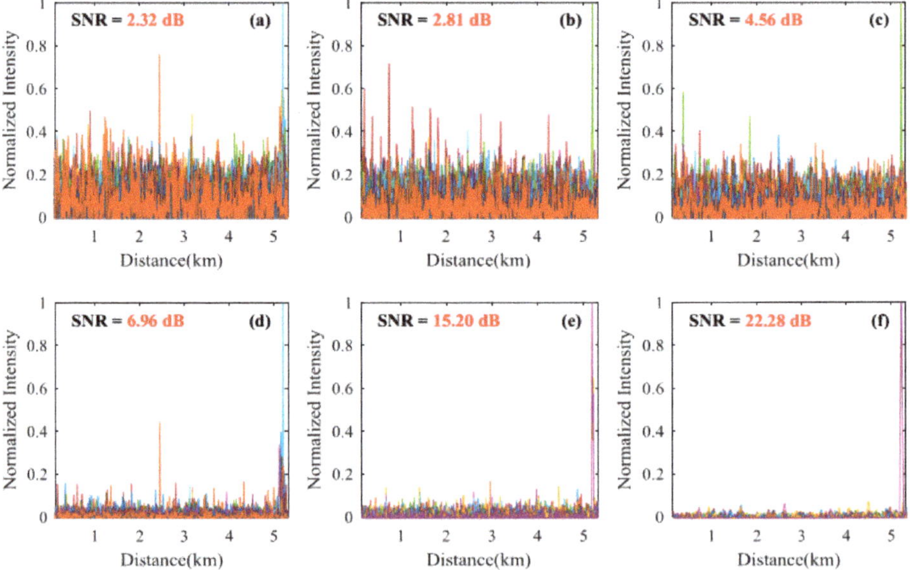

Figure 6. The locating results of pencil-break under 500 ns pulse width. (**a**) The result of MAD method; (**b**) The result of TED method; (**c**) The result of MWD method; (**d**) The result of WD method; (**e**) The result of SK method; (**f**) The result of SAK method.

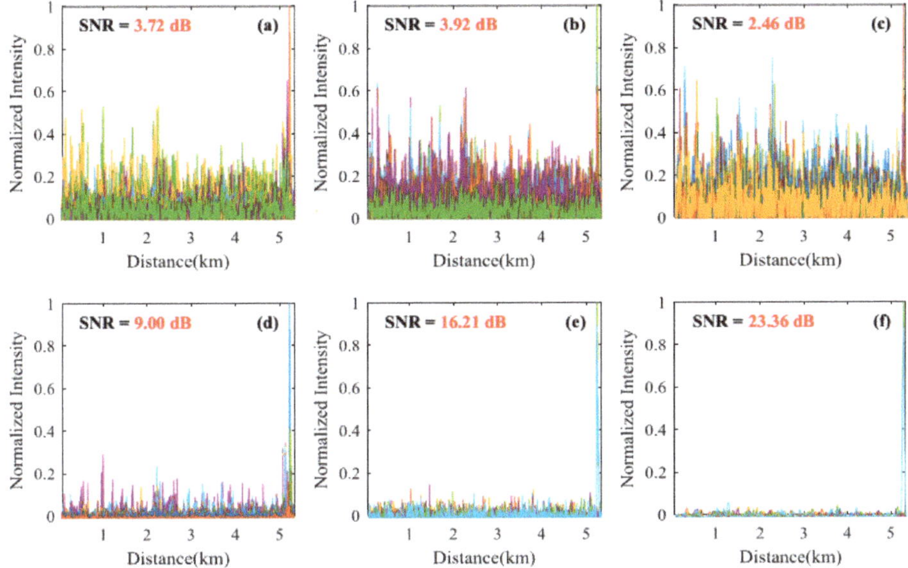

Figure 7. The locating results of digging event under 500 ns pulse width. (**a**) The result of MAD method; (**b**) The result of TED method; (**c**) The result of MWD method; (**d**) The result of WD method; (**e**) The result of SK method; (**f**) The result of SAK method.

The average locating SNRs of each method on the 200 locating tests are shown in Table 1. We can see that, comparing with the average of the MAD, TED, WD, and MWD methods, the SAK method improves the pencil-break locating SNR by 13.95 dB (under 100 ns pulse width) and 16.6 dB (under 500 ns pulse width); and the digging locating SNR by 10.60 dB (under 100 ns pulse width) and 17.3 dB (under 500 ns pulse width).

Table 1. The average locating SNRs of each method on the 200 locating tests.

Event	Pulse Width	MAD	TED	WD	MWD	SK	SAK
Pencil-break (dB)	100 ns	9.14	10.07	14.72	9.75	19.79	24.87
	500 ns	5.12	5.03	6.86	5.39	14.82	22.20
Digging (dB)	100 ns	8.70	11.57	14.31	11.18	16.22	22.04
	500 ns	4.56	6.42	8.93	6.12	15.72	23.81

3.2.2. Locating Accuracy

The box plot of locating results is shown in Figure 8. We can see that for both pencil-break and digging, the range of the locating results gets wider when the pulse width increases from 100 ns to 500 ns. Thus the increasing pulse width may degrade the locating accuracy. We can also see that the SAK method has more consistent results than other methods for both pencil-break vibrations locating and digging events locating. Quantitatively, the averages and standard deviations of pencil-break and digging locating results by the 6 methods under two different pulse widths are shown in Table 2. We can see that the averages of the locating results by these methods show no significant differences, which is because for all of the methods the locating results may fluctuate above and below the identical real locations. However, the standard deviations of the locating results differ. We can see that the SAK method has the lowest standard deviation for both pencil-break and digging locating. When the pulse width is 100 ns, the standard deviation of the pencil-break and digging locating results by SAK method reduce from 1.7 m and 1.4 m (the average locating standard deviation of the previous methods) to 0.8 m and 0.7 m, respectively. Two-sample F-tests show that this improvement is statistically significant for all of the previous methods ($p < 0.0056$). When the pulse width is 500 ns, the pencil-break and digging locating standard deviation of SAK method are only 1.9 m and 1.8 m respectively, while the average locating standard deviation of the previous methods are as high as 9.2 m and 10.9 m respectively. Two-sample F-tests show that this improvement is also statistically significant for all of the previous methods ($p < 10^{-19}$). Comparing with SK method, the SAK method decreases the localization standard deviation of pencil-break and digging under 500 ns pulse width by 8.4 m and 10.4 m, respectively, which means that moving average on the spatial dimension can effectively improve the locating accuracy.

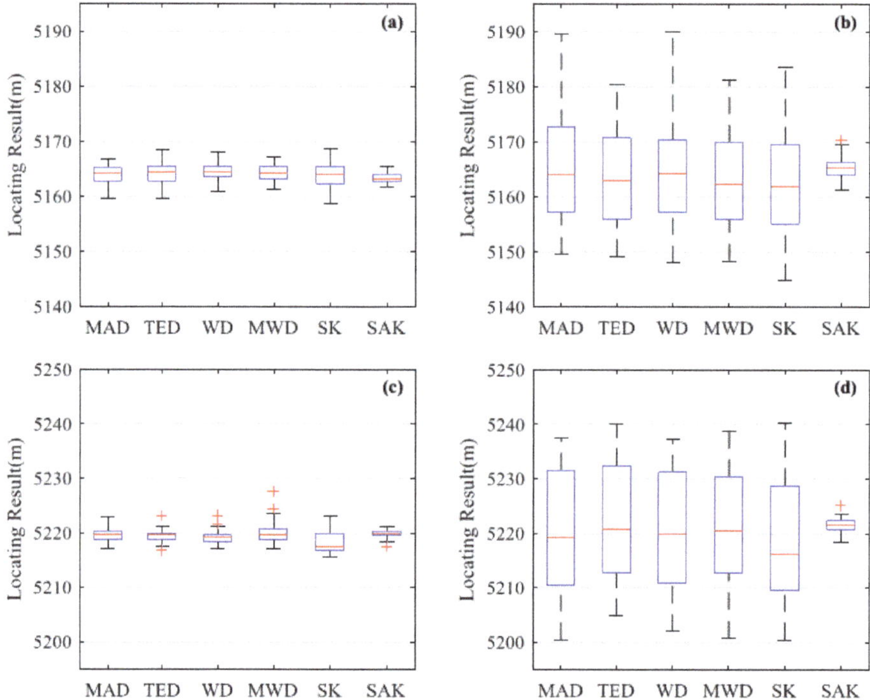

Figure 8. The statistic results of locating results. (**a**) The result of pencil-break vibration under 100 ns pulse width; (**b**) The result of pencil-break vibration under 500 ns pulse width; (**c**) The result of digging under 100 ns pulse width; (**d**) The result of digging under 500 ns pulse width.

Table 2. The average and standard deviation of the locating results under different pulse widths.

Event	Pulse Width	Statistics	MAD	TED	WD	MWD	SK	SAK
Pencil-break (m)	100 ns	Avg.	5163.9	5164.2	5164.3	5164.3	5163.6	5163.4
		Std.	1.7	1.9	1.7	1.6	2.3	**0.8**
	500 ns	Avg.	5165.8	5163.7	5164.6	5162.9	5162.4	5165.4
		Std.	9.6	8.6	9.9	8.7	10.3	**1.9**
Digging (m)	100 ns	Avg.	5219.7	5219.5	5219.3	5220.0	5218.5	5219.9
		Std.	1.3	1.2	1.1	2.0	2.1	**0.7**
	500 ns	Avg.	5220.4	5221.9	5221.3	5221.2	5219.0	5221.5
		Std.	11.6	11.0	10.9	10.3	11.6	**1.2**

3.2.3. Time Consumption

For a real-time Φ-OTDR system, the locating speed should be faster than the data generating speed. Therefore, the time consumption of the localization algorithm should be as less as possible. We compared the time cost of SAK method with the previous locating methods under different fiber length, shown in Figure 9. The methods are performed under 5000 Rayleigh traces (corresponding to 5 s long data in our system). The CPU used in our PC is an Intel Core i7-4930K. The algorithms are executed by MATLAB R2017a (MathWorks, Inc., Natick, MA, USA). We can see that the time consumption of SAK method is close to that of the MAD and TED methods and it outperforms both the WD and MWD methods. When the fiber length is 30 km (the spatial sampling interval is 1 m), the SAK method takes only 1.77 s. Consequently, SAK method can meet the time consumption requirement of the real-time Φ-OTDR system.

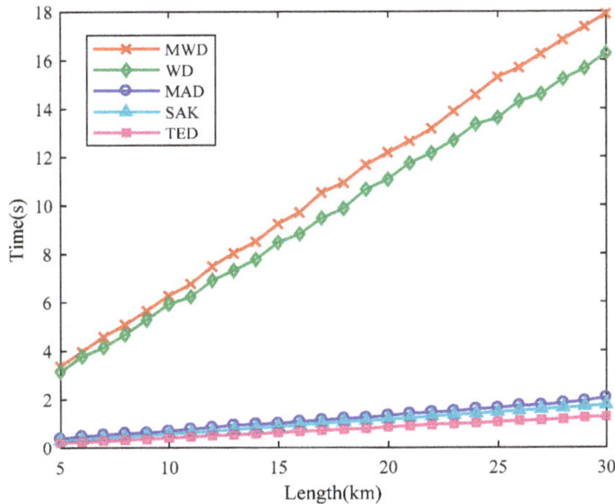

Figure 9. The time consumption of various locating methods under different fiber length.

3.3. Discrimination of Perturbations

In Section 2.1, we analyzed the kurtosis of different types of signals in Φ-OTDR. The simulation results indicated that we may discriminate the random noise, instantaneous destructive perturbations, and stationary environmental interferences by kurtosis in simple scenarios. In this section, we performed two experiments to validate the simulation results. In addition to using pencil-break vibration and digging as the instantaneous destructive perturbations, we drove the PZT with various frequencies to simulate the interference signals.

Figures 10 and 11 show the locating results of various tests using SAK method. Figure 10a shows the SAK curves of 5 Hz PZT vibration plus pencil-break vibration. The SAK trough at ~5075 m and the SAK peak at ~5164 m locate the PZT vibration and the pencil-break vibration respectively. The SAK image of PZT plus pencil-break is shown in Figure 10b. One conspicuous dark stripe can be observed at ~5075 m, which indicates the PZT vibration. One bright point is observed at ~5164 m, which indicates the pencil-break vibration. Figure 10c,d show the locating results of 200 Hz PZT vibration plus five digging events. The SAK trough at ~5075 m and the SAK peak at ~5220 m in Figure 10c correspond to the PZT vibration and digging events respectively. The dark stripe at ~5075 m and 5 bright points at ~5220 m shown in Figure 10d, respectively, locate the PZT vibration and digging events as well, which can also be seen clearly in the three-dimensional SAK image (shown in Figure 11). It should be noted that when the PZT starts to work and stops working, the SAK may be high as well. Thus such "onset" and "offset" events could also be regarded as instantaneous destructive perturbations in that moment.

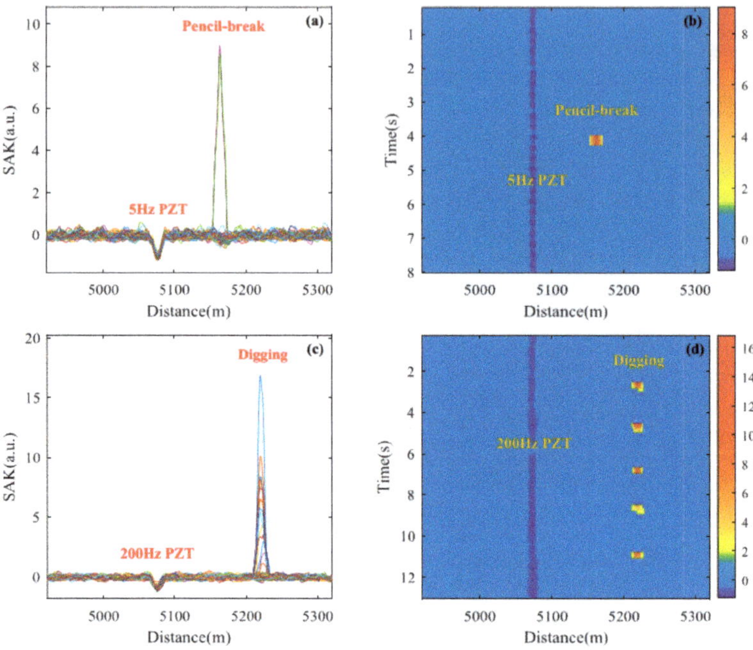

Figure 10. The location results of piezoelectric transducer (PZT) and instantaneous destructive perturbations (pencil-break and digging) through proposed method. (**a**) The SAK curves of 5 Hz PZT vibration plus pencil-break vibration; (**b**) The SAK image of 5 Hz PZT vibration plus pencil-break vibration; (**c**) The SAK curves of 200 Hz PZT vibration plus digging; (**d**) The SAK image of 200 Hz PZT vibration plus 5 digging events.

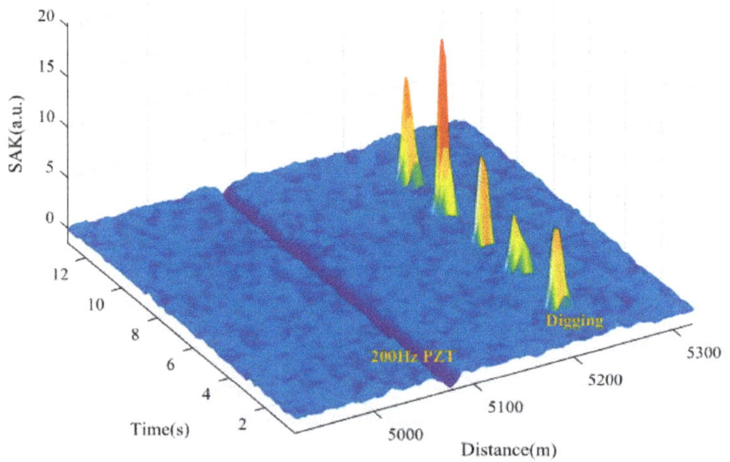

Figure 11. The three-dimensional SAK image of 200 Hz PZT vibration plus 5 digging events.

In order to generally evaluate the discrimination performance of proposed method, we computed and analyzed the SAK value of a set of various tests (including 126 pencil-break vibrations, 168 digging events, 294 PZT vibrations with various frequencies, and 302 environmental noises). The statistic SAK result of various Φ-OTDR signals is shown in Figure 12a, where IDP means the instantaneous destructive perturbations (pencil-break vibration and digging event); N_{max} and N_{min} are the SAK maximum and SAK minimum of the noises; IF means interference (PZT vibration with different frequencies including 5, 50, 100 and 200 Hz). We can see that different types of signals have different SAK value interval.

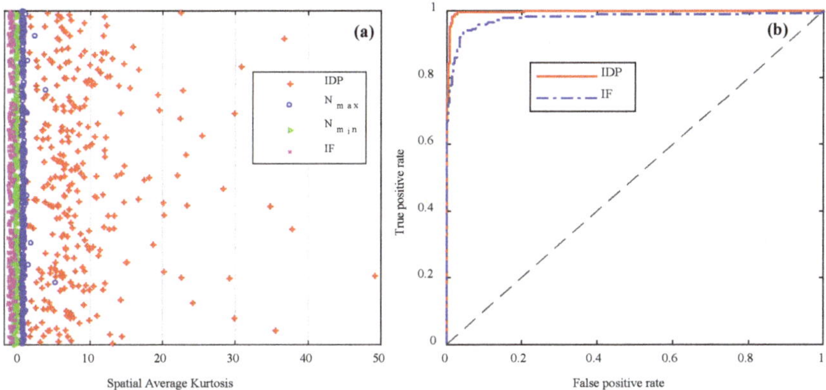

Figure 12. The statistic results of perturbation discrimination by SAK method. (a) The SAK distribution of different types of signals; (b) The receiver operating characteristic (ROC) curves of perturbation discrimination. IDP: instantaneous destructive perturbations; IF: interference.

We used true positive rate (TPR) and false positive rate (FPR) to evaluate the identification ability of the SAK method, where the TPR and FPR are defined as:

$$\text{TPR} = \frac{\text{TP}}{\text{TP} + \text{FN}}, \ \text{FPR} = \frac{\text{FP}}{\text{TN} + \text{FP}}. \tag{8}$$

Here, TP is the count of correct positive prediction, FP is the count of wrong positive prediction, TN is the count of correct negative prediction, and FN is the count of wrong negative prediction. The TPR is also known as the probability of detection, while the FPR is also known as the probability of false alarm. By plotting the TPR against FPR at various threshold settings, the receiver operating characteristic (ROC) curve is obtained, shown in Figure 12b. Four-fold cross-validation was used to evaluate the TPR and FPR of the perturbation discrimination by SAK. The SAK thresholds of discriminating the pencil-break (or digging) and PZT vibration were fitted on the training data of each fold using a logistic regression model. The evaluation results on test data of each fold are show in Table 3.

We can see that the average TPR, FPR and accuracy of discriminating pencil-break or digging from noise are 95.57%, 1.02%, and 97.32%, respectively, while the average TPR, FPR, and accuracy of discriminating PZT vibration from noise are 92.54%, 4.38%, and 94.13%, respectively.

Table 3. The true positive rate (TPR), false positive rate (FPR), and accuracy (Acc.) of signal discrimination using SAK method.

Events	Fold Index	TPR	FPR	Acc.
Pencil-Break or Digging	1	94.12%	1.23%	96.64%
	2	97.26%	1.32%	97.99%
	3	95.65%	0.00%	97.99%
	4	95.24%	1.54%	96.64%
	Avg.	**95.57%**	**1.02%**	**97.32%**
PZT Vibration	1	93.15%	6.58%	93.29%
	2	90.41%	1.32%	94.63%
	3	94.20%	2.50%	95.97%
	4	92.41%	7.14%	92.62%
	Avg.	**92.54%**	**4.38%**	**94.13%**

4. Discussion

Compared with the previous locating methods for Φ-OTDR systems, the SAK method has the following advantages:

(1) It can locate the instantaneous destructive perturbations with a higher SNR. This is because that the Φ-OTDR signals caused by instantaneous destructive perturbations generally have extreme value and result in large positive kurtosis, while noises in Φ-OTDR basically obey normal distribution and result in zero-kurtosis.

(2) It has better locating accuracy. By using a moving average on the spatial dimension, the SAK method can accurately locate the center of the vibration segment, even if the subsection is in an insensitive state, thus leading to more accurate locating results. It should be noted that "moving average on the spatial dimension" can be also used in other locating methods to fast figure out the accurate vibration segment center.

(3) It can simultaneously locate the perturbations and to some degree evaluate whether they are destructive or just interference. Because instantaneous destructive perturbations generally have higher positive kurtosis, while stationary interferences usually have lower negative kurtosis. The pencil-break, digging, and PZT discrimination results show that the SAK might be a promising feature for further event recognition.

(4) The time consumption of SAK method is short enough for a long distance real-time Φ-OTDR system. It should be noted that here the time consumption include both the locating time and discriminating time.

The SAK method has two main defects. One is that the localization time interval is directly related to the time step s in Equation (5). When the time step is 150 and the sampling rate is 1 kHz, the time interval of localization will be 0.15 s. However, for an Φ-OTDR threat alarm system this is totally acceptable. The other is that the SAK method only captures the short-time characteristics of the perturbations. Instantaneous destructive perturbations and stationary interferences might be discriminated by SAK, whereas long-term perturbations still require more features to determine whether they are a threat.

5. Conclusions

In this paper, we proposed a new perturbation localization and discrimination method for Φ-OTDR systems. Pencil-break vibration and digging events are used as the instantaneous destructive perturbations while PZT vibration is used to simulate interference. Experimental results show that, comparing with previous methods (MAD, TED, WD, MWD), the proposed method can locate the instantaneous destructive perturbations with a higher SNR. We also introduced moving average on the spatial dimension to improve the locating accuracy when coherent fading problem occur, thus leading to the smaller standard deviation of locating results. In addition, the SAK method can discriminate the instantaneous destructive perturbations from system noise and stationary environmental interference to some extent. Due to the little time consumption, SAK can be a promising real-time locating and perturbation discrimination method for Φ-OTDR systems.

Author Contributions: Conceptualization, F.J. and H.L.; Formal analysis, F.J.; Funding acquisition, H.L., Z.Z. and X.Z.; Methodology, F.J.; Resources, Z.Z. and Y.Z.; Software, F.J.; Supervision, H.L.; Validation, F.J.; Writing-original draft, F.J.; Writing-review & editing, F.J. and Y.Z.

Funding: This work is funded by National Natural Science Foundation of China under Grant No. 61627816 and No. 61773059.

Conflicts of Interest: The authors declare no conflict of interest.

References

1. Juarez, J.C.; Taylor, H.F. Distributed fiber optic intrusion sensor system. *J. Light. Technol.* **2005**, *23*, 2081–2087. [CrossRef]
2. Juarez, J.C.; Taylor, H.F. Field test of a distributed fiber-optic intrusion sensor system for long perimeters. *Appl. Opt.* **2007**, *46*, 1968–1971. [CrossRef] [PubMed]
3. Martins, H.F.; Martin-Lopez, S.; Corredera, P.; Filograno, M.L.; Frazão, O.; González-Herráez, M. High visibility phase-sensitive optical time domain reflectometer for distributed sensing of ultrasonic waves. *Proc. SPIE* **2013**, *8794*, 87943F-1–87943F-4. [CrossRef]
4. Peng, F.; Wu, H.; Jia, X.H.; Rao, Y.J.; Wang, Z.N.; Peng, Z.P. Ultra-long high-sensitivity Φ-OTDR for high spatial resolution intrusion detection of pipelines. *Opt. Express* **2014**, *22*, 13804–13810. [CrossRef] [PubMed]
5. Lu, Y.; Zhu, T.; Chen, L.; Bao, X. Distributed vibration sensor based on coherent detection of phase-OTDR. *J. Light. Technol.* **2010**, *28*, 3243–3249. [CrossRef]
6. Qin, Z.; Chen, L.; Bao, X. Wavelet denoising method for improving detection performance of distributed vibration sensor. *IEEE Photon. Technol. Lett.* **2012**, *24*, 542–544. [CrossRef]
7. Yi, S.; Hao, F.; Zhoumo, Z. A long distance phase-sensitive optical time domain reflectometer with simple structure and high locating accuracy. *Sensors* **2015**, *15*, 21957–21970. [CrossRef]
8. Qin, Z.; Chen, H.; Chang, J. Signal-to-noise ratio enhancement based on empirical mode decomposition in phase-sensitive optical time domain reflectometry systems. *Sensors* **2017**, *17*, 1870. [CrossRef]
9. Qin, Z.; Chen, H.; Chang, J. Detection performance improvement of distributed vibration sensor based on curvelet denoising method. *Sensors* **2017**, *17*, 1380. [CrossRef]
10. Ölçer, İ.; Öncü, A. Adaptive temporal matched filtering for noise suppression in fiber optic distributed acoustic sensing. *Sensors* **2017**, *17*, 1288. [CrossRef] [PubMed]
11. Zhu, T.; Xiao, X.; He, Q.; Diao, D. Enhancement of SNR and spatial resolution in Φ-OTDR system by using two-dimensional edge detection method. *J. Light. Technol.* **2013**, *31*, 2851–2856. [CrossRef]
12. Wang, Y.; Jin, B.; Wang, Y.; Wang, D.; Liu, X.; Bai, Q. Real-time distributed vibration monitoring system using Φ-OTDR. *IEEE Sens. J.* **2017**, *17*, 1333–1341. [CrossRef]
13. He, H.; Shao, L.; Li, H.; Pan, W.; Luo, B.; Zou, X.; Yan, L. SNR enhancement in phase-sensitive OTDR with adaptive 2-D bilateral filtering algorithm. *IEEE Photon. J.* **2017**, *9*, 1–10. [CrossRef]
14. Qin, Z.; Chen, L.; Bao, X. Continuous wavelet transform for non-stationary vibration detection with phase-OTDR. *Opt. Express* **2012**, *20*, 20459–20465. [CrossRef] [PubMed]
15. Hui, X.; Ye, T.; Zheng, S.; Zhou, J.; Chi, H.; Jin, X.; Zhang, X. Space-frequency analysis with parallel computing in a phase-sensitive optical time-domain reflectometer distributed sensor. *Appl. Opt.* **2014**, *53*, 6586–6590. [CrossRef] [PubMed]
16. Yue, H.; Wu, Y.; Zhao, B.; Ou, Z. Simultaneous and signal-to-noise ratio enhancement extraction of vibration location and frequency information in phase-sensitive optical time domain reflectometry distributed sensing system. *Opt. Eng.* **2015**, *54*, 047101:1–047101:6. [CrossRef]
17. Shi, Y.; Feng, H.; Huang, Y.; Zeng, Z. Correlation dimension locating method for phase-sensitive optical time domain reflectometry. *Opt. Eng.* **2016**, *55*, 091402:1–091402:5. [CrossRef]

18. Wu, H.; Xiao, S.; Li, X.; Wang, Z.; Xu, J.; Rao, Y. Separation and determination of the disturbing signals in phase-sensitive optical time domain reflectometry (Φ-OTDR). *J. Light. Technol.* **2015**, *33*, 3156–3162. [CrossRef]
19. Sun, Q.; Feng, H.; Yan, X.; Zeng, Z. Recognition of a phase-sensitivity OTDR sensing system based on morphologic feature extraction. *Sensors* **2015**, *15*, 15179–15197. [CrossRef] [PubMed]
20. Tejedor, J.; Martins, H.F.; Piote, D.; Macias-Guarasa, J.; Pastor-Graells, J.; Martin-Lopez, S.; Guillén, P.C.; Smet, F.D.; Postvoll, W.; González-Herráez, M. Toward prevention of pipeline integrity threats using a smart fiber-optic surveillance system. *J. Light. Technol.* **2016**, *34*, 4445–4453. [CrossRef]
21. Tejedor, J.; Macias-Guarasa, J.; Martins, H.; Piote, D.; Pastor-Graells, J.; Martin-Lopez, S.; Corredera, P.; Gonzalez-Herraez, M. A novel fiber optic based surveillance system for prevention of pipeline integrity threats. *Sensors* **2017**, *17*, 355. [CrossRef] [PubMed]
22. Jiang, F.; Li, H.; Zhang, Z.; Zhang, X. An event recognition method for fiber distributed acoustic sensing systems based on the combination of MFCC and CNN. *Proc. SPIE* **2017**, *10618*, 1061804-1–1061804-7. [CrossRef]
23. Izumita, H.; Furukawa, S.I.; Koyamada, Y.; Sankawa, I. Fading noise reduction in coherent OTDR. *IEEE Photon. Technol. Lett.* **1992**, *4*, 201–203. [CrossRef]

Article

Distributed Temperature Sensing Monitoring of Well Completion Processes in a CO_2 Geological Storage Demonstration Site

Dasom Sharon Lee [1,2], Kwon Gyu Park [1,*], Changhyun Lee [1,*] and Sang-Jin Choi [1]

[1]   Petroleum & Marine Division, Korea Institute of Geoscience and Mineral Resources, 124 Gwahak-ro, Yuseong-gu, Daejeon 34132, Korea; dslee@kigam.re.kr (D.S.L.); sang-jin@kigam.re.kr (S.-J.C.)

[2]   Department of Energy System Engineering, Seoul National University, 1 Gwanak-ro, Gwanak-gu, Seoul 08826, Korea

*   Correspondence: kgpark@kigam.re.kr (K.G.P.); jetlee@kigam.re.kr (C.L.); Tel.: +82-42-868-3250 (K.G.P.); +82-42-868-3945 (C.L.)

Received: 8 November 2018; Accepted: 2 December 2018; Published: 3 December 2018

Abstract: The Distributed Temperature Sensing (DTS) profiles obtained during well completion of a CO_2 monitoring well were analyzed to characterize each well completion process in terms of temperature anomalies. Before analysis, we corrected the depth by redistributing the discrepancy, and then explored three temperature calibration methods. Consequently, we confirmed the depth discrepancy could be well corrected with conventional error redistribution techniques. Among three temperature calibration methods, the conventional method shows the best results. However, pointwise methods using heat coil or in-well divers also showed reliable accuracy, which allows them to be alternatives when the conventional method is not affordable. The DTS data revealed that each well completion processes can be characterized by their own distinctive temperature anomaly patterns. During gravel packing, the sand progression was monitorable with clear step-like temperature change due to the thermal bridge effect of sand. The DTS data during the cementing operation, also, clearly showed the progression up of the cement slurry and the exothermic reaction associated with curing of cement. During gas lift operations, we could observe the effect of casing transition as well as typical highly oscillating thermal response to gas lifting.

Keywords: distributed temperature sensing; temperature calibration; depth correction; fiber optic sensing; well completion monitoring

---

1. Introduction

The most distinctive advantage of fiber optic Distributed Temperature Sensing (DTS) technology over other point sensing and wireline logging technologies is its 'distributed sensing' feature. The distributed sensing, as the fiber itself is the sensing medium, enables obtaining a temperature profile for the entire fiber length with much higher spatial density than point sensors in a period. It also enables continuous real-time temperature profiling at any time throughout the well, once it is permanently installed inside the well [1]. Unlike the production logging run, which obtains temperatures after the event, the DTS can identify in-well performance changes as the events occur during production and shut-in [2]. In addition, fiber optic distributed sensing has many more advantages compared to conventional point sensors or wireline logging as it can be deployed in any harsh or unusual environment and it is small, light, corrosion–resistant, and has a long sensing range, good sensitivity, and electromagnetic immunity.

In May of 2017, as a part of an onshore pilot-scale CO_2 sequestration demonstration project in Korea, multifunctional monitoring wells with discrete sensors and fiber optic cable for DTS and

Distributed Acoustic Sensing (DAS) have been completed. A key consideration factor in developing injection and/or monitoring wells for CO_2 geological storage is the integrity of the well that prevents the wells from becoming a conduit for CO_2 leakage. Generally, many sensors and pieces of equipment are installed in CO_2 monitoring well depending on the aim of the monitoring. Therefore, more care is needed for well design, installation, and completion process in order to ensure that the many sensors and the equipment perform at their best as well as to maintain integrity. Since we applied a permanent method using gravel packing and cementing for the completion, which is irreversible unlike a non-permanent method that uses a packer, monitoring each well completion process is critical to determine whether each process has gone as planned.

For the purpose of monitoring each well completion process in terms of temperature anomaly patterns, we monitored DTS data during whole completion processes without any intervention. However, on-site analysis was limited to qualitative interpretation because temperature pre-calibration and depth correction were not possible due to the field schedule and conditions.

In this paper, we applied a depth correction method and three temperature post-calibration methods for more quantitative and accurate analysis. Then, we characterized the thermal characteristics of each well completion process in order to explore the feasibility of DTS as a well completion monitoring and control tool.

The uncertainty of depth accuracy is inevitable in practical application of DTS to boreholes because it is difficult to detect any depth errors due to stretching and wrapping of the cable during installation. In this paper, conventional depth correction is applied by considering both stretching and wrapping of cable [3]. Then, we confirmed the accuracy of depth correction with our gas lift temperature profile using the depth of casing transition as a priori information.

The conventional way of temperature calibration is submerging both ends of the fiber in ice or hot water to calibrate the temperature offset to an absolute value and its rate of change along the fiber length [3]. It is commonly recommended to apply temperature calibration before installation. In many cases, however, calibration before installation is not possible or limited due to practical factors like site schedule and conditions. This was our case, therefore, we corrected temperature after installation. In addition to a conventional temperature calibration method of submerging the long length in water, we also explored the possibility of two types of pointwise post-calibration methods that may overcome the practical limitation of the submerging method on the site.

During well completion, real-time DTS data were analyzed to monitor the entire well completion process in terms of temperature changes and to control or optimize the process. However, it was limited to a qualitative analysis because the depth correction and temperature calibration were not done. After the depth correction and temperature calibration, thermal events of each well completion process could be characterized more precisely, and each well completion process could be identified with distinctive temperature anomaly pattern. The progress up the well of sand and cement slurry were monitored with clear step-like temperature change. Thermal events due to exothermal reaction during cement curing process was also clearly observed. Consequently, we could conclude that observing the entire well completion process with pre-calibrated DTS data can be a promising tool of real-time control and optimization of well completion process.

2. Materials and Methods

2.1. Site Description

The thermal data were collected from the monitoring well, JG-M, which was completed at the end of May 2017 as a part of onshore pilot-scale CO_2 geological storage demonstration program at Janggi field. The Janggi field site is located at Pohang city which is the southeastern part of Korean Peninsula as illustrated in Figure 1. The total depth (TD) of the monitoring wells is 1092.44 m, and the well is slanted by $2°$ to the southeast direction.

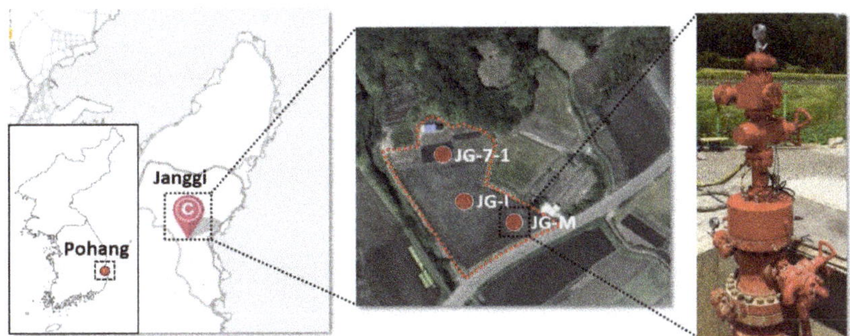

Figure 1. Location of Janggi field site including relative position of auxiliary monitoring well (JG-7-1), monitoring well (JG-M) with its wellhead, and injection well (JG-I).

2.2. Well Schematic and Completion Process

The monitoring sensors and equipment deployed in JG-M include 32-level electrode array, 24-level 3C geophone strings, a U-tube [4] fluid sampler, two piezo pressure/temperature sensors (P/T), an experimental pH sensor, and a fiber cable for DAS, DTS, and heat-pulse monitoring [5]. The schematic of the Janggi JG-M monitoring well and the completion assembly employed is presented in Figure 2 to help understand the thermal events associated with completion, equipment and materials.

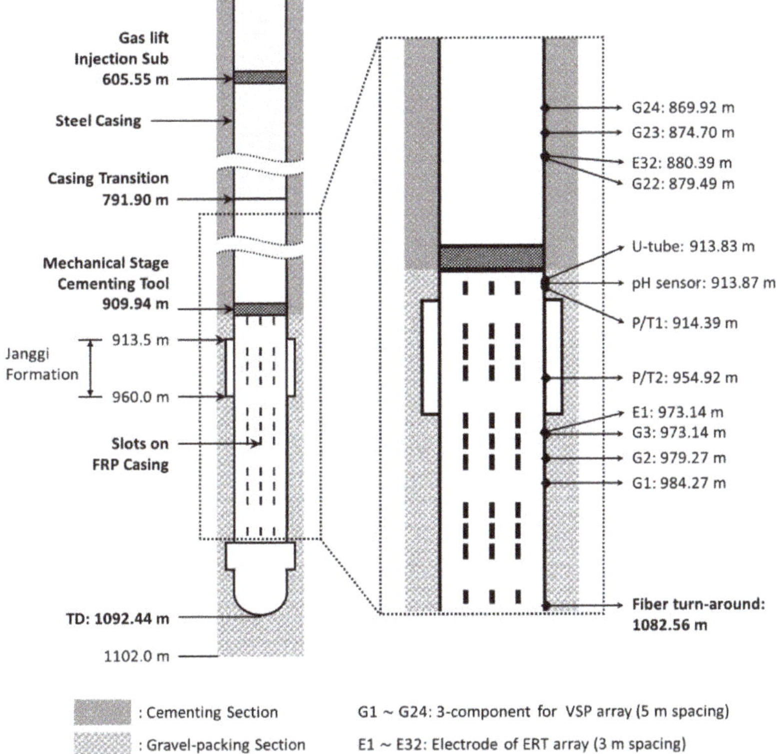

Figure 2. Schematic of the monitoring well, JG-M and the completion assembly employed with gravel packing and cementing.

The 32-level electrodes were installed for surface-to-borehole and cross-well Electrical Resistivity Tomography (ERT) to image the distribution and migration of the CO_2 plume. Due to fact the 32 electrodes cover a 93 m interval around the reservoir formation (JGF = Janggi conglomerate formation), almost all materials below 803 m should be non-metallic or electrically insulated. Therefore, instead of steel casing, 3-1/2″ Fiberglass Reinforced Plastic (FRP) casing was used at 300.54 m interval below 791.90 m, where the slotted (depicted by dashed line in Figure 2) and non-slotted casing alternated to guarantee fluid circulation and to maintain strength against pressure. The 3C geophones were installed for the purpose of acquiring Vertical Seismic Profiling (VSP) and passive micro-seismic monitoring data. The P/T sensors were installed at the top and bottom boundary of JGF reservoir, which were also used for pressure and temperature during the well completion. The fiber cable turnaround which protect spliced part of fiber core for dual-ended mode measurement was installed at 1082.56 m. The gas-lift manifold for fluid production and cementing stage tool for cementing were installed at 605.55 m and 909.94 m, respectively.

All sensors and instruments are installed in the tubing by the convey method by attaching sensors and control line to 3-1/2″ steel and FRP casing. A series of centralizers and clamps were used to protect the cables during installation and cement jobs. For the FRP casing section, we used nylon decentralizers which were attached to the casing with 3/8-inch set screws. The mid joint clamp was used to hold the stainless-steel tubes. For the steel casing section, cannon cross coupling protectors with fins provide centralization during cementing.

The two different patterns in the well schematic shown in Figure 2 depict the well completion schemes behind the casing. The bottom section below 910 m indicates the gravel packing section. The basic requirement of the reservoir section is to guarantee the circulation of pore fluid and CO_2. The ERT and VSP sensors should be well coupled with the surrounding formations for sensing, which is the reason for choosing the gravel packing over a packer. Above 910 m of the well was completed with cementing to ensure the pressure isolation of the reservoir section and the mechanical integrity of well that prevents the well from becoming a conduit for CO_2 migration.

After landing the tubing hanger, DTS system and permanent pressure temperature gauge was used to monitor the process of gravel packing and cementing operation in May 2017. We directly injected 12/20 sand into the annulus and recirculated the fluid up with the 1.9-inch IJ tubing. After the gravel packing procedure, cementing is conducted by circulating slurry into the well with a stage cementing tool at 909.94 m.

2.3. Fiber Optic and DTS Interrogator Configuration

The fiber optic control line used in JG-M consists of four optical fiber lines for DTS and DAS and two copper conduct lines for heat-pulse monitoring [5], which are protected by stainless steel sheaths and polypropylene jackets to prevent them from being damaged during installation. Among the four optical fiber lines two single-mode fibers are for DAS and the other two are multi-mode fibers for DTS. Temperature profiles are obtained in dual-ended mode by splicing two multi-mode fibers at the end of the control line. The schematic of optical fiber control line and specification, and turnaround sub are as shown in Figure 3.

Silixa's XT-DTS DTS interrogation system was used to monitor the temperature changes for five days during completion and for a month during curing. The data was obtained using a passive mode where no active heating was done to the cable itself. Our interrogation unit can measure temperature profiles up to 5 km, with minimum spatial and temporal resolution of 0.25 m and 5 s, respectively. The temperature profiles were collected with a spatial sampling interval of 0.25 m, with a temporal interval of 600 s during well completion monitoring from May to June of 2017 and 60 s during background monitoring and temperature calibration tests from November, 2017 to April, 2018.

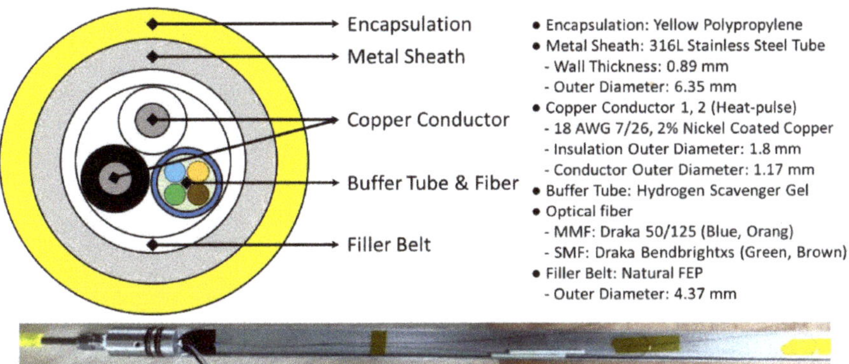

Figure 3. Schematic of fiber optic control line and specifications (**above**), and photo of turnaround sub for dual-ended configuration (**below**).

3. Depth and Temperature Correction

There are many potential factors that can cause errors when using the DTS system. The errors which arise from the instrumentation, the fiber, the nature of the installation, and other sources show up as depth discrepancies and temperature errors [3]. In order to specify the exact location of thermal events and to analyze the events more quantitatively, we applied depth correction and temperature calibration before interpreting thermal events for the well completion processes.

3.1. Depth Correction

Depth mismatch is an unavoidable issue in the field. Unlike temperature correction, depth discrepancies cannot be pre-calibrated before installation. Even though the cables were carefully handled during the installation, many factors such as cable coiling up around the tubing, stretching of fiber optic cable itself, the usage of a decentralizer, and inner fiber optic twisting may inevitably introduce depth discrepancies. Although it may depend on the application, Smolen and van der Spek [3] stated that generally a depth mismatch of $\pm$ 0.3 m is realistic for most applications. With the depth issue comes the issue of determining the end of the fiber depth. The length measured by DTS should be mapped for depth correction to a real physical profile length of interest. In dual-ended or partially returned configurations in borehole applications, the end of the fiber can be easily identified with the peak temperature due to ambient geothermal gradient and it can be confirmed in the raw data by a sudden intensity drop around the splicing point.

In our case, the peak temperature was observed mostly around 1197.114 m in fiber length. On the other hand, the maximum intensity drop occurred between 1196.606 m and 1196.860 m with the difference of 21.17 A.U. (relative decrease of 10.4%) and the second maximum drop occurred between 119.860 m to 1197.114 m with the difference of 18.62 A.U. (relative decrease of 9.3%). Although there is a difference of 0.5 m between the peak temperature point and the maximum intensity drop point, we used the peak temperature point as the fiber-end to determine total fiber length because such a difference is negligible considering our spatial sampling interval of 0.25 m and the width of the probe pulse. With respect to this fiber end, the fiber cable length (D_f) that corresponds to total casing length of 1082.56 m is 1095.599 m. Therefore, in our case, a depth mismatch (ΔD) of 13.039 m was identified, which is 1.2% longer than the total casing length.

A simple way to match depth discrepancies is to redistribute the amount of depth mismatch to the total length. Thus, we redistributed total depth error throughout the total length under the assumption that the rate of depth mismatch is constant throughout the total length.

For the confirmation of depth correction accuracy, we compared the temperature profiles before and after correction for gas lift process in May and November, 2017 as shown in Figure 4. Since the casing material is changed from steel to FRP casing at 791.90 m as shown in Figure 2, any temperature anomaly associated with the heat conductivity difference of two casing material should be identified around this depth. The effect of casing transition is clearly identifiable with the temperature increase pattern from data in both May and November, regardless of depth correction.

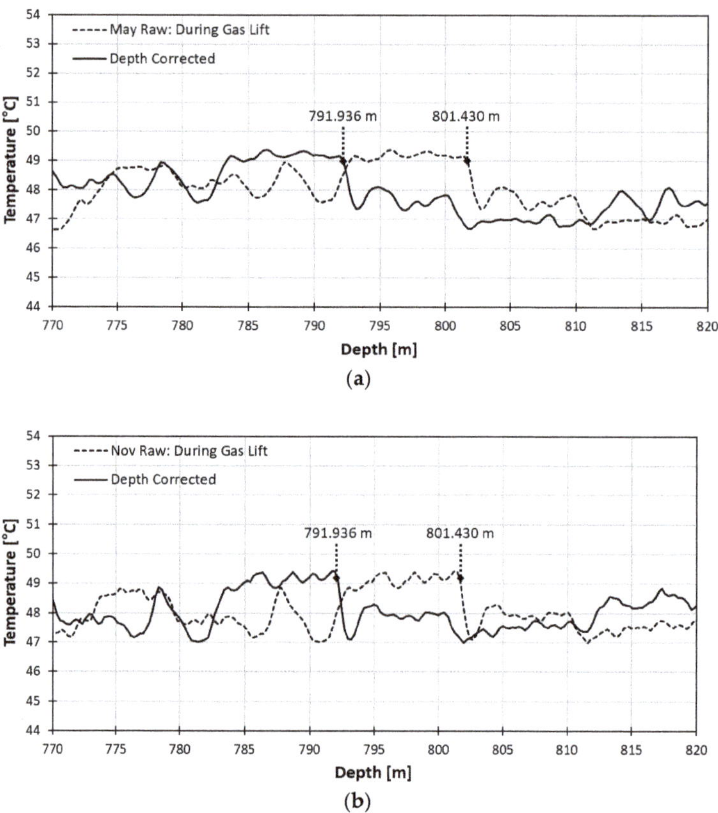

Figure 4. Comparison of depth correction effect with gas lift data from May (**a**) and November (**b**).

After correcting the depth, the depth of thermal changes due to casing transition moved up 9.494 m from 801.430 m to 791.936 m. The depth after correction, 791.936 m matches well with the true depth of the casing transition point (791.90 m) within one sampling-point difference, which is 0.25 m in terms of the DTS system's sampling interval. In both May and November, the difference to the true depth was 0.036 m.

Considering the true depth of casing transition point is known and the position of the thermal anomaly due to casing transition can be clearly identifiable, we may redistribute total depth error to each section above and below the transition point section by section under the assumption that the rate of depth mismatch is constant but different in each section. However, the recalculated sampling interval turns out to be the same to the third decimal place regardless of the section division. Thus, no major improvement in depth calibration using sections was seen.

The new sampling interval was applied to the cement injection data for the verification of the depth correction effect again as shown in Figure 5. Although there is no known absolute position of the top of the gravel pack and the bottom of cementing, the top of the gravel pack should be near 911 m and bottom of cement injection near 910 m according to the completion schematic shown in Figure 2.

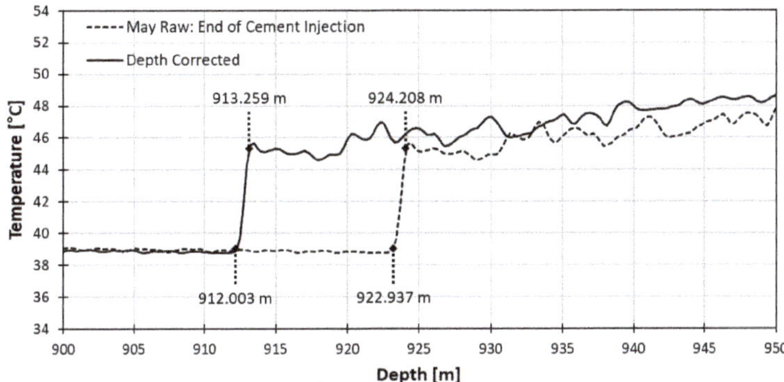

Figure 5. Temperature profiles before and after applying depth correction to data at the end of cement injection in May.

Once the cement slurry meets the gravel pack, it will permeate slowly into the gravel pack. This process can be delineated by a temperature transition between the low temperature of the cooler cement slurry and the high temperature caused by the warmer gravel pack as shown in Figure 5, and this thermal transition remains until the temperature goes back to the normal geothermal gradient.

Considering this, we can interpret the start and the end of the thermal transition zone to the top of gravel pack and the bottom of cement slurry, respectively. The top numbers and the bottom numbers in the figure indicate the depth of the bottom of cement and the depth of the top of gravel pack, respectively. After correcting the depth, the top of gravel pack moved up 10.934 m from 922.937 m to 912.003 m that is 2.003 m difference with expected depth of 910 m. The bottom of cement moved up 10.949 m from 924.208 m to 913.259 m that is 2.259 m with expected depth of 911 m.

3.2. Temperature Correction

Many DTS acquisition systems have an internal calibration system, however higher accuracy is needed for geoscience operating environments where the fiber optic cable may be under conditions of rapid change and large temperature fluctuations. Under static conditions, the accuracy of calibrated temperatures is generally within ± 0.3 °C, however under rapid operating conditions, the accuracy range becomes widened to ± 2 °C [6]. The temperature measurement accuracy also declines as the length of the fiber is increased due to the power loss of the laser pulse [3]. In our experiment, pre-calibration could not be done before installation, therefore we suggest a post-calibration method that is within the acceptable range of error, ± 0.3 °C.

Three calibration methods were investigated in the JG-M borehole in April 2018 and the experimental setting is displayed in Figure 6. In a dual-ended configuration, only one absolute temperature point is needed because knowing the temperature at a point results in having two points in this configuration. The "water tub" is the conventional method of correction by submerging about one or more pulse width lengths of fiber cable into water to find the stable absolute temperature. However, submerging enough length of fiber longer than a pulse width length into water may not be possible once the cable is installed. Therefore, two pointwise temperature correction methods are investigated. The 'heat coil' heats a portion of fiber cable near the well with a heat coil. The "in-well" is another pointwise method of correction by using the position of a heat coil to apply a slope correction and using the absolute temperature value inside the well to apply the offset correction. For all cases, the slope correction was applied to temperature profile with no events in the well, and background temperature data was collected for 15 min before the calibration method experiments.

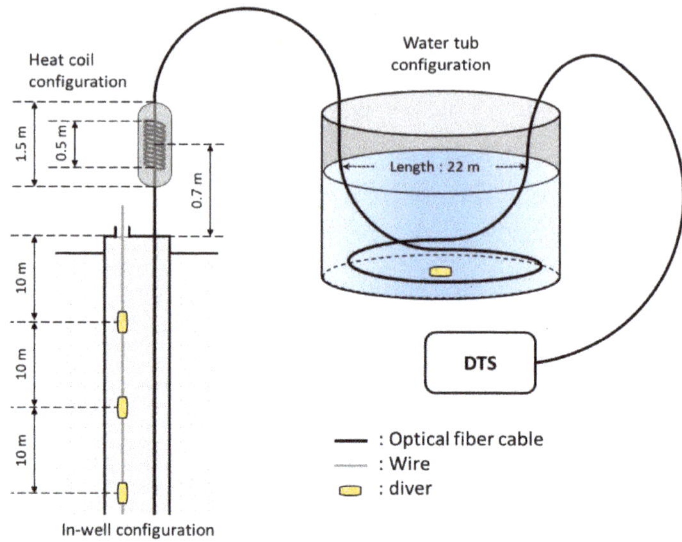

Figure 6. Schematic diagram and setting of three temperature calibration experiments.

3.2.1. Long Point Calibration Method with Mater Tub

In the water tub experiment, 22 m of fiber optic cable, which is longer than a pulse width length of a laser, was submerged in water. Figure 7 shows the process of the temperature being stabilized in the calibration bath. The temperature fluctuation became smaller as time went by, and about 1 h later, finally stabilized to that of water inside the calibration bath.

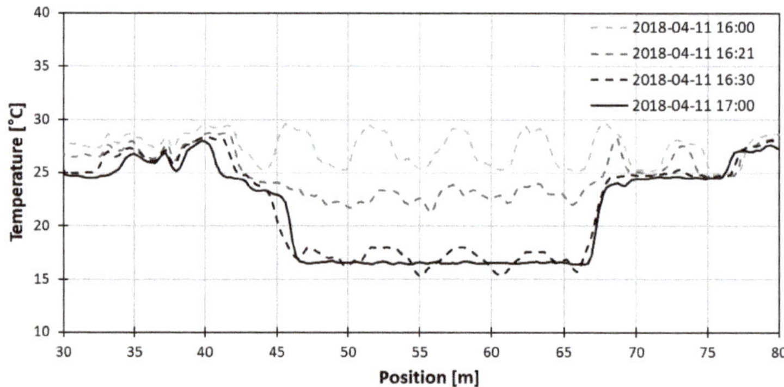

Figure 7. Temperature change with time at the fiber optic cable submerged in water.

The position of the cable submerged in water could be clearly seen both on the left and right side of the DTS data, due to the dual-ended installation of the fiber. The stable range on the left side of the data ranges from 48.896 m to 64.148 m, and the stable range on the right side of the data, though it is not displayed in the figure, ranges from 2329.826 m to 2345.078 m. Since the submerged part of the fiber optic cable must have the same temperature, the inner 60 points were averaged to eliminate fluctuation of the temperature value on each side. The midpoint on the left side was at 56.522 m, and midpoint on the right side was at 2337.706 m. To make sure the right midpoint matched the left midpoint, same number of 4887 points to max peak were used.

In order to find a better slope correction factor, the temperature change of water within an hour was considered. Even though the specific heat of water is lower than air, the minimum difference of average temperature between the left side and the right side was considered for better estimation of the reference temperature. The ten-minute range which gave the minimum difference between the left and the right side is from 17:53 to 18:02. From the best ten-minute range, the minimum, maximum, and average slope value was used to find the slope equation. The maximum slope gave the minimum difference between the left and the right value. The minimum slope gave the maximum difference between the left and the right value. The slope curve of minimum, maximum and average slope is shown in Figure 8a. In the figure, "all" represents using all the data points without considering the temperature change of water, "DTS" represents the best ten-minute data in terms of minimum difference between the left and the right endpoints. The "Diver" represents the most stable temperature measured with the diver in the water tub for ten minutes.

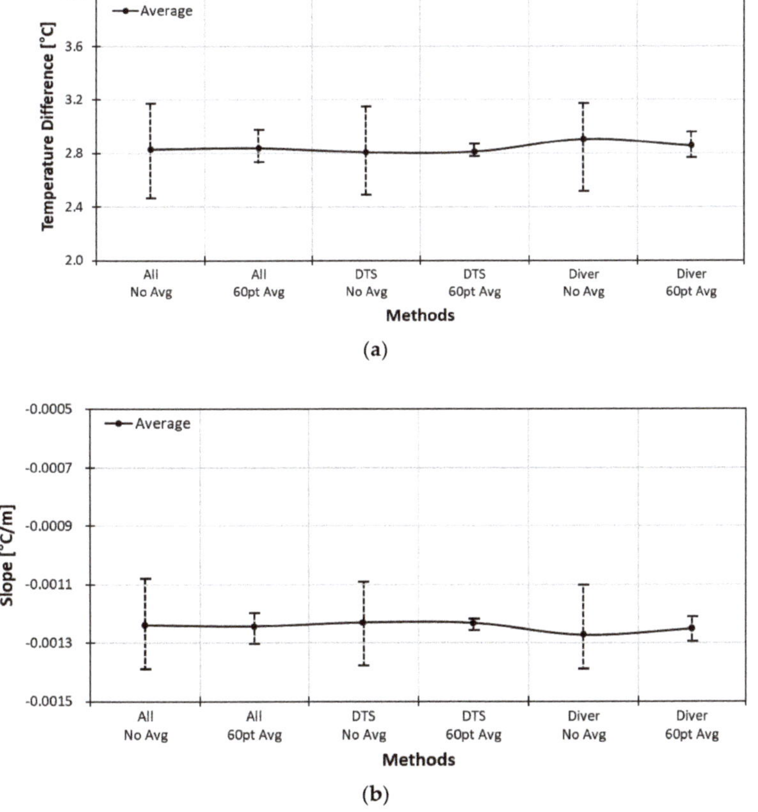

Figure 8. Comparison of long point water tub methods in terms of temperature difference (**a**) and slope range (**b**).

The "DTS" method gave the best result when 60 points were averaged to eliminate fluctuations of the reference temperature as shown in Figure 9a. This was confirmed again by comparing all three methods as displayed in Figure 8. The averaging of 60 points was the most effective in all three cases as it had smaller slope and difference range.

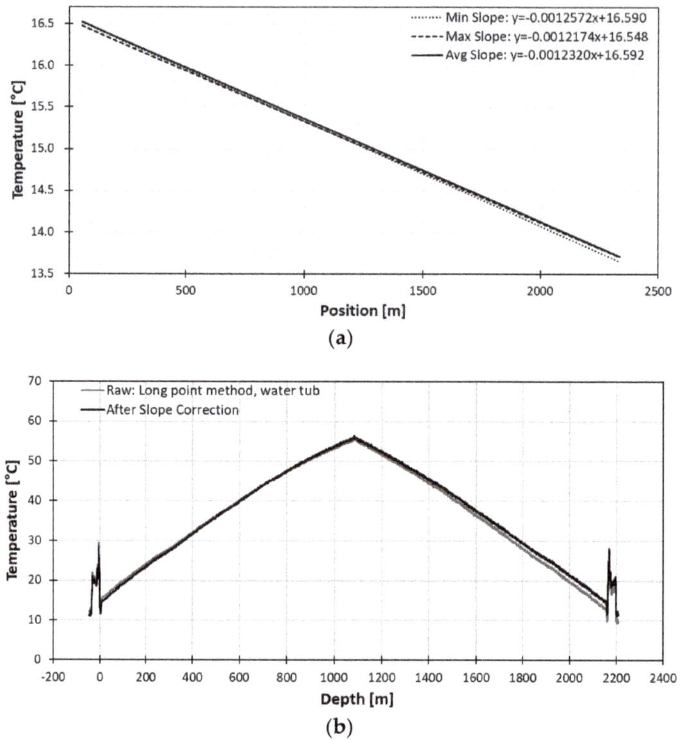

Figure 9. (a) Slope curve of long point method with best 10-minute range of DTS 60-point average from 17:53 to 18:02 and (b) heat profile after long point water tub temperature correction.

For all three cases, the slope equation was applied to the background data from April 12 before any experiment because it is best to apply to the data without any events. The temperature accuracy improved after temperature correction, as the difference to true temperature decreased from 2.6 °C before correction to 0.02 °C, as illustrated in Figure 9b.

3.2.2. Pointwise Calibration Method with Heat Coil

The heat-coil was used to locate the position on the DTS data for slope correction. A heat coil of 0.5 m in length was wrapped around the fiber optic cable near the well and insulated with a sponge tube of 1.5 m and wrapped in tape to prevent the temperature from fluctuating due to the wind as shown in Figure 6. The distance from the middle of heat coil to the well was 70 cm. In the heat coil experiment, obtaining a stabilized absolute temperature was difficult because the heat coil was shorter than the pulse width of interrogation system and air temperature varied too quickly. Therefore, the ambient temperature value at the heat coil position was used for slope correction. On April 12, the fiber optic cable was heated with heat coil to about 50 °C for about 10 minutes and the cooling observed for 27 min.

The slope curves are displayed in Figure 10a, which shows very different slope difference from the long point water tub method. In this case, only slope correction was applied because the absolute temperature was not determined.

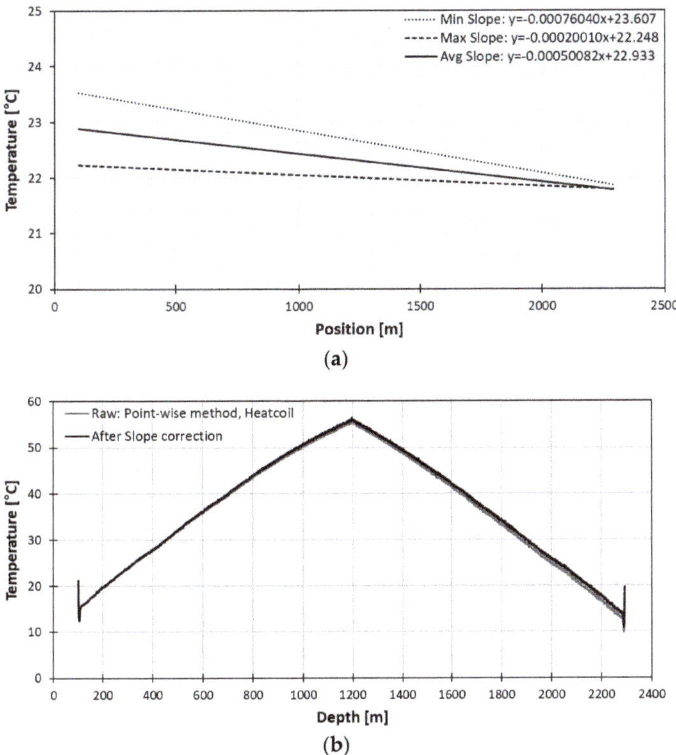

Figure 10. (**a**) Slope curve of pointwise method with heat coil and (**b**) heat profile after pointwise heat coil slope correction.

Even with only slope correction, we could evaluate if the calibration is done correctly by comparing the two endpoints because two endpoints on the left and the right should have the same temperature value after correction. However, we can identify in Figure 10b, that temperature at the right endpoint, even with an increase of 1.1 °C after correction, still has a 1.5 °C difference with that of the left endpoint.

3.2.3. In-Well Pointwise Calibration Method

To compensate the difficulty of obtaining a stable absolute temperature with the heat coil method, three divers were lowered into the well at 10 m, 20 m, 30 m below groundwater which has lower specific heat constant than the air. The diver position on the DTS could be located by counting from the known heat-coil position. The position could also be confirmed by temperature change when the fiber optic meets groundwater, which was only 1 m to in-well water from the heat-coil position.

The slopes found with the pointwise in-well method are presented in Figure 11. Compared to the long point water tub method, the variation ranges are wider than the long point method but the average slopes are similar to that of the long point method. The slope and its range did not vary greatly with the depth of diver. The temperature after correction with pointwise in-well diver at 10 m is presented in Figure 12. The temperature accuracy improved as the difference to true temperature is only 0.06 °C compared to the difference of 2.8 °C before correction.

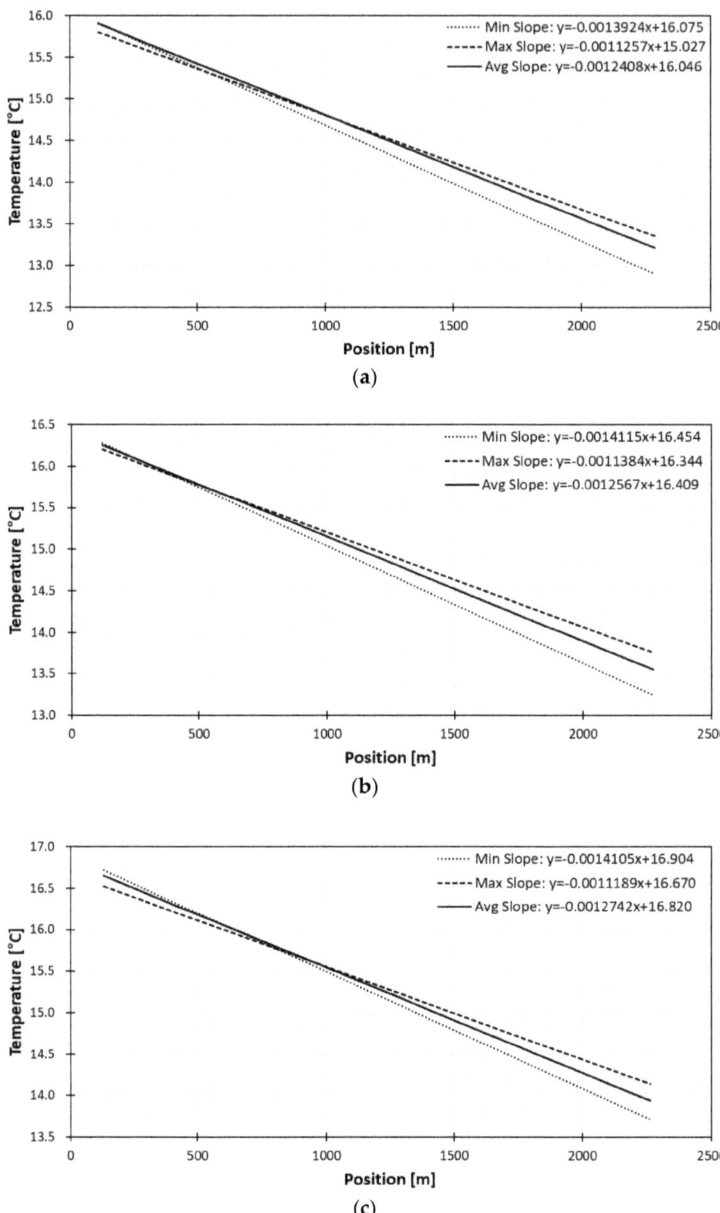

Figure 11. Slope curves of pointwise calibration method by in-well diver at (**a**) 10 m, (**b**) 20 m, (**c**) 30 m.

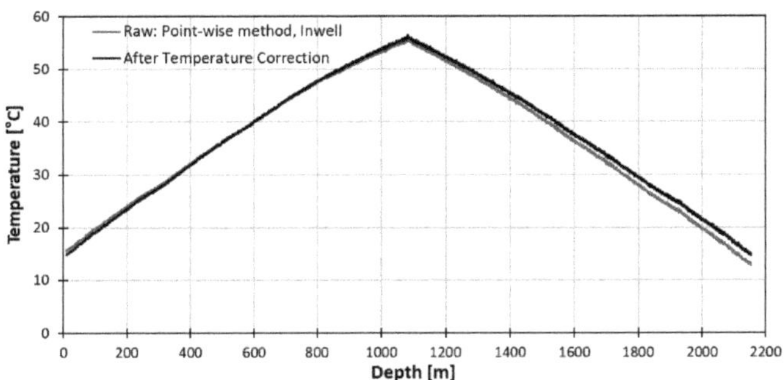

Figure 12. Temperature profile after correction by pointwise method with 10 m in-well diver.

3.2.4. Discussion

The slope and temperature difference curves of three calibration methods are summarized in Figure 13. The long point calibration method with water tub gave the best result. Although the pointwise calibration method using the diver showed a larger range than the long point method, it gave an average slope and the average temperature difference comparable with that of the long point method. Comparing the variation range of the heat coil method and the in-well method, we can conclude that it is crucial to obtain accurate and stable absolute temperatures.

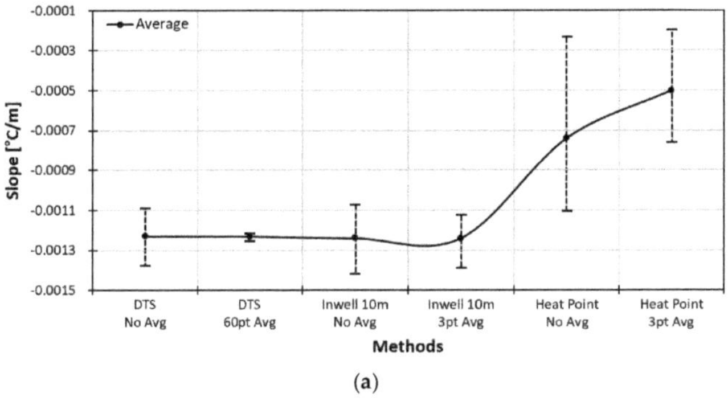

(a)

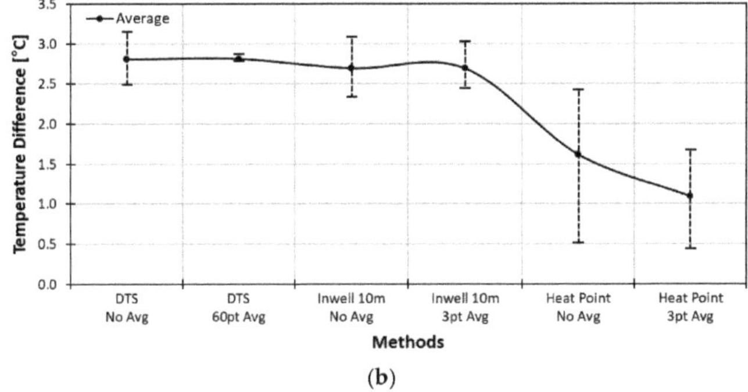

(b)

Figure 13. Slope curves (a) and temperature difference (b) of post-calibration method.

The stability analysis was carried by dividing the total round-trip length of the fiber into 10 sections in terms of the same number of sample points per section. The downward trip sections, from wellhead to well bottom, are marked as L5–L1 and the upward trip sections are marked as R1–R5. The number of samples per each section is 897 points for the long point method, 855 points for the pointwise method with in-well diver, and 862 points the pointwise method with heat coil.

The stability curve of all three calibration methods is as shown in Figure 14, which displays the stabilities at 13 points including two top points (Endpt), a bottom point (Maxpeak), and ten midpoints of each section. The slope stability curve of long point calibration method has a range of +0.06 °C and −0.03 °C. The slope stability curve of pointwise calibration method with in-well of 10 m has a range of +0.35 °C and −0.25 °C. The slope stability curve of pointwise calibration method with heat coil has a range of +0.7 °C and −0.6 °C. The temperature difference increases as the length of fiber increase in all three calibration methods as expected due to the power loss that increases as the traveling distance become longer.

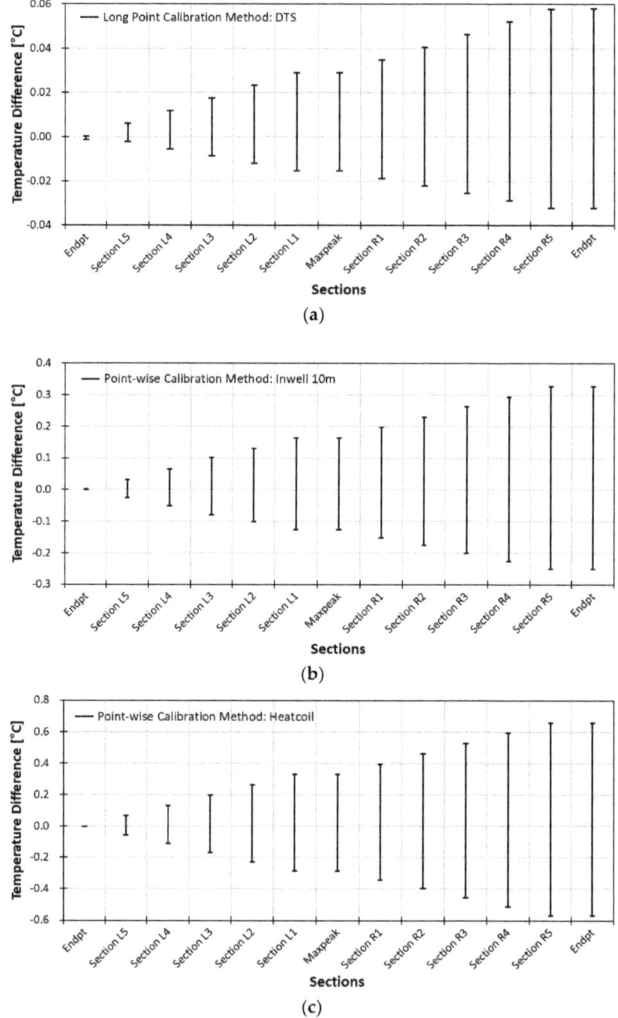

Figure 14. Slope stability of (**a**) long point calibration with water tub (**b**) pointwise calibration with in-well and (**c**) pointwise calibration with heat coil.

The appropriate temperature error range is within 0.3 °C, and both water tub and in-well data fall within the range. However, the error range with water tub slope correction is within 0.3 °C with stability of 0.05 °C and the temperature error range with diver slope correction is within 0.1 °C with a stability of 0.3 °C. Therefore, even though the long point calibration works best, the pointwise calibration method can be a useful alternative where long point calibration cannot be done.

4. DTS Monitoring of Well Completion Processes

The uniqueness of the Janggi field DTS data is the continuous real-time monitoring of the whole well completion process from gravel packing to complete curing without well intervention. The heat map obtained from May 11th to June 4th is shown in Figure 15. Anomalies numbered from 1 to 4 correspond to gravel packing, cement injection, gas lift, and cement curing, respectively.

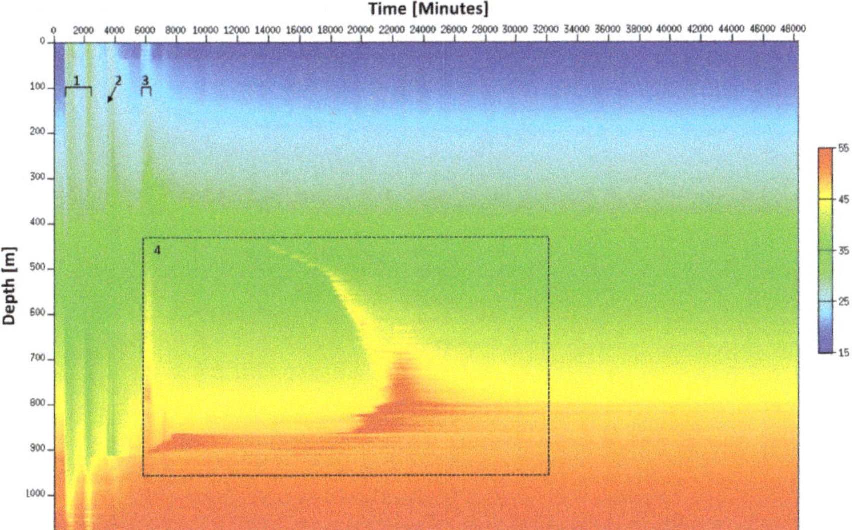

Figure 15. Heat map obtained during well completion where the gravel packing is denoted by 1, the cement injection by 2, the gas lift operation by 3 and the cement curing process by 4.

4.1. Gravel Packing

There has been monitoring of production in gravel packed completions but monitoring the gravel packing process itself was not yet done [7]. At the Janggi field site, fiber optic cable was connected directly after casing installation to the DTS unit, which allowed real-time monitoring of the gravel packing process throughout the entire length of the well. The detailed temperature profile during gravel packing can be seen in Figure 16 which outlines the major events during the gravel packing process.

The reference (black line) refers to the temperature profile at the end of well completion. The gravel packing process can be characterized by the decrease of temperature below the reference line. The temperature decreases because of the cooling effect due to the injection of sand mixed with fluid that is cooler than the borehole fluid inside the well. In the deeper parts of the well, high temperature anomalies with oscillating pattern can be seen due to water circulating but leaving sand to be accumulated. The sand deposited in the annulus plays a role of thermal bridge between the surrounding formations and fiber cable on tubing. Because sand has higher heat conductivity than water, such a thermal bridging effect explains the high temperature anomalies. The section where sand is not accumulated does not have this connectivity between formations and cable still shows a low temperature trend. The point where the temperature clearly drops indicate the top level of the gravel pack. The progression up of this anomaly along the well shown in the figure (May 11 19:39 and May 12 14:51) clearly shows that the top of gravel pack can be monitored well by DTS. After gravel packing, the top of gravel pack estimated from DTS data (May 12) matches closely with the planned depth of 910 m as mentioned in Section 3.2.1.

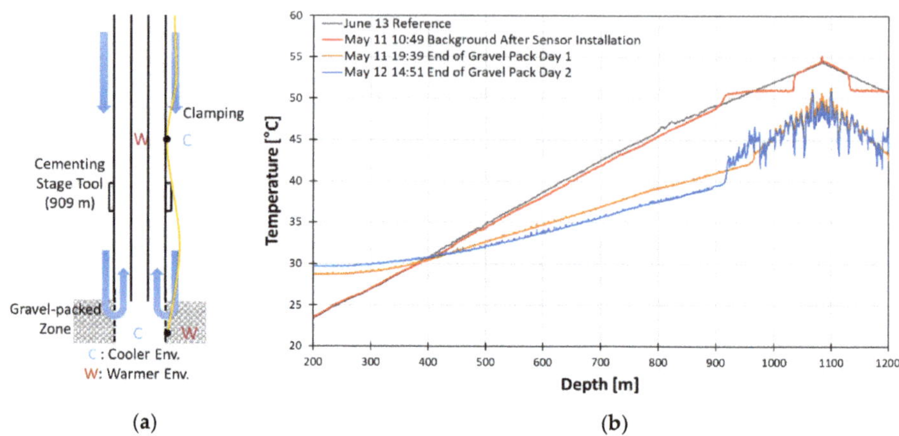

Figure 16. Schematic of gravel packing operation (**a**) and temperature profile during the gravel packing operation (**b**).

In the low temperature anomaly section, small peaks that seem like noise are identified. However, they are anomalies due to the clamps being spaced at approximately every 9 m. The clamps provide more direct connection between inside and outside of tubing wall. The effect of clamps appears as high temperature anomalies in the unpacked section while it appears as low temperature anomalies in the gravel-packing zone. In the case of the high temperature anomalies, inside the casing is warmer than the annulus because the sand was directly injected into the annulus. Consequently, the clamped point of the DTS cable is directly connected to the warmer environment while the other parts are not. In the case of low temperature anomalies, inside the casing is cooler in the packed region. Thus, the clamping point has the reverse effect compared to the previous case. The anomalies associated with clamps indirectly shows high resolution of DTS and its usefulness in dynamic condition.

From the heat map in Figure 15, we can identify that two gravel packing processes that were carried out over the two days have similar patterns to each other. The temperature is increasing above the gravel-packed zone and decreasing in the unpacked zone during the operation. The short burst of cooling at the top may be explained due to either flow change or colder fluid injection or breaks during gravel packing operation.

The ability to monitor the progression of sand volume allows the possibility to control the amount of sand volume to reach the top of the intended depth and the capability to check for possible plugging as the sand is poured. All these factors help optimize the gravel packing process at the field level as the change in temperature is displayed without interrupting the process. The gravel packing process also showed potential for detecting sand production and the response to sand production can be optimized to diminish the risk of reduced well productivity.

4.2. Cement Injection and Curing Process

After the gravel packing operation, the cement slurry was injected to the casing to complete the well. The temperature profile of the cement injection process is displayed in Figure 17. The temperature before cement injection has a lower temperature than the reference, which indicates the effect from gravel packing operation still remains and needs more time to recover to an undisturbed normal geothermal gradient. As the slurry is injected, the temperature is being gradually lowered until the end of the injection because the cement slurry is cooler than the borehole fluid. The top of the gravel pack and the bottom of cement can be determined as mentioned in Section 3.2.1.

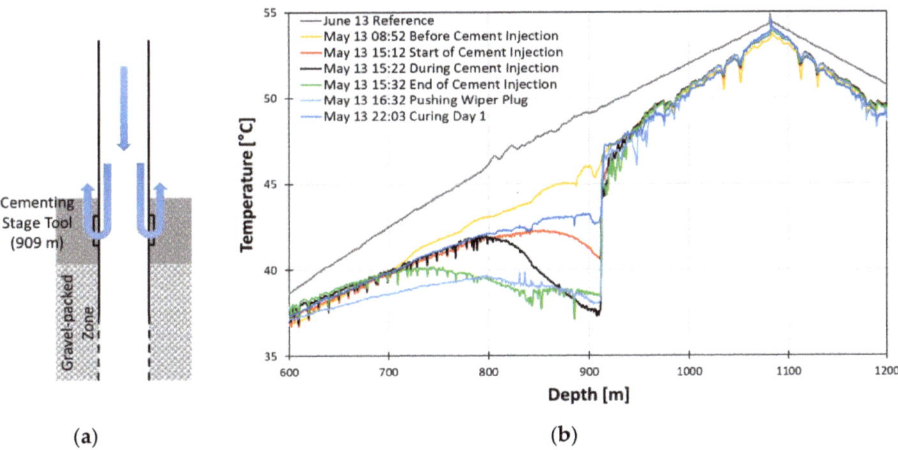

(**a**)　　　　　　　(**b**)

Figure 17. Schematics of cementing injection process (**a**) and temperature profiles during the cement injection process (**b**).

The major events during curing process from right after injection to a month after injection are shown in Figure 18.

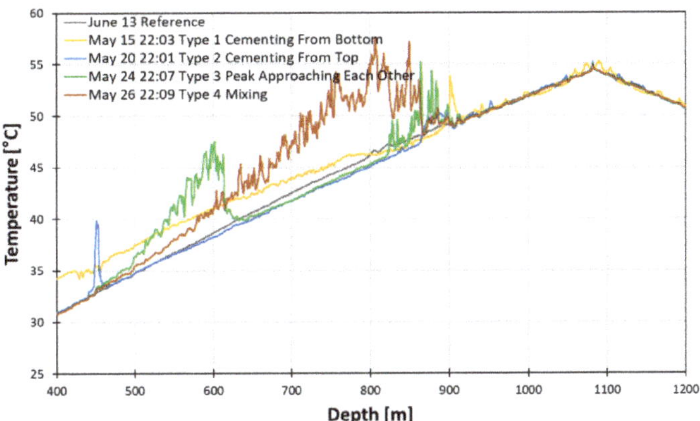

Figure 18. Temperature profiles during the curing process in detail.

In the yellow line denoted as type 1, though the high temperature anomaly due to gas lifting remains at depths shallower than 850 m, we can identify the peak near the stage cement tool which can be interpreted as the start of the exothermic process of the curing. The effect of the gas lift anomaly is gone in type 2 blue line; however, another peak appears near 450 m. The peak indicates the start of curing from the upper part. The type 3 green line displays two peaks from the top and bottom boundary. These boundaries move toward each other, indicating the curing process propagates from the boundary to the intermediate. The type 4 red line displays the two peaks finally meet and start settling down.

From the heat map in Figure 15, the major trends of the whole cementing operation can be observed. We can identify when and where the curing starts, when the curing finally done, and how long it took for complete curing from the heat map provide by constant DTS monitoring.

The cementing is the final critical steps to ensure the integrity of the well. Therefore, monitoring the entire process from cement slurry injection to curing in real-time can provide key information in order to control and optimizes the process. For example, the progression of injection outlines both injection rate and volume of cement, which offers the ability to check for possible plugging during injection. The constant monitoring of the entire well also allows correlating unusual anomaly to exothermic reaction due to the settling of the cement, which can provide better insights to the different stages of exothermic reactions. As the time of complete curing could be accurately determined by the homogeneity in temperature, monitoring curing process help move on to the next stage with confidence.

4.3. Gas Lift

The gas lift process was carried out after cement injection but during cement curing. During the gas lift, the N_2 gas was released, and the fluid was pulled upwards due to pressure. The major events of one cycle of gas lift operation is displayed in Figure 19. The temperature increased during gas lift operation in the data below 791.90 m. The increase of temperature is due to pulling up of the deeper and warmer reservoir fluid. The high temperature anomaly associated with gas lift operation has a highly oscillating pattern. In our case, however, its amplitude is different at two regions above and below 791.90 m. The difference in amplitude of the oscillation is due to heat conductivity difference of the casing materials. The 791.90 m is the transition point of the two casing materials from steel to FRP. Since steel has higher heat conductivity than FRP, we can observe a higher amplitude of oscillation in the steel casing region than the FRP casing region.

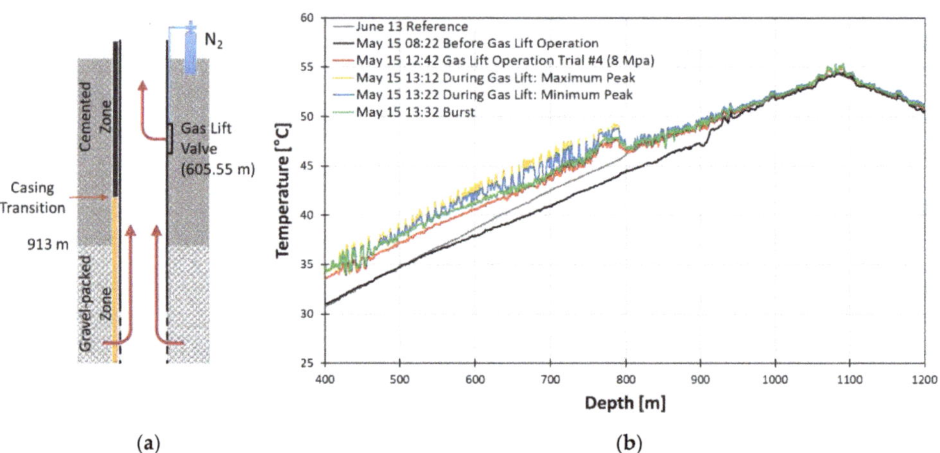

Figure 19. Schematics of gas lift process (**a**) and temperature profiles during the gas lift in May (**b**).

The heat map during the gas lift cycles which shows the global trend during the gas lift is displayed in Figure 20. The number of repeated patterns during gas lift matches the number of gas lift cycles, which was eight cycles. The width of each repeated pattern is the duration of each cycle. Even after the gas lift operation was finished, its effect lasted a day to return to temperature before the gas lift operation began. The temperature did not completely recover due to the starting of curing process near 900 m during the middle of gas lift operation.

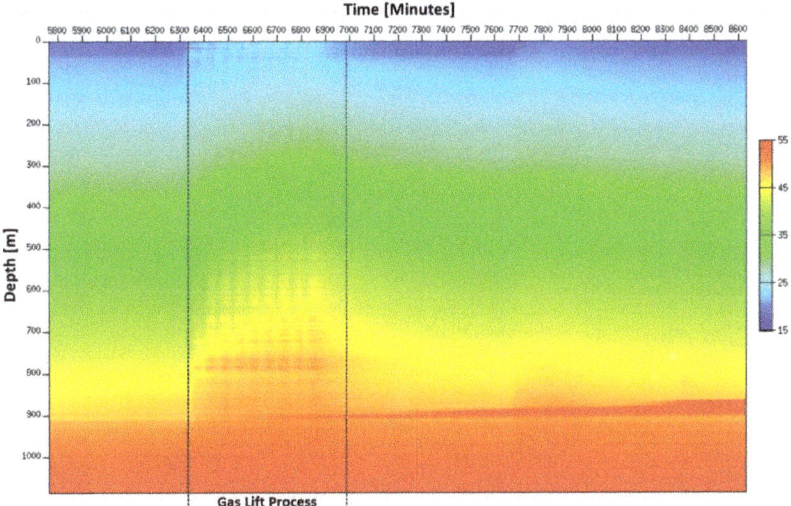

Figure 20. Heat map of entire gas lift operation in May

As mentioned in Section 3.1, our gas lift data shows not only a typical highly oscillating thermal anomaly pattern of gas lifting event but also the effect of casing transition, which supports the confirmation of the accuracy of depth correction. The DTS data has the potential to detect multiple points of injection and determine gas rate through each injection point. All these are possible as the event is happening and require no shut-in when measuring DTS temperature data [8]. Therefore, DTS helps capture the temperature response throughout the entire length of the well simultaneously as the event is happening.

5. Conclusions

We have presented continuous real-time DTS monitoring data of the whole well completion process from gravel packing to complete curing without well intervention, and discussed the observed thermal anomaly patterns during each well completion process.

The real-time DTS monitoring data revealed that each well completion process can be characterized by its own distinctive temperature anomaly patterns. In addition, it is remarkable that the upward progression of sand and cement slurry and the thermal anomaly due to curing of cement can be monitored. This clearly shows that pre-calibrated DTS, due to its continuous real-time and high spatial resolution nature compared to conventional point sensors and wireline tools, can be a powerful tool not only for monitoring, but also for controlling and optimizing well completion processes without intervention in each process. However, because the thermal event characteristics associated with each well completion process are site- and condition-dependent, further studies to quantify the temperature anomaly depending on factors like injection rate, hydraulic and thermal properties of the formation and fluid etc. are required for quantitative control and optimization of well completion process by DTS.

For depth calibration, the conventional method distributing the error to the total length is enough. The accuracy could be enhanced more if more points where the depth are known are available in addition to the top and the bottom of well. In the case of temperature calibration, pre-calibration using the conventional method is best. However, our investigation shows that the pointwise method can be an alternative with acceptable accuracy of less than 0.3 °C and stability less than 0.3 °C.

Author Contributions: Data curation, D.S.L.; Formal analysis, D.S.L., S.-J.C., K.G.P. and C.L.; Funding acquisition, K.G.P.; Investigation, D.S.L., S.-J.C. and K.G.P.; Methodology, K.G.P.; Supervision, K.G.P.; Validation, C.L. and K.G.P.; Writing – original draft, D.S.L.; Writing – review & editing, K.G.P.

Funding: This work was supported by the Korea CCS R&D Center (Korea CCS 2020 Project) grant funded by the Korea government (Ministry of Science and ICT) in 2017 (KCRC-2014M1A8A1049287).

Conflicts of Interest: The authors declare no conflict of interest.

References

1. Hurtig, E.; Großwig, S.; Kühn, K. Fiber optic temperature sensing: application for subsurface and ground temperature measurements. *Tectonophysics* **1996**, *257*, 101–109. [CrossRef]
2. Schlumberger. The Essentials of Fiber optic Distributed temperature Analysis. Sugar land: Schlumberger Educational services. 2009. Available online: https://www.slb.com/resources/publications/books/fiber_optic_distributed_temperature_analysis_book.aspx (accessed on 7 November 2018).
3. Smolen, J.; van der Spek, A. *Distributed Temperature Sensing—A DTS Primer for Oil & Gas Production*, 1st ed.; Shell International Exploration and Production, Shell: Hague: Netherlands, 2003; pp. 1–50.
4. Freifeld, B.M.; Trautz, R.C.; Yousif, K.K.; Phelps, T.J.; Myer, L.R.; Hovorka, S.D.; Collins, D. The U-Tube: A novel system for acquiring borehole fluid samples from a deep geologic CO_2 sequestration experiment. *J. Geophys. Res.* **2005**, *110*, B10203. [CrossRef]
5. Freifeld, B.M.; Finsterle, S.; Onstott, T.C.; Toole, P.; Pratt, L.M. Ground Surface temperature reconstruction: Using in situ estimates for thermal conductivity acquired with a fiber-optic distributed thermal perturbation sensor. *Geophys. Res. Lett.* **2008**, *35*, L14309. [CrossRef]
6. Hausner, M.; Suárez, F.; Glander, K.; Giesen, N.; Selker, J.; Tyler, S. Calibrating Single-Ended Fiber optic Raman Spectra Distributed Temperature Sensing Data. *Sensors* **2011**, *11*, 10859–10879. [CrossRef] [PubMed]
7. Brown, G.A.; Pinzon, I.D.; Davies, J.E.; Mammadkhan, F. Monitoring Production from Gravel-Packed Sand-Screen Completions on BP's Azeri Field Wells Using Permanently Installed Distributed Temperature Sensors. In Proceedings of the SPE Annual Technical Conference and Exhibition, Anaheim, CA, USA, 11–14 November 2007.
8. Hemink, G.; van der Horst, J. On the Use of Distributed Temperature Sensing and Distributed Acoustic Sensing for the Application of Gas Lift Surveillance. *SPE Prod. Oper.* **2018**, *33*, 896–912. [CrossRef]

Article

Research on the Earth Pressure and Internal Force of a High-Fill Open-Cut Tunnel Using a Bilayer Lining Design: A Field Test Using an FBG Automatic Data Acquisition System

Tianyuan Xu *, Mingnian Wang, Li Yu *, Cheng Lv, Yucang Dong and Yuan Tian

Key Laboratory of Transportation Tunnel Engineering, Ministry of Education, Southwest Jiaotong University, Chengdu 610031, Sichuan, China; 19910622@163.com (M.W.); lvchengaust@163.com (C.L.); yucang_dong@my.swjtu.edu.cn (Y.D.); ytian_ty@163.com (Y.T.)
* Correspondence: x_tianyuan@outlook.com (T.X.); yuli_1026@swjtu.edu.cn (L.Y.)

Received: 30 January 2019; Accepted: 20 March 2019; Published: 27 March 2019

Abstract: When there are railway tunnels on both sides of a valley, a bridge is usually built to let trains pass. However, if the valley is very close to an urban area, building an open-cut tunnel at the portal and then backfilling it to create available land resources for the city and to prevent excavation slag from polluting the environment would be a wise choice. This has led to the emergence of a new type of structure, namely, the high-fill open-cut tunnel. In this paper, by performing an automatic long-term field test on the first high-fill open-cut tunnel using a bilayer design in China, the variations of earth pressure and structural internal force during the backfilling process were obtained, and different tunnel foundation types were studied. The results showed that the earth pressure significantly exceeded the soil column weight, with a maximum earth pressure coefficient between 1.341 and 2.278. During the backfilling process, the earth pressure coefficient increased at first and then decreased slowly to a relatively stable value, and a stiffer foundation would make the structure bear higher earth pressure (1.69 times the normal one observed during monitoring). The change of internal force had two stages during backfilling: before the backfill soil reached the arch crown, the internal force of the lining changed slowly and then grew linearly as the backfill process continued. Moreover, the axial force ratio of the inner and outer linings was close to their thickness proportion, and the interaction mode between the two layers was very similar to the composite beam.

Keywords: high-fill open-cut tunnel; bilayer lining; FBG sensors; automatic monitoring system; earth pressure of open-cut tunnel

---

1. Introduction

In China, the high-speed railway network is rapidly extending into the mountainous area of the southwest. Due to the geological conditions of the southwest region, there has been a substantial increase in the number of tunnels needed, some of which are near cities. There is a special situation for some of these cities: sometimes, there is a valley between two tunnels that is very close to urban areas. In a normal situation, a bridge must be built above the valley to let trains pass. However, if an open-cut tunnel were built at the portal and then backfilled with the engineering spoils from the tunnel excavation, available land resources can be created for cities, which would prevent waste from polluting the environment and lead to the emergence of high-fill open-cut tunnels, as shown in Figure 1. Compared with the traditional open-cut tunnels, a high-fill open-cut tunnel has the following two characteristics:

(1) The height of backfill layer can reach up to 30–40 m, which is almost five times larger than a common one;

(2)   In order to bear such high earth pressure, the thickness of the tunnel lining can reach up to 2 m or more, which is 4–5 times thicker than those with shallow overburden. Details of existing high-fill open-cut tunnels in China are shown in Table 1.

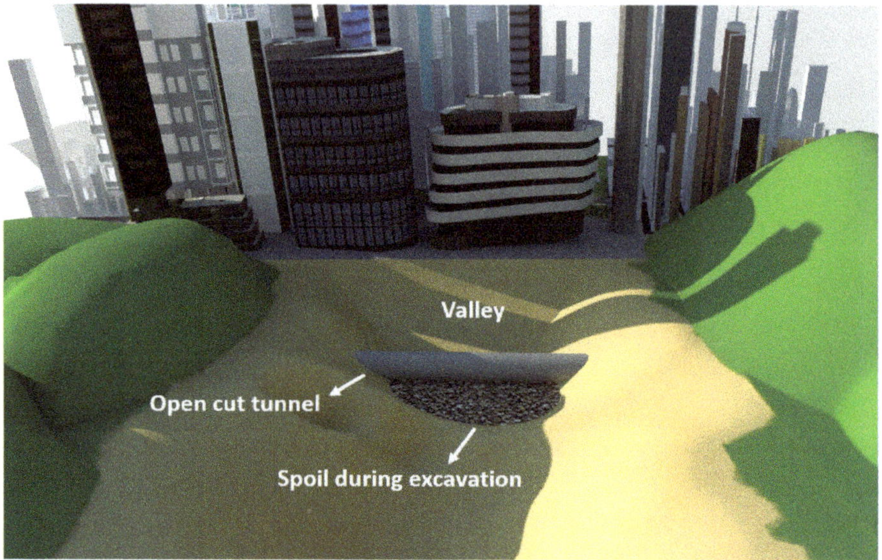

Figure 1. Open-cut tunnel built in a valley close to a city.

Table 1. Details of three existing open-cut tunnels in China.

Project Name	Backfill Height—From Tunnel Crown to Backfill Surface (m)	Lining Thickness (m)
Lanyu open-cut tunnel	40	1.8–2.4
Longdongbao open-cut tunnel	33	2.8–3.0
Fengdu open-cut tunnel	28	1.9–3.3

These two characteristics present two research difficulties:

(1) A large amount of hydration heat can be generated after construction if a monolithic design is adopted for these superthick concrete linings, causing a significant temperature difference between the internal and external surfaces of the lining, which would result in structural shrinkage cracks and safety issues. Therefore, a more reasonable structure design must be adopted.

In order for the lining structure to meet the bearing capacity requirement for such a high backfill layer and to minimize the shrinkage cracks at the same time, a bilayer design is used for high-fill open-cut tunnels which divides the overthick lining into two parts—outer and inner linings. Two layers are concreted separately at 14-day intervals, resulting in a significant decrease of hydration heat generated by each layer of lining. However, the bilayer lining design initially appeared in shield tunnels, which consist of segmental and secondary linings, and has never been used in open-cut tunnels.

Yang et al. [1] carried out a numerical study on the performance of segmental and secondary linings as well as the stress transmission between them. They reported that linings with rebar had a combined bearing capacity, while linings with membranes had separate bearing capacities. Vogel et al. [2] examined double-shell linings in respect to direct shear stress capacity. Their results revealed that a spray-applied waterproofing membrane was able to transfer stresses between both concrete linings but could not transfer any shear stress. Su and Bloodworth [3] presented laboratory tests on beam samples cut from bilayer linings. They compared the influence of different membrane thicknesses and substrate roughness and provided parameters based on test results for further research and design. Their subsequent research [4] developed a composite mechanical behavior quantification method. Chuan et al. [5] presented several load tests on bilayer shield tunnel models, the results of which revealed that the secondary lining bore most of the bending moment. Their following study focused on a calculation model for bilayer linings [6].

The abovementioned studies provide a better understanding of the mechanical characteristics of the bilayer lining tunnel design, as well as the calculation model and load shearing ratio needed to design the structure. However, the construction sequence of open-cut tunnels is completely the opposite of that of shield tunnels, in which the lining is built at last. Moreover, the size and shape of open-cut tunnels are also quite different from shield tunnels (circular design), as shown in Figure 2.

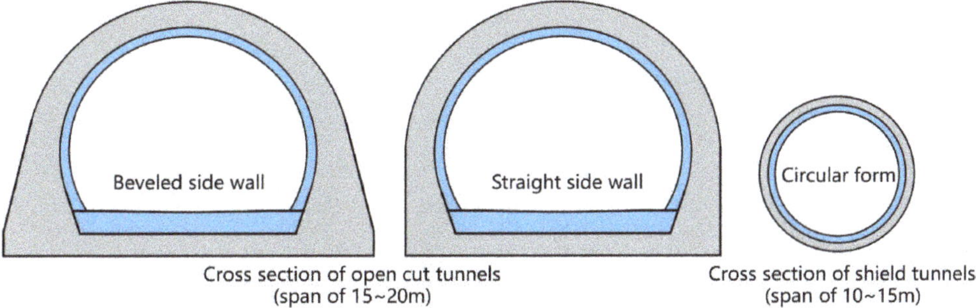

Figure 2. Comparison of size and shape of open-cut and shield tunnels using the bilayer design.

(2) Under such a high backfill layer, it still remains unclear whether the earth pressure on the tunnel crown equals the overburden pressure. Moreover, a deep foundation has been adopted for high-fill structures in some situations, which can also affect the earth pressure of the structure.

A high-fill situation usually occurs in culverts and pipelines in mountainous areas and has never been adopted in railway open-cut tunnels. Qiang et al. [7] analyzed several calculation methods of earth pressure on a culvert crown. The results showed that under a high embankment fill condition, the calculated results by the "neutral point" method were closer to the values measured in situ. Li et al. [8] carried out a field monitoring and numerical simulation to investigate the state and distribution of stress on the exterior surface of slab-culverts under high embankments. The results showed that the earth pressure on culverts was larger than the self-weight of filling, and the pressure distribution was uneven. Chen et al. [9] presented a new formula to calculate the vertical earth pressure on a culvert, the calculation of which results were compared with field tests. Zhang et al. [10] carried out a centrifugal experiment on the earth pressure distribution of a culvert top. The results showed that a box culvert with a pile foundation bore higher earth pressure than a normal one, and they suggested that soil could be backfilled first on both sides of the culvert before it was constructed. Some researchers developed a load calculation method for deeply buried culverts by summarizing the displacement distribution characteristics on the culvert crown [11,12]. Additionally, controlling displacement was considered to reduce the vertical pressure on culverts. Moreover, backfill materials and some engineering measures also affect the earth pressure of high-fill structures. Meguid [13] presented an experimental investigation to measure the earth pressure distribution on a rigid pipe backfilled with tire-derived aggregate. The average measured earth pressure above the crown of the pipe was found to be as low as 30% of the overburden pressure. An optimum soft zone geometry for imperfect trench installation was proposed to maximize the reduction of the earth pressure on buried corrugated steel pipes [14], and the maximum wall stress was reduced by 69%.

According to previous research, the earth pressure of an open-cut structure is closely related to the side slope angle and foundation type. As a more rigid foundation usually results in a higher earth pressure, it is suggested to avoid using a deep foundation in the design of culverts. However, some culverts still use a strengthened foundation to achieve better performance in particular situations [15]. For railway open-cut tunnels, the track surface subsidence must be strictly controlled to meet the operation requirements of high-speed trains, making it necessary to use deep foundations for railway open-cut tunnels in poor geological conditions. In conclusion, a high-fill open-cut tunnel using a bilayer lining design is very different from the current culvert or shield tunnel in terms of construction method, shape, size, and foundation type, and its mechanical characteristics are not clear yet. Besides, current research rarely contains long-term field measurement data, especially measurements conducted when the structure is built on different types of foundations.

In geotechnical and structural domains, fiber Bragg grating (FBG) sensors are applied to measure structural strain [16–19], seepage pressure [20,21], temperature [22], and vibration [23]. However, the monitoring work for tunnels does not continue after the lining has stabilized. Therefore, vibrating wire sensors are more widely used to reduce unnecessary costs, and these sensors typically require manual data acquisition. However, for dynamic or long-term data acquisition from tunnels, such as the real-time safety monitoring of tunnels in special geological conditions and measuring the vibrations generated by train travel [24,25], it is more sensible to use FBG sensors, as they have strong anti-interference and long-term stability advantages. In our research, FBG pressure cells and strain gauges were adopted to acquire the soil pressure and structural internal force for further analysis.

Moreover, for an open-cut tunnel, once the backfilling process begins, personnel have to enter the tunnel from the undercut portal, which sometimes is more than 10 km from the test section. Therefore, an automatic data acquisition system was adopted to provide more timely and accurate data. These test results can be a useful complement to both high-fill open-cut and bilayer structures to help provide a deeper understanding of them.

2. Field Test Procedure

2.1. Tunnel Description

The field test was carried out at the first high-fill open-cut tunnel in Fengdu, China, where the bilayer lining design was firstly adopted, with a 0.5-m-thick C35 concrete inner lining covered by a 1.4–2.8-m-thick C35 concrete outer lining. Further, a waterproof layer consisting of nonwovens and a polyethylene sheet was applied between the inner and outer linings as well as on the surface of the outer lining. The outer lining was constructed first. When the outer lining was stabilized, the inner lining was constructed, and finally, the backfilling process was carried out.

This open-cut tunnel is 373-m long with a backfill height up to 22–28 m (from tunnel arch crown to the backfill surface) and a 100-m horizontal backfill range on both sides of the tunnel axis. Both sides of the tunnel were filled symmetrically with 0.5-m-thick earth in each cycle, which took 6–7 days, and the whole backfill process lasted about 9 months. Also, a C30 concrete dam was adopted as the foundation for some part of the open-cut tunnel to control the subsidence of the tunnel bottom. The remaining parts of the open-cut tunnel were set on bedrock with weatherproof protection. More design details are shown in Figures 3–5.

2.2. Automatic Data Acquisition System

The application of intelligent test systems and methods in tunnels has been a growing trend in recent years [26–29]. In our field test, an automatic data acquisition system was developed and adopted, using FBG sensors to measure the structural internal force and soil pressure. The system was developed by the research team in collaboration with Sensorlead Technology Co., Ltd, Shanghai, China. It can capture the wavelength data of the sensors at 100 Hz and allows users to input different formulas to convert wavelength data into stress, pressure, displacement, etc. Moreover, it has a remote operating subsystem to allow users to view the data anytime by personal computers or even smartphones.

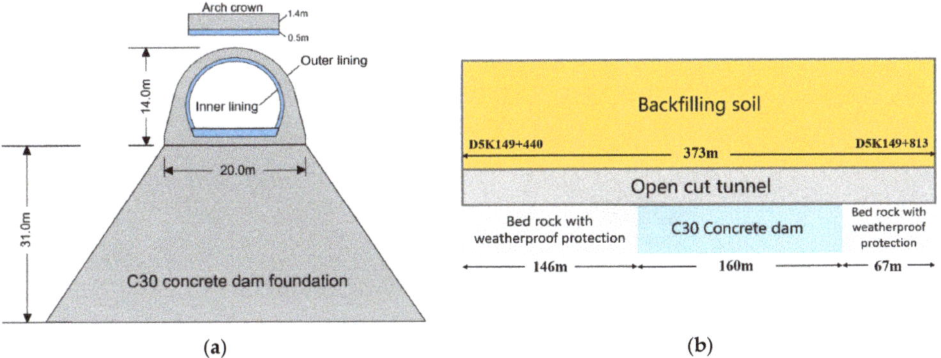

Figure 3. Design details of the tunnel. (**a**) Cross section of the tunnel and the concrete dam foundation; (**b**) Profile view of the whole project.

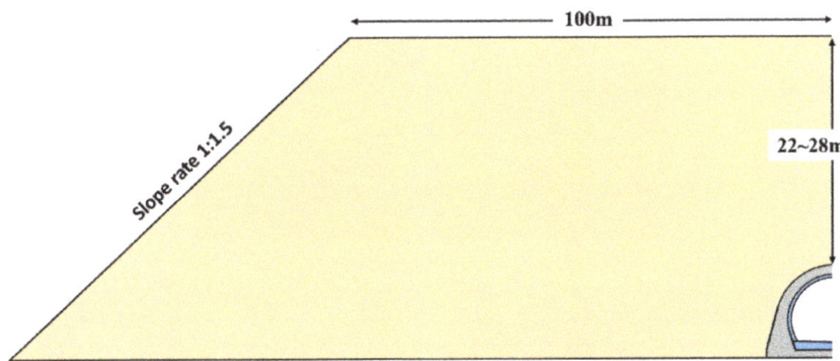

Figure 4. Backfilling range (half side).

Figure 5. Open-cut tunnel in the backfill process.

Step 1: Adjacent sensors were connected by optical fibers at both ends and were finally connected to a fiber optic cable closure, as shown in Figure 6.

Step 2: All the closures were connected in a series and were finally connected to one main fiber cable.

Step 3: The main fiber cable was connected to FBG demodulators (Figure 7), which transmitted the data to a remote-control computer, as shown in Figures 8 and 9.

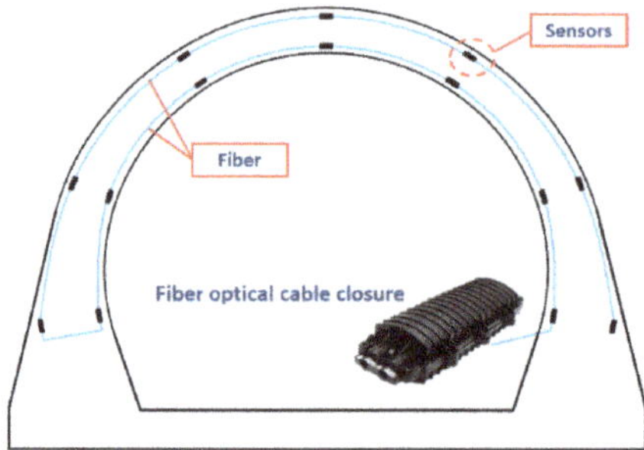

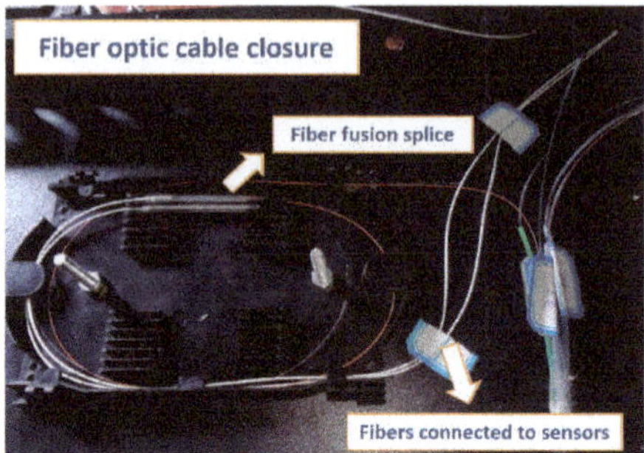

Figure 6. Fiber optic cable closure.

Figure 7. FBG demodulators.

Figure 8. Remote-control computer.

Figure 9 shows the flow chart of the whole data acquisition system. In order to reduce the test error caused by construction disturbance, the data acquired at 4:00 a.m. was used for subsequent analysis.

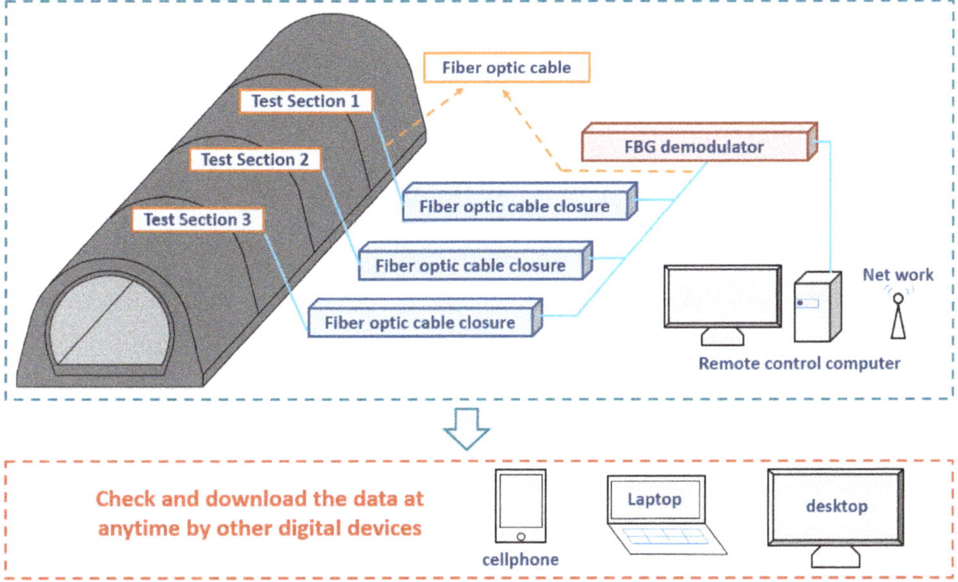

Figure 9. Automatic data collecting and delivering system.

2.3. Sensor Layout and Basic Structure

Two sections were chosen to test the earth pressure and structural internal force, using bedrock and the concrete dam as their foundation, respectively, as shown in Table 2.

Table 2. The measuring sections.

Type	Backfill Height	Foundation Form
Sample A1	28.0 m	Bedrock with weatherproof protection
Sample A2	22.0 m	C30 Concrete dam

In order to obtain the variation law of earth pressure and structural internal force during the backfilling process, earth pressure cells were installed at the arch crown and arch rib on the outer lining surface, and strain gauges were installed in the inner and outer linings, respectively, as shown in Figure 10. For those pressure cells installed on the outer lining surface, metal cable boxes were used to protect the fiber cables from being damaged by the falling backfill soil, as shown in Figure 11.

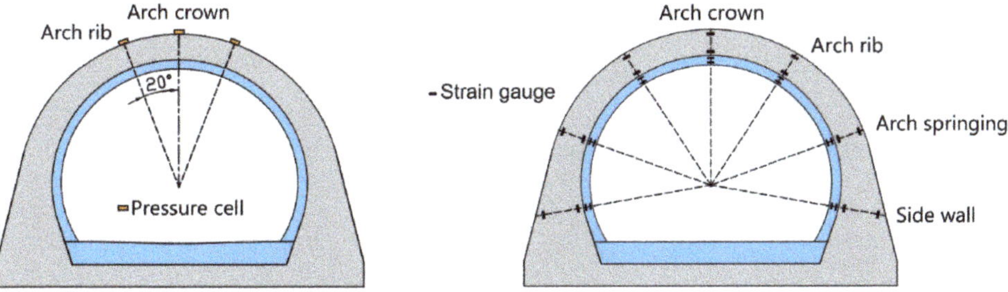

Figure 10. The layout of pressure cells and strain gauges.

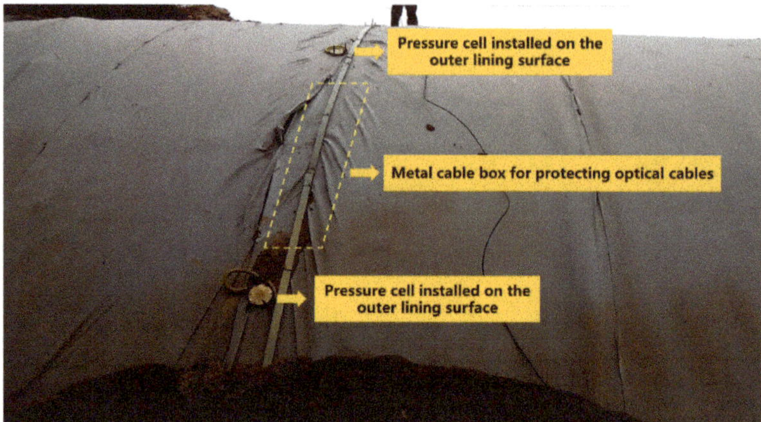

Figure 11. Pressure cells on the outer lining surface with metal cable boxes to protect the optical cables.

The gauges and cells adopted in this research were made by Sensorlead Technology Co., Ltd. Shanghai, China. Their model and basic structure are shown in Figures 12 and 13. Changes in stress and pressure can cause changes in the wavelength of the sensor, which could be measured by the gratings. Further, real-time ambient temperature could be read by the temperature compensation gratings in the sensors, correcting errors due to seasonal temperature differences and providing more accurate data.

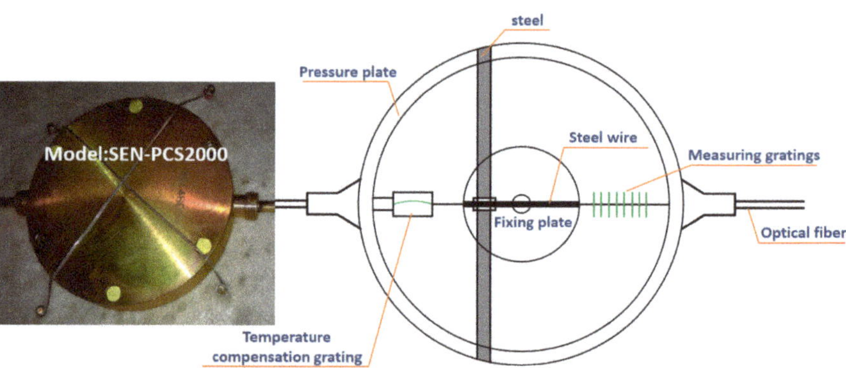

Figure 12. Model and basic structure of pressure cells.

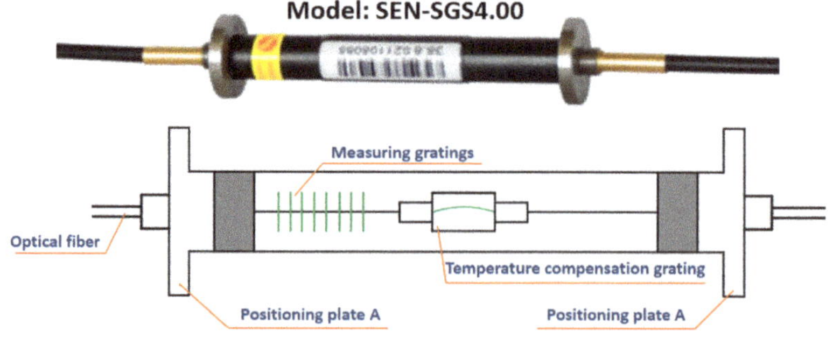

Figure 13. Model and basic structure of gauges.

3. Numerical Model

For further study and design improvement of a high-fill open-cut tunnel using bilayer linings, a numerical model was built in this study, as shown in Figure 14. The elastic model was adopted for the lining and concrete foundation of the open-cut tunnel, and the Drucker–Prager elastic–plastic model was used for the bedrock and backfill soil. The contact elements were added between the open-cut tunnel and the backfill to simulate the mutual squeezing and slip between them (Figure 15). According to the site situation, a waterproof layer consisting of nonwovens and a polyethylene sheet was set between the outer lining and backfill soil as well as between the outer and inner linings. The friction coefficient between the nonwovens and the polyethylene sheet was 0.23. Moreover, the backfilling process of the open-cut tunnel was simulated by the method of layered activation, that is, finite element units for every single backfilling layer (0.5 m) were activated sequentially. The calculation parameters of the model are shown in Table 3.

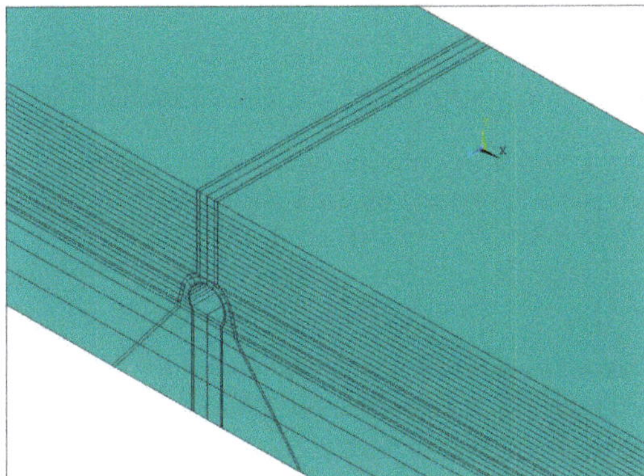

Figure 14. Three-dimensional finite element model for a numerical simulation from ANSYS 12.0 software.

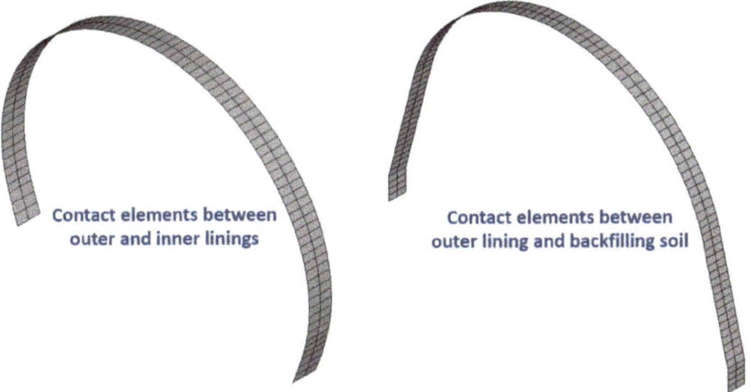

Figure 15. Contact elements between different structures.

Table 3. The physical and mechanical parameters of the model.

Object	Elastic Modulus	Density	Poisson Ratio	Cohesion	Internal Friction Angle
Backfill	12.6 MPa	2100 kg/m^3	0.4	0.1 MPa	22°
Bedrock	7.25 GPa	2300 kg/m^3	0.32	0.2 MPa	31°
Lining	33.5 GPa	2500 kg/m^3	0.2		
Dam	31.5 GPa	2500 kg/m^3	0.2		

In the 3-D numerical model, the node stress in the original Cartesian coordinate system was transformed into the tangential stress of the lining by the coordinate transformation of elastic mechanics. As a result, the bending moment and axial force of the lining were obtained and used for further analysis. The transformation method is described below (Figure 16).

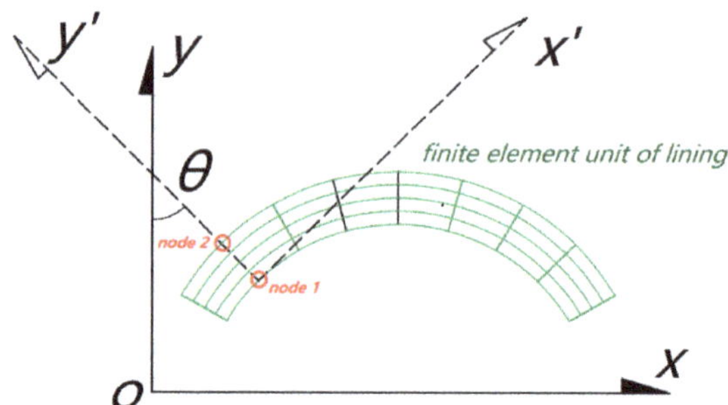

Figure 16. Coordinate transformation for node stress of lining unit.

Taking nodes 1 and 2 on the outer and inner sides of a section, and assuming that the angle between the line of the two nodes above and the vertical plane is θ, the tangential stress of each node on the lining section can be calculated from Equation (1) [30]:

$$\sigma_n = \sigma_x \cos^2 \theta + \sigma_y \sin^2 \theta + \tau_{xy} \sin 2\theta \tag{1}$$

where σ_x, σ_y, and τ_{xy} are node stress components in the original coordinate system, respectively, and θ is the angle between the outer normal of the section and the y axis.

Therefore, the tangential stress of nodes 1 and 2 was obtained, and then the axial force N and bending moment M of the lining section could be deduced according to Equations (2) and (3) [30]:

$$M = bh^2 \frac{\sigma_1 - \sigma_2}{12} \tag{2}$$

$$N = bh \frac{\sigma_1 + \sigma_2}{2} \tag{3}$$

4. Earth Pressure Test Results and Discussion

4.1. Earth Pressure Test Results

The in situ earth pressure was obtained in the backfilling process of the open-cut tunnel. The measured results are represented by the earth pressure coefficient (c), which was defined as earth pressure divided by bulk density, as described in Equation (4). The average earth pressure of the left and right arch rib acted to reduce the testing error. The testing results are shown in Figure 17:

$$c = \frac{p}{\rho g h} \tag{4}$$

where p is the earth pressure, ρ is the soil density, and h is the backfilling height.

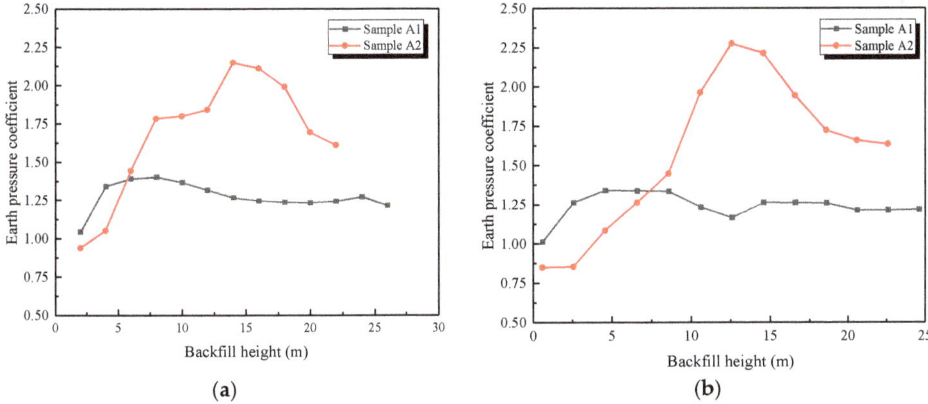

Figure 17. Variation of earth pressure coefficient during the backfilling process. (**a**) Arch crown; (**b**) Arch rib.

As shown in Figure 17:

(1) The earth pressure coefficient firstly increased and then decreased slowly with the increasing backfilling height. The earth pressure coefficient of sample A1 reached its peak value when the backfilling height was around 8 m and then decreased slowly until stabilizing. However, the maximum coefficient of sample A2 appeared at the stage when the backfilling height was about 14 m and then rapidly decreased until stabilizing. The different change trends of the earth pressure coefficient between samples A1 and A2 means that the interaction process in the soil was gradual during the backfilling phase.

(2) At same backfilling height, the earth pressure coefficient at the arch crown and arch rib of sample A2 was obviously higher than that of sample A1, respectively, as listed in Table 4. The earth pressure coefficient at the arch crown of sample A2 was 1.53 times that of sample A1, and sample A2 was 1.69 times that of sample A1 at the arch rib. The root cause of this phenomenon was the additional shear stress difference between samples A1 and A2. The additional shear stress was caused by the settlement difference between the inside and outside soil columns, as shown in Figure 18. The inside soil column consisted of lining and the above soil. Its stiffness was much higher than that of the outside backfilled soil, which led to the settlement difference between the outside and inside soil columns. So, the outside soil column pulled the inside soil column down, resulting in additional shear stress, as shown in Figure 18. Therefore, compared with the stiffness of sample A1, that of sample A2 was much higher, resulting in greater differential settlement and, finally, a significant increase in earth pressure.

Table 4. The earth pressure coefficients of different samples.

Position	Sample A1	Sample A2	Times (A2/A1)
Arch crown	1.404	2.15	1.53
Arch rib	1.341	2.278	1.69

In summary, at the same backfilled height, the increasing stiffness and thickness of the base caused there to be higher earth pressure acting on the open-cut tunnel, which increased the bearing capacity requirement for the lining. Nevertheless, rail surface settlement must be strictly controlled to satisfy the operation requirements of railway trains. Thus, a deep foundation is always applied in open-cut railway tunnels which cross through weak and soft ground. So, to avoid excessive loading of the lining, load shedding measures can be used which have already been adopted in high-fill culverts and pipes [31]. For instance, the load shedding layer was installed above the structure to decrease the settlement difference between the inside and outside soil columns, which reduced the loading of the open-cut structure [32].

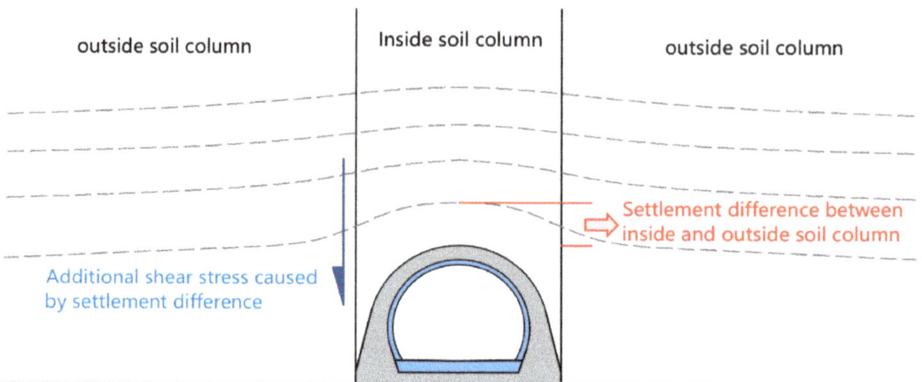

Figure 18. Additional shear stress caused by the settlement difference between inside and outside soil columns.

4.2. The Comparative Analysis of In Situ Testing and Numerical Simulation

In order to validate the abovementioned conclusion, a comparative analysis of the in situ test and numerical simulation is shown in Figure 19.

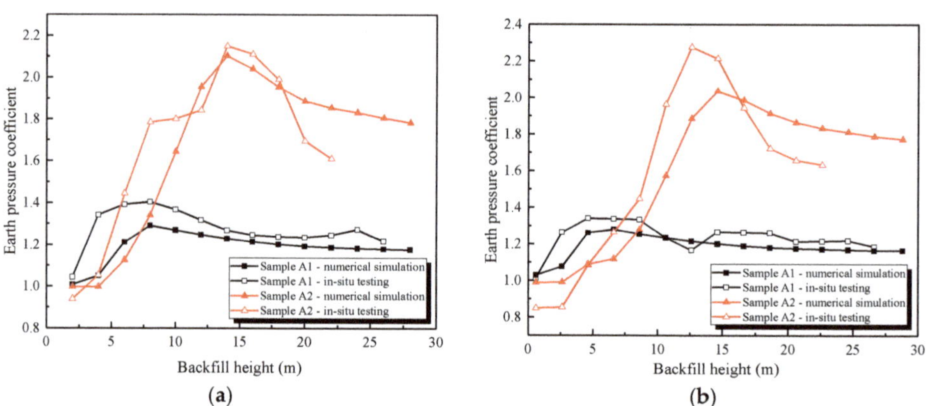

Figure 19. Earth pressure coefficient (comparison of numerical simulations and in situ testing). (**a**) Arch crown; (**b**) Arch rib.

As shown in Figure 19, the earth pressure coefficient change trend of the numerical simulation was similar to that of the in situ test, which firstly increased and then decreased until stabilizing. Moreover, the vertical earth pressure coefficient of sample A2 was obviously higher than that of sample A1, as listed in Table 5. The earth pressure coefficient at the arch crown of sample A2 was 1.63 times that of sample A1, and sample A2 was 1.59 times that of sample A1 at the arch rib.

Table 5. The earth pressure coefficients of different samples in numerical simulations.

Position	Sample A1	Sample A2	Times (A2/A1)
Arch crown	1.292	2.104	1.63
Arch rib	1.281	2.037	1.59

The in situ testing results were in good agreement with the numerical simulation results, and the maximum error was 21.7% at the beginning phase of the backfilling process. The reason for this could be that the backfill had not been completely compacted and stabilized during the actual construction.

In summary, both the in situ testing and the numerical simulation demonstrated that the total loading that acted on the high-fill open-cut tunnel consisted of two parts: the additional loading and the soil column loading. Moreover, the value of additional loading was closely related to the stiffness of the base. In order to more accurately determine the earth pressure of a high-fill open-cut tunnel, the lining and foundation below must be considered as a whole structure. The results also verified the accuracy of the numerical model, which can be used to analyze the settlement difference between the inside and outside soil columns, as shown in Section 4.3.

4.3. The Settlement Difference Analysis between the Inside and Outside Soil Columns

The key factor for the additional loading was the settlement difference between the inside and outside soil columns. However, the change trend or results of this difference could not be obtained exactly through in situ measurement, making it difficult to explain the gradual decreasing phenomenon of the earth pressure coefficient. Therefore, the relationship between earth pressure and settlement difference was analyzed via numerical simulation. The settlement difference between the inside and outside soil columns of two base types is illustrated in Figure 20.

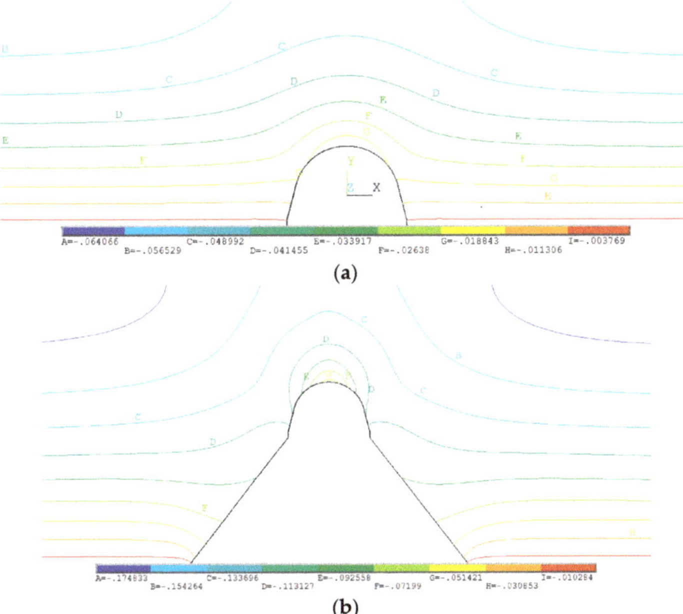

Figure 20. Settlement difference between the inside and outside soil columns. (**a**) Settlement difference for sample A1; (**b**) Settlement difference for sample A2.

As presented in Figure 20, the settlement difference of sample A2 was obviously higher than that of sample A1. The settlement difference results from the arch crown in the 2–28-m range are shown in Figure 21, where the settlement difference on the horizontal plane near the arch crown for sample A1 was 18 mm, while that of sample A2 was 107 mm, about six times that of sample A1, which means the stiffness of the base had a significant influence on the settlement difference. Moreover, the settlement difference gradually decreased to a stable value with the increasing backfill height, which clearly explains the change trend of the earth pressure coefficient.

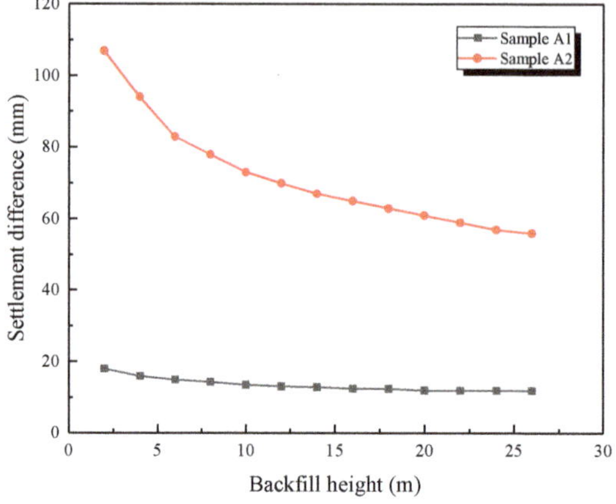

Figure 21. Settlement difference variation law as the backfill height increased.

5. Internal Force Test Results and Discussion

The structural strain data were firstly transformed into tangential stress by multiplying the elastic modulus of the material. The test data of the measuring points in the symmetrical position were averaged to reduce the error. The dynamic change of axial force and bending moment of the bilayer lining during the backfilling process is shown in Figures 22 and 23.

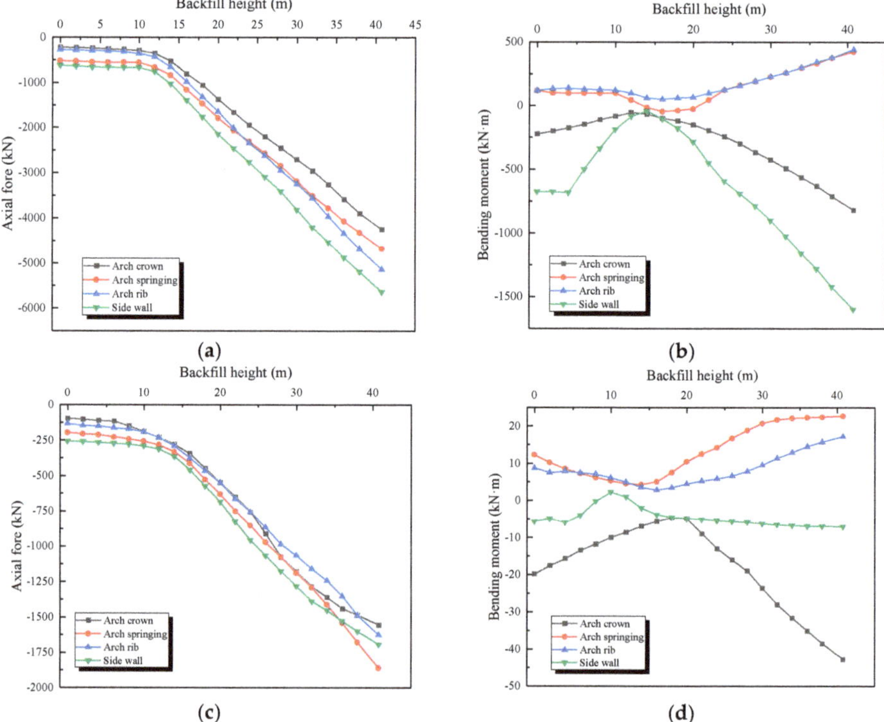

Figure 22. Variation of internal force during the backfilling process for sample A1. (**a**) Axial force of outer lining (sample A1); (**b**) Bending moment of outer lining (sample A1); (**c**) Axial force of inner lining (sample A1); (**d**) Bending moment of inner lining (sample A1).

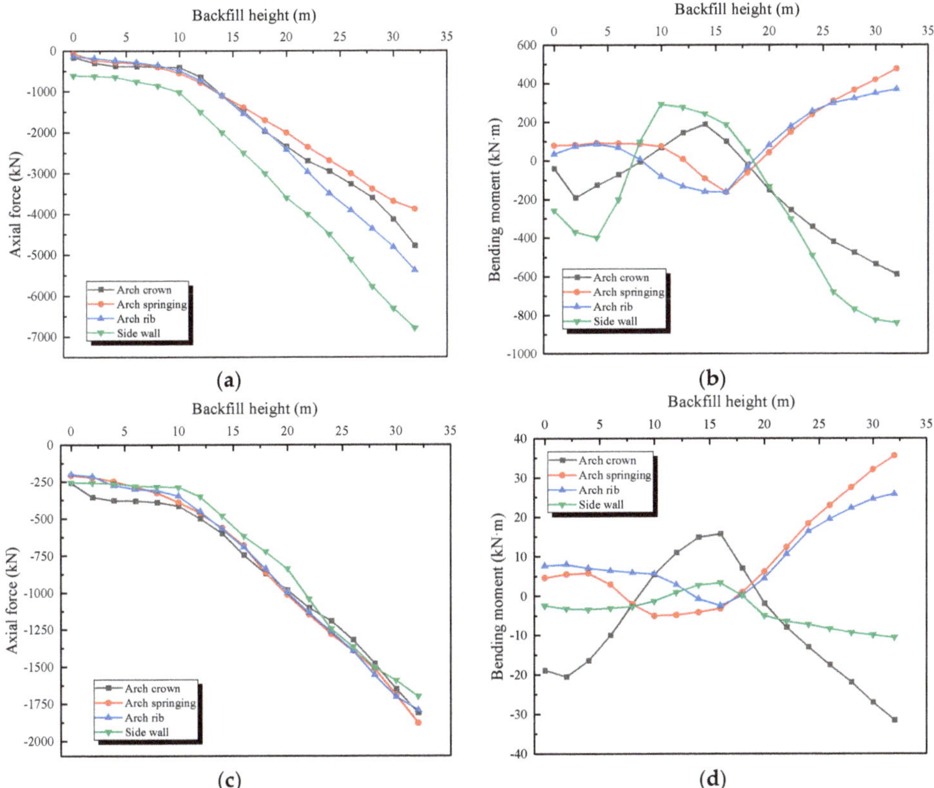

Figure 23. Variation of internal force during the backfilling process for sample A2. (**a**) Axial force of outer lining (sample A2); (**b**) Bending moment of outer lining (sample A2); (**c**) Axial force of inner lining (sample A2); (**d**) Bending moment of inner lining (sample A2).

The change trend of axial force and bending moment can be represented as two stages:

Stage 1: Before the backfilling soil reached the arch crown, the structure only bore lateral pressure, resulting in a very small increase of the axial force and bending moment.

Stage 2: When the backfilling soil exceeded the arch crown, the lining bore both lateral and vertical pressure. The axial force showed linear growth, while the bending moment linearly decreased to a negative value. Moreover, the internal force change trend of samples A1 and A2 during the backfilling process was quite similar.

However, the absolute value of the internal force of the tunnel on the concrete dam foundation was much higher than the one without the concrete dam. The axial force of sample A1 was 1.2–1.6 times that of the other, and the bending moment was 1.6–2.4 times that of the other, as seen in Table 6. This further proves that an open-cut tunnel using a concrete dam foundation will bear greater earth pressure at the same backfilling height, and the internal force of the structure will also increase significantly.

In addition, the axial force of the outer lining was about 2.6–3.2 times that of the inner lining, which was very close to their thickness ratio (3:1), as shown in Table 6. This was because the waterproof layer transition between the two linings reduced the friction between the outer and inner linings. The transmission of shear force between the outer and inner linings was blocked, while only the axial force could be transmitted. That is, the mechanical behavior of the double lining could be analyzed according to the combined beam model.

Table 6. Relationship between the internal force and the tunnel foundation.

Object	Outer/Inner Lining	Foundation Type	Arch Crown	Arch Rib	Arch Springing	Side Wall
Axial force (kN)	Outer lining	Bedrock	−2960.13	−3552.58	−3504.25	−4200.27
		Concrete dam	−4772.29	−5367.43	−5807.45	−6781.54
	Inner lining	Bedrock	−1286	−1158	−1289	−1386
		Concrete dam	−1810	−1789	−1878	−1698
Bending moment (kN·m)	Outer lining	Bedrock	−492.94	262.17	512.23	−1025.67
		Concrete dam	−987.53	547.12	918.45	−1775.46
	Inner lining	Bedrock	−28	11.42	21.8	−6.5
		Concrete dam	−68.52	25.89	35.6	−10.53

6. Conclusions

A field test study was carried out at the first high-fill open-cut tunnel using the bilayer design in China. An automatic data acquisition system using FBG sensors was adopted to obtain the earth pressure and internal force of the tunnel, and numerical models were created for further analysis as well. The following major conclusions can be drawn:

(1) The earth pressure coefficient increased first and then decreased during the backfilling process, and the earth pressure value was significantly higher than the soil column weight. This was because the difference in settlement between the inner and outer soil columns could produce shear forces downward, and the settlement value slowly decreased as the backfill height increased, which was proved by the numerical model above.

(2) At the same backfilling height, the open-cut tunnel on the concrete dam foundation bore greater earth pressure than the one on bedrock, and the internal force of the tunnel on the concrete dam also significantly increased compared with the other one. The lining and the foundation below must be considered as a whole structure to more accurately determine the earth pressure.

(3) The change in axial force and bending moment had two stages, and the boundary point was when the soil reached the arch crown. Before that, the axial force and bending moment increased very slowly. When the soil exceeded the arch crown, an obvious linear growth of the absolute value of the internal force was observed.

(4) Because of the low friction coefficient between the inner and outer linings, the transmission of shear force between them was blocked, while the axial force could be transmitted smoothly, indicating that the mechanical behavior of the double lining was quite similar to the combined beam model.

Author Contributions: Conceptualization, M.W.; Methodology, T.X. and L.Y.; Software, T.X. and C.L.; Validation, M.W. and T.X.; Investigation, T.X. and Y.D.; Data Curation, T.X. and C.L.; Writing-Original Draft Preparation, T.X.; Writing-Review & Editing, T.X., L.Y. and M.W.; Visualization, Y.D., T.Y. and T.X.; Supervision, M.W.

Funding: This research was funded by the National Natural Science Foundation of China, grant number 51878568, and the Chongqing Development Plan Project, grant number cstc2014yykfB30003.

Acknowledgments: The authors would like to thank Rui Xu, Xin Zheng, Hao Cai and Hanbo Chen for their help in the field test. The constructive comments from the reviewers are greatly appreciated.

Conflicts of Interest: The authors declare no conflict of interest.

References

1. Yang, F.; Cao, S.; Qin, G. Mechanical behavior of two kinds of prestressed composite linings: A case study of the Yellow River Crossing Tunnel in China. *Tunn. Undergr. Space Technol.* **2018**, *79*, 96–109. [CrossRef]
2. Vogel, F.; Sovják, R.; Pešková, Š. Static response of double shell concrete lining with a spray-applied waterproofing membrane. *Tunn. Undergr. Space Technol.* **2017**, *68*, 106–112. [CrossRef]

3. Su, J.; Bloodworth, A. Interface parameters of composite sprayed concrete linings in soft ground with spray-applied waterproofing. *Tunn. Undergr. Space Technol.* **2016**, *59*, 170–182. [CrossRef]

4. Su, J.; Bloodworth, A. Numerical calibration of mechanical behavior of composite shell tunnel linings. *Tunn. Undergr. Space Technol.* **2018**, *76*, 107–120. [CrossRef]

5. He, C.; Guo, R.; Xiao, M.; Zhou, J.; He, Y. Model Test on Longitudinal Mechanical Properties of Single and Double Layered Linings for Railway Shield Tunnel. *China Railw. Sci.* **2013**, *34*, 40–46.

6. Yao, C.; Yan, Q.; He, C. An improved analysis model for shield tunnel with double-layer lining and its applications. *Chin. J. Rock Mech. Eng.* **2014**, *33*, 80–89.

7. Ma, Q.; Ku, Z.; Xiao, H.; Hu, B. Calculation of earth pressure on culvert underlying flexible subgrade. *Results Phys.* **2019**, *12*, 535–542. [CrossRef]

8. Li, G.; Ou, J.; Qiu, H.; Wu, J. Mechanism of stress characteristics on surface of slab-culverts under high embankments. *Chin. J. Geotech. Eng.* **2018**, *40*, 1152–1160.

9. Chen, B.; Song, D.; Jiao, J.; Zhang, J. Vertical earth pressure on high fill culverts under load reduction condition. *J. Huazhong Univ. Sci. Technol.* **2015**, *43*, 112–116.

10. Zhang, J.; Zheng, J.; Zhang, T. Experimental Research of Centrifuge Model of Soil Pressure on Positive-buried Box Culvert on Soft Foundation. *J. Highw. Transp. Res. Dev.* **2014**, *07*, 53–59.

11. Shukla, S.K.; Sivakugan, N. Analytical expression for geosynthetic strain due to deflection. *Geosynth. Int.* **2009**, *16*, 402–407. [CrossRef]

12. Kim, K.; Yoo, C.H. Design loading on deeply buried box culverts. *J. Geotech. Geo-Environ. Eng.* **2005**, *131*, 20–27. [CrossRef]

13. Meguid, M.A.; Youssef, T.A. Youssef. Experimental investigation of the earth pressure distribution on buried pipes backfilled with tire-derived aggregate. *Transp. Geotech.* **2018**, *14*, 117–125. [CrossRef]

14. Kang, J.; Parker, F.; Yoo, C.H. Soil–structure interaction for deeply buried corrugated steel pipes Part II: Imperfect trench installation. *Eng. Struct.* **2008**, *30*, 588–594. [CrossRef]

15. Kheradi, H.; Ye, B.; Nishi, H.; Oka, R.; Zhang, F. Optimum pattern of ground improvement for enhancing seismic resistance of existing box culvert buried in soft ground. *Tunn. Undergr. Space Technol.* **2017**, *69*, 187–202. [CrossRef]

16. Kaklauskas, G.; Sokolov, A.; Ramanauskas, R. Ronaldas Jakubovskis Reinforcement Strains in Reinforced Concrete Tensile Members Recorded by Strain Gauges and FBG Sensors: Experimental and Numerical Analysis. *Sensors* **2019**, *19*, 200. [CrossRef] [PubMed]

17. Tsuda, H.; Kumakura, K.; Ogihara, S. Ultrasonic sensitivity of strain-insensitive fiber Bragg grating sensors and evaluation of ultrasound-induced strain. *Sensors* **2010**, *10*, 11248–11258. [CrossRef]

18. Xu, D. A new measurement approach for small deformations of soil specimens using fiber bragg grating sensors. *Sensors* **2017**, *17*, 1016. [CrossRef]

19. Her, S.C.; Huang, C.Y. Effect of coating on the strain transfer of optical fiber sensors. *Sensors* **2011**, *11*, 6926–6941. [CrossRef] [PubMed]

20. Campopiano, S.; Cutolo, A.; Cusano, A.; Giordano, M.; Parente, G.; Lanza, G.; Laudati, A. Underwater acoustic sensors based on fiber Bragg gratings. *Sensors* **2009**, *9*, 4446–4454. [CrossRef]

21. Ho, S.C.M.; Ren, L.; Li, H.N.; Song, G. A fiber Bragg grating sensor for detection of liquid water in concrete structures. *Smart Mater. Struct.* **2013**, *22*, 055012. [CrossRef]

22. Costa, B.J.; Figueiras, J.A. Figueiras, Fiber optic based monitoring system applied to a centenary metallic arch bridge: Design and installation. *Eng. Struct.* **2012**, *44*, 271–280. [CrossRef]

23. Xiao, F.; Chen, G.; Hulsey, J. Monitoring Bridge Dynamic Responses Using Fiber Bragg Grating Tiltmeters. *Sensors* **2017**, *17*, 2390. [CrossRef] [PubMed]

24. Li, Z.; Xu, T.; Wu, Q.; Yu, L.; Wang, M. Field experimental study of basement structural dynamic properties of the heavy-haul railway tunnel in broken surrounding rock. *Rock Soil Mech.* **2018**, *39*, 949–956.

25. Li, Z.; Wang, M.; Yu, L.; Zhao, Y. Study of the basement structure load under the dynamic loading of heavy-haul railway tunnel. *Int. J. Pavement Eng.* **2018**. [CrossRef]

26. Dong, L.; Tong, X.; Li, X.; Zhou, J.; Wang, S.; Liu, B. Some developments and new insights of environmental problems and deep mining strategy for cleaner production in mines. *J. Clean. Prod.* **2019**, *210*, 1562–1578. [CrossRef]

27. Rodrigues, C.; Cavadas, F.; Félix, C.; Figueiras, J. FBG based strain monitoring in the rehabilitation of a centenary metallic bridge. *Eng. Struct.* **2012**, *44*, 281–290. [CrossRef]

28. Shi, S.; Xie, X.; Wen, Z.; Zhou, Z.; Li, L.; Song, S.; Wu, Z. Intelligent Evaluation System of Water Inrush in Roadway (Tunnel) and Its Application. *Water* **2018**, *10*, 997. [CrossRef]
29. Dong, L.; Li, X.; Xu, M.; Li, Q. Comparisons of Random Forest and Support Vector Machine for Predicting Blasting Vibration Characteristic Parameters. *Procedia Eng.* **2011**, *26*, 1772–1781.
30. Wang, G. *Elastic Mechanics*, 3rd ed.; Higher Education Press: Beijing, China, 2015; pp. 106–107.
31. Jiang, C.; Chen, B.; Mao, X.; She, M. Stress characteristics of high fill load-shedding culvert on flexible foundation. *Rocks Soil Mech.* **2019**, *01*, 1–6.
32. Chen, B.; Song, D.; Mao, X.; Chen, E.J.; Zhang, J. Model Test and Numerical Simulation on Rigid Load shedding Culvert Backfilled with Sand. *Comput. Geotech.* **2016**, *79*, 31–40. [CrossRef]

Article

Field Performance of Open-Ended Prestressed High-Strength Concrete Pipe Piles Jacked into Clay

Hai-Lei Kou [1,2,*], Wen-Zhou Diao [1], Tao Liu [3], Dan-Liang Yang [1] and Suksun Horpibulsuk [4]

[1]   College of Engineering, Ocean University of China, Qingdao 266100, China;
      dwzouc@163.com (W.-Z.D.); danliang1995@163.com (D.-L.Y.)
[2]   Cooperative Innovation Center of Engineering Construction and Safety in Shandong Blue Economic Zone,
      Qingdao 266033, China
[3]   College of Environmental Science and Engineering, Ocean University of China,
      Qingdao 266100, China; ltmilan@ouc.edu.cn
[4]   School of Civil Engineering, and Center of Excellence in Innovation for Sustainable Infrastructure
      Development, Suranaree University of Technology, Nakhon Ratchasima 30000, Thailand; suksun@g.sut.ac.th
*   Correspondence: hlkou@ouc.edu.cn; +86-137-3098-7254

Received: 31 October 2018; Accepted: 29 November 2018; Published: 1 December 2018

Abstract: The behavior of open-ended pipe piles is different from that of closed-ended pipe piles due to the soil plugging effect. In this study, a series of field tests were conducted to investigate the behavior of open-ended prestressed high-strength concrete (PHC) pipe piles installed into clay. Two open-ended PHC pipe piles were instrumented with Fiber Bragg Grating (FBG) sensors and jacked into clay for subsequent static loading tests. Soil plug length of the test piles was continuously measured during installation, allowing for calculation of the incremental filling ratio. The recorded data in static loading test were reported and analyzed. The distribution of residual forces after installation and the effect on the bearing capacity were also discussed in detail. The test piles were observed to be in partially plugged condition during installation. The measured ultimate shaft resistance and total resistance of the test piles were 639 and 1180 kN, respectively. The residual forces locked in the test piles after installation significantly affected the evaluation of the axial forces, and thus the shaft and end resistances. It tended to underestimate the end resistances and overestimate the shaft resistances if the residual forces were not considered in the loading test. However, the residual forces did not affect the total bearing capacity of open-ended PHC pipe piles in this study.

Keywords: open-ended pipe piles; field test; static loading test; soil plugging; residual forces

---

1. Introduction

Prestressed high-strength concrete (PHC) pipe piles are commonly adopted as deep foundations in the coastal areas of Southeast China. The installation methods usually involve jacking and driving. A vibration and predrilling method is also utilized in the field. Compared with other installation methods, the jacking method is preferred as it avoids noise, vibration and slurry pollution. Hence, it is more suitable to install piles in urban areas. The performance of jacked pipe piles during and after installation has becomes a focus of attention in recent years [1,2], but many issues remain to be solved, in particular for open-ended pipe piles, such as the termination criteria, the bearing capacity, the design methods and so on [3,4]. This is mainly because the physical properties involved in open-ended pipe piles are extremely complicated [5,6].

Many references have reported on the behavior of jacked steel H-piles, including the effect of excess pore water pressure, the set-up, the estimation of shaft and end resistances and so on [7]. As full displacement piles, the steel H-piles have no soil plug effect during installation [8,9]. However, for non-displacement or partial displacement piles, soil plugging was identified as an important factor in pile behavior [10,11]. A number of field and model tests were conducted to study the effect of soil plugging on the behaviour of open-ended piles jacked into sand [12]. Kou et al. [13,14] revealed that the loading-settlement behavior was closely related to the development of soil plugs. Yu and Yang [15] compared major open-ended pipe pile design methods and proposed the Hong Kong University (HKU) method to estimate the end bearing capacity of open-ended steel piles in sand based on the Imperial College Pile (ICP) and the University of Western Australia (UWA) methods [16,17]. The proposed HKU method was capable of producing satisfactory predictions over a wide range of pile lengths, diameters and slenderness. Moormann et al. [18] adopted Coupled Eulerian-Lagrangian (CEL) method to simulate the soil plug behavior in open-ended steel piles in loose sand to estimate the current design methods. For silty clay, soil plug behavior research focuses on open-ended steel piles with model tests [19] or field tests [20,21]. Few studies were conducted to investigate the soil plugging effect to the performance of jacked open-ended concrete pipe piles. Liu et al. [22] investigated the formation of soil plug in open-ended concrete pipe piles in the field, however, the effect of soil plugging on pile behavior was not discussed.

Further, the conventional strain gauges used in the monitoring of axial strains has some disadvantages [23]. Fiber optic sensors are becoming a well-established technology for a variety of geophysical and civil engineering applications [24]. Schmidt-Hattenberger et al. [25] conducted strain measurements by fiber Bragg grating sensors for in situ pile loading tests. The monitored strain results of fiber Bragg grating sensors agree fairly well with conventional concrete strain gages. Authors have also examined the skin friction between piles and subsoils as well as the settlement behavior of the pile based on the measured strain data. Klar et al. [26] conducted a static loading test on bored piles based on fiber optic technology and discussed the differences between fiber measurements and conventional discrete measurements. Doherty et al. [27] used a fiber sensing system to monitor the axial loads of soil-cement injected precase piles and steel pipe piles in the field. The advantages of fiber optic sensors compared to conventional strain gauges are long-term performance, resistance to corrosive environments, immunity against electromagnetic interferences, array-capability by wavelength-demultiplexing, and miniaturized size. Hence, there is a practical demand to investigate the behaviour of open-ended concrete pipe piles jacked into silty clay considering soil plugging using novel fiber optic sensors [28].

In this study, Fiber Bragg Grating (FBG) sensors were adopted to instrument two open-ended PHC pipe piles to study the soil plugging effect on pile behaviour when jacked into clay. The performance of test piles was continually monitored during the installation and the loading tests. The main objectives of this study were to investigate: (1) the behaviour of open-ended PHC pipe piles jacked into clay; (2) the soil plugging during and after jacking; (3) the residual forces after installation and the effect to the evaluation of total bearing capacity; (4) the loading transfer and settlement mechanisms of the test piles in loading test.

2. Site Conditions and Test Program

2.1. Site Description

Full-scale tests were conducted at a site in Hangzhou, China. Before the test pile installation, two auger borings B_1 and B_2, and one cone penetration test (CPT) were conducted to investigate the soil information, as shown in Figure 1. Figure 2a,b show the soil profiles from boreholes with the cone resistance q_c, sleeve friction f_s and friction ratio R_f. It is indicated that the test site is in sequence by 2.1 to 2.4 m of fill layer, 2.1 m of silty clay, 3.5 m of sandy clay, 2.0 m of muddy-silty clay, 2.5 m of silty clay, and 9.2 to 9.5 m thick alluvium layer. The soil properties in embedded length of test piles are summarized in Table 1.

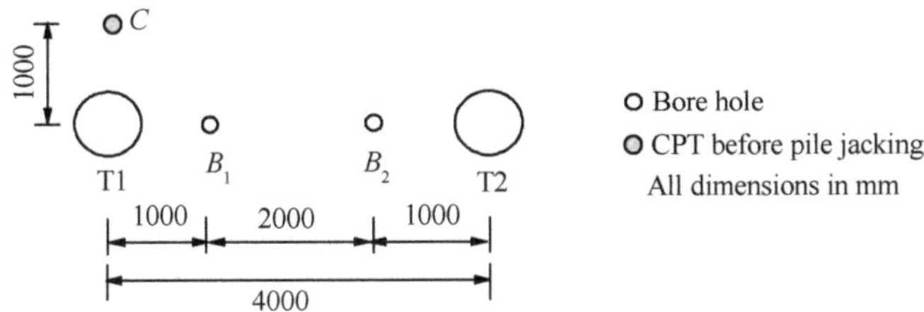

Figure 1. In-situ test layout.

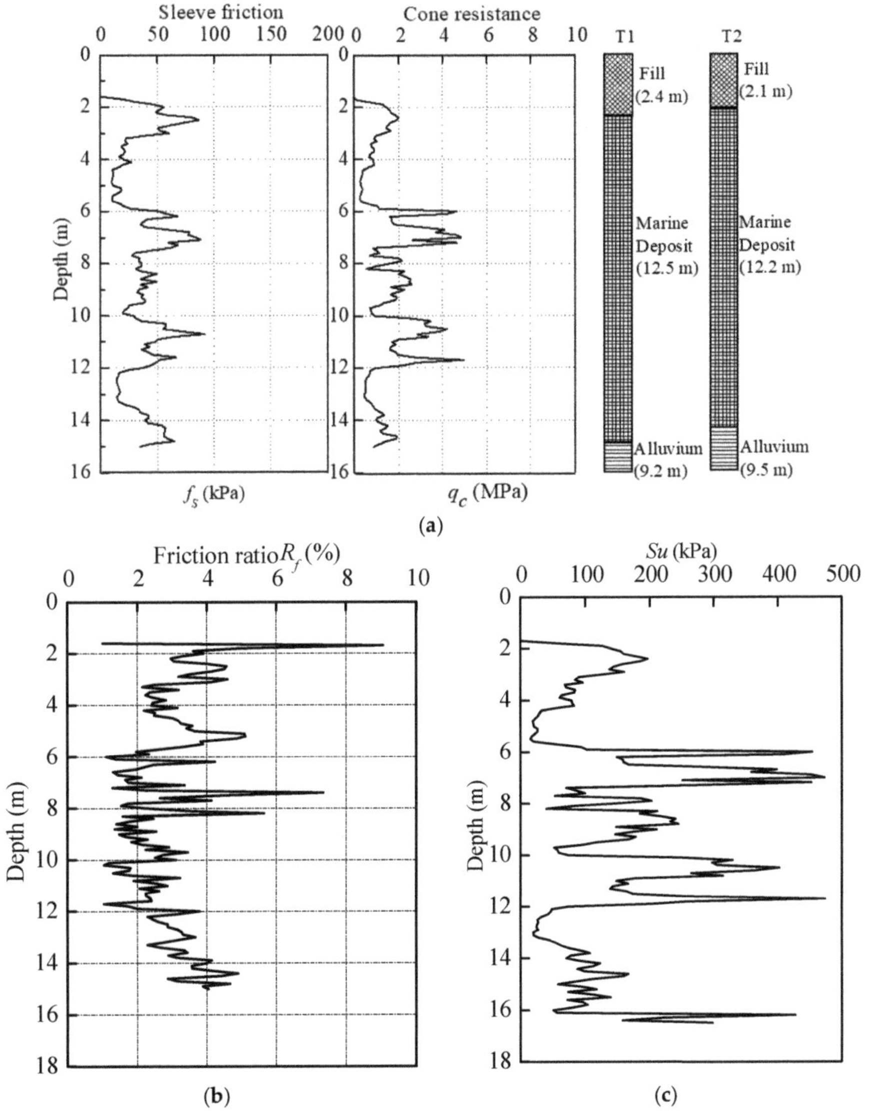

(a)

(b)   (c)

Figure 2. CPT data and Su values in test site. (**a**) Soils profiles and CPT values along depth; (**b**) Friction ratio with depth; (**c**) Undrain shear strength with depth.

Table 1. Subsoil Properties.

Soils Type	Depth (m)	W (%)	γ (kN/m^3)	e_0 (%)	G_s	LL (%)	PL (%)	$\alpha_{1\text{-}2}$ (MPa^{-1})	c (kPa)	φ (°)
Silty clay	2.1~5.1	25.7	19.36	0.73	2.72	30.1	17.9	0.15	14.0	21.5
Sandy clay	5.1~7.6	31.2	18.52	0.87	2.69	33.3	26.4	0.18	7.1	29.4
Muddy-silty clay	7.6~10.4	44.8	17.07	1.28	2.74	37.0	19.7	0.87	15.8	8.0
Silty clay	10.4~13.0	23.4	19.77	0.67	2.72	28.6	15.8	0.22	28.5	22.8

Note: c and φ were determined using the quick shear tests; $\alpha_{1\text{-}2}$ was determined using one-dimensional compressibility test.

The coefficient of compressibility $\alpha_{1\text{-}2}$, defined as the decrease in volume per unit volume produced by a unit change of pressure, given in the table was determined through one-dimensional compressibility test. The cohesion c and friction angle φ were determined using direct shear test. The undrained shear strength S_u is calculated from CPT results according to Equation (1) proposed by Schmertmann [29]:

$$S_u = (q_t - \sigma_{v0})/N_{kt} \tag{1}$$

where q_t is the modified cone resistance; σ_{v0} is the total overburden pressure; N_{kt} is a cone bearing capacity factor. A N_{kt} value of 10 was used here.

2.2. Test Piles Details and Instrumentation

In order to separate all resistance components for partially plugged open-ended piles, the instrumented double-walled pile system is usually used in model tests [30,31] or full-scale field tests [32]. In this paper, the FBG sensors are installed along test piles using embedded methods. However, the survival rate of embedded FBG system in pile instrumentation is lower than that of conventional sensors because the exposed grating points are very fragile and require further cable protection [33]. Hence, an epoxy resin was used in this test to seal the FBG sensors and cables.

The test piles used in this study are two open-ended PHC pipe piles with 400 mm outer diameter, 75 mm wall thickness, and 13.0 m length. Details are given in Table 2.

Table 2. Details of Test Piles

Pile number	Size (mm)	Embedded Length (m)	Elastic Modulus (GPa)	Axial Compressive Strength (MPa)
T1	400 (75)	13.0	36.0	35.9
T2	400 (75)	13.0	36.0	35.9

Note: In size, 400 mm is the outer diameter of the test piles, 75 mm is the wall thickness; 36.0 GPa is the elastic modulus of the test piles material.

Fourteen FBG sensors were installed in each test pile at seven levels to monitor the axial strain along the test piles and then separate the end and shaft resistances. The detailed configuration of FBG sensors instrumentation is shown in Figure 3a. The sensors were installed about 2.5 m intervals and closer together near the pile toe. After two shallow grooves with 50 mm width and 15 mm depth were created, all FBG sensors were attached on the groove bottom according to pre-designed intervals. Then the FBG sensors were sealed with epoxy resin to prevent any damage caused by groundwater and the surrounding soils during jacking, as shown in Figure 3b,c. The axial strain recorded by FBG sensors could be transformed into axial forces. The Young's modulus of the test piles material was taken as 3.6×10^4 N/mm^2 in this study without considering pre-stressing. The end resistance was estimated with linear extrapolation method using the measured data. As discussed above, the shaft resistances between soil plug and inner surface were not measured in this program. In other words, the estimated end resistances include annulus and plug resistances.

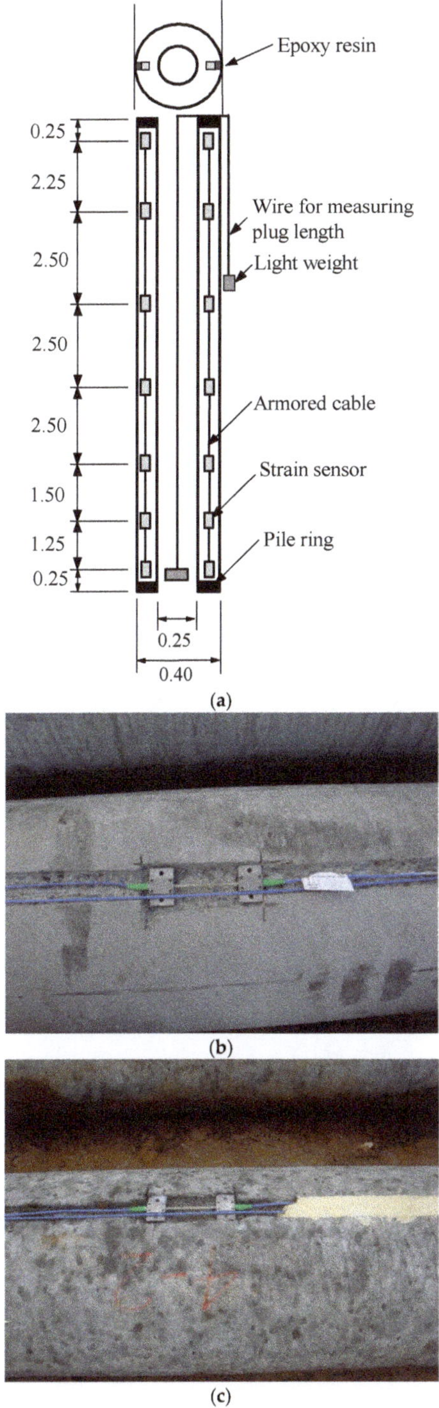

Figure 3. Strain gauges installation on test piles. (**a**) Schematic arrangement for strain gauges (Unit: m); (**b**) Installation of strain gauges along test piles; (**c**) Strain gauges sealed with epoxy resin in field.

2.3. Test Program

Test piles were installed using a ZYJ680A jacking machine (Figure 4), which has a loading capacity of 6.8 tons. The jacking machine includes two components of clamps and hydraulic jacks. In jacking, the clamp with four arc pieces was used to stabilize against the pile vertically. Then a hydraulic jacking force was applied to push the piles. After the hydraulic clamp reached the maximum travel range, the hydraulic clamp was released and moved up to the top position. The step was repeated to push the piles down until the test piles were installed into the ground. Two test piles were both jacked into 13.0 m depth and founded in the layer of marine deposits. The ultimate bearing capacity of the test piles was estimated as 1200 kN from the calculation before loading test.

Figure 4. Jacking machine used in the test.

The jacking forces were recorded to assess the drivability of the test piles. The soil plug length during jacking was continuously measured using two different weights, as shown in Figure 5.

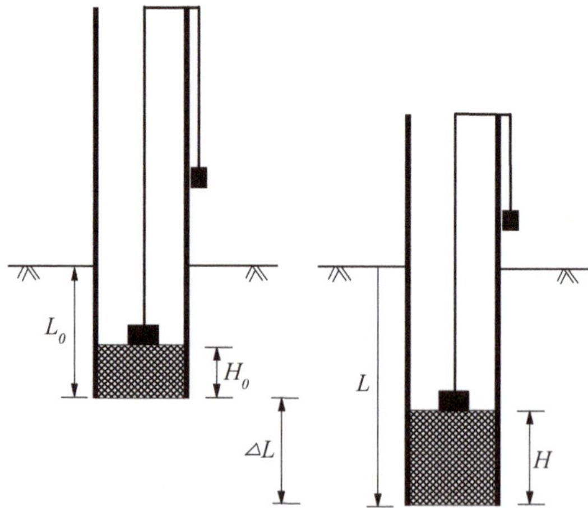

Figure 5. Measurement of soil plug length during pile jacking.

The loading test was conducted on test piles of T2 at 17 days after jacking, as illustrated in Figure 6. The total applied loads to the piles head were controlled by a hydraulic pump. The vertical displacement was monitored using two dial gauges mounted in the reference beams. The values of all FBG sensors attached to T2 were set to be zero before loading in order to measure the axial forces induced by applied loads. The static loading test was slowly maintained loading test and conducted in accordance with the Chinese technical code for testing of building foundation piles [34], with no unload-load loops.

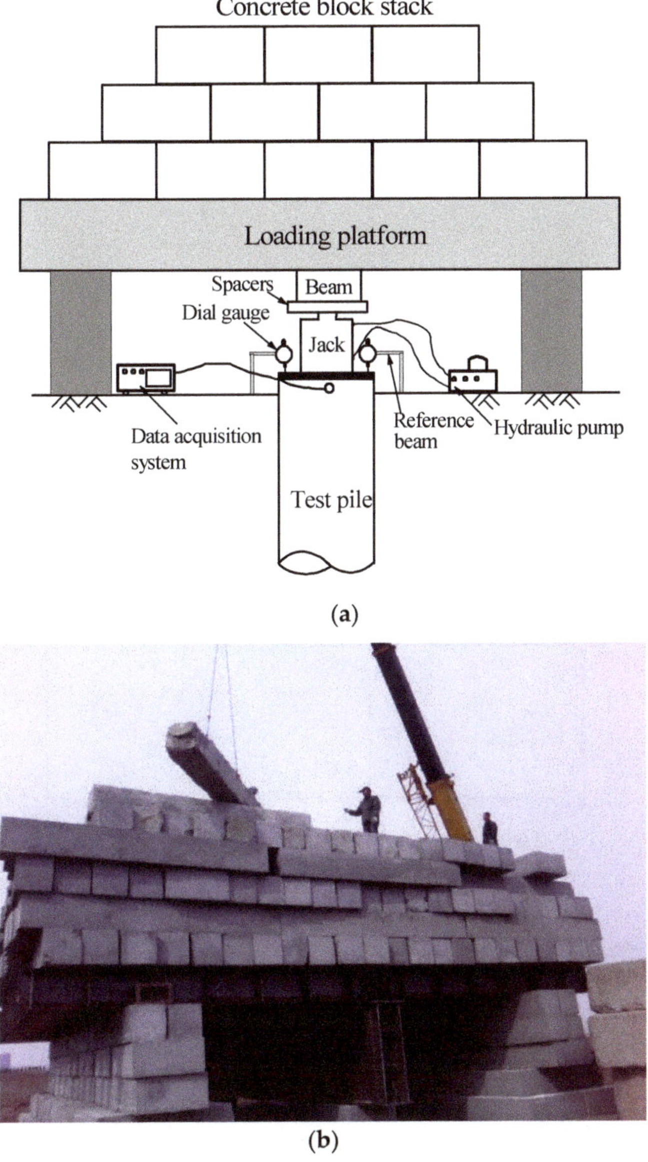

Figure 6. Static loading test of T2. (**a**) Schematic view; (**b**) Static loading test in field.

3. Test Resultis and Discussion

3.1. End and Shaft Resistance during Jacking

The jacking forces of test piles during installation are plotted in Figure 7. It can be seen from the figure that the jacking forces increased with penetration depth. At final depth of 13.0 m, the jacking forces were 806 and 815 kN for T1 and T2, respectively. The induced axial forces along test piles are illustrated in Figure 8. It is indicated that the axial forces increased with jacking depth and the distribution shown similar characteristics to each other. It also should be noticed that the induced axial forces in Figure 8 are different from the jacking forces in Figure 7. The induced axial force on the piles head is the same as the jacking forces at the corresponding depth in Figure 7.

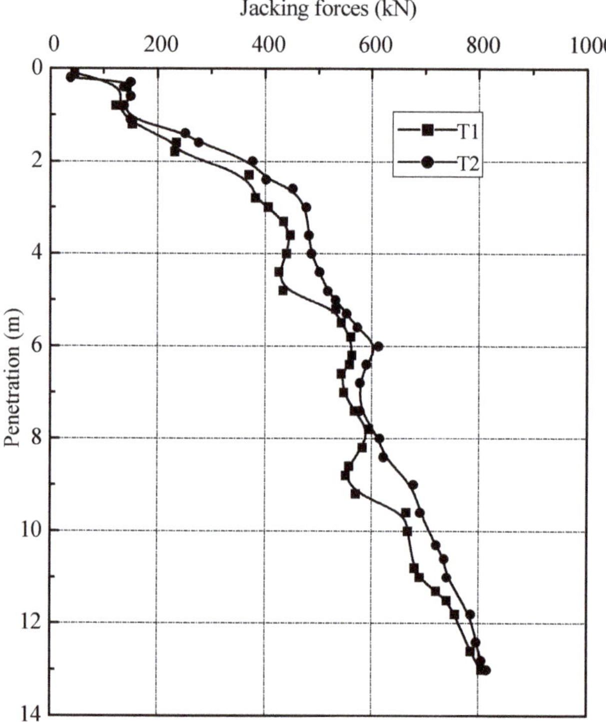

Figure 7. Jacking forces versus penetration depth.

For a better view of the loading transfer, the end and shaft resistances are separated and shown in Figure 9.

The shaft resistances of T1 and T2 consistently increased with installation, while the end resistances were in a range from 21 to 151 kN. At a penetration depth of 5.5 m, the shaft resistances of test piles reached a level of 457.3 and 471.2 kN, taking on 94.8 and 95.7% of the jacking forces, respectively. At 13.0 m, the shaft resistance for T1 was about 771.6 kN, amounting to 95.8% of jacking forces. For T2, the shaft resistance reached a value of 793.1 kN, while the end resistance was about 22.7 kN. The corresponding ratios of the shaft and end resistances to the jacking forces were 97.2 and 2.7%, respectively. The average shear stresses (τ_{av} = shaft load/shaft area) and unit end resistance (q_b) measured during installation are plotted against depth, as shown in Figure 10.

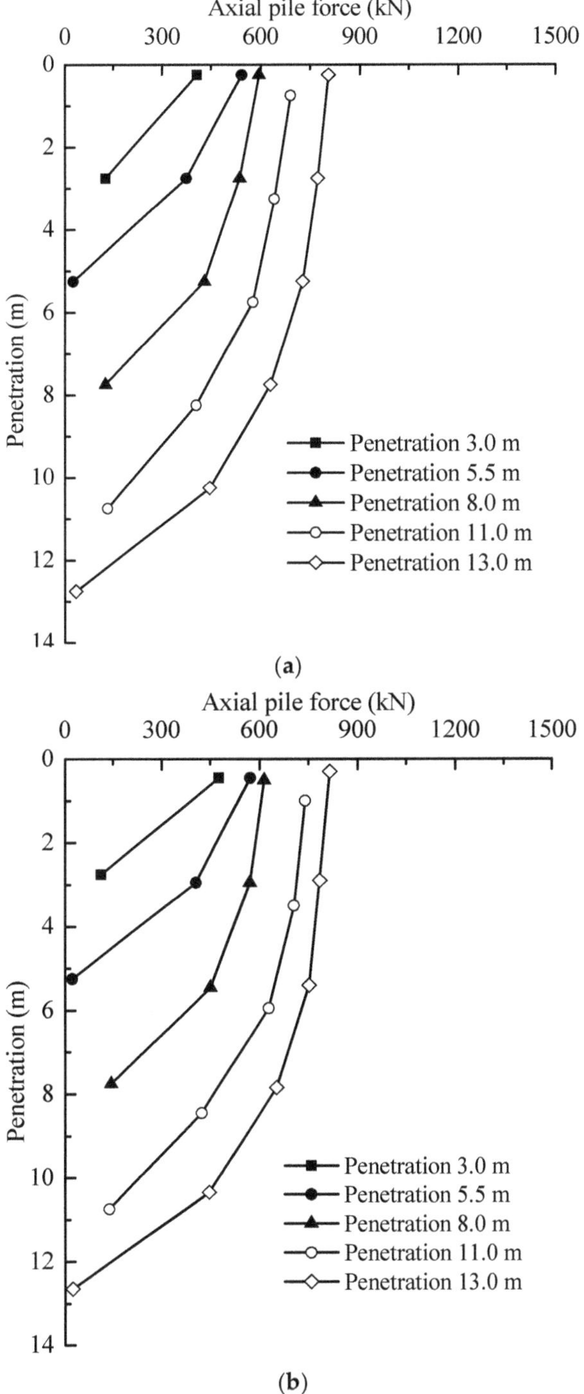

Figure 8. Variation of axial pile forces during jacking: (**a**) T1; and (**b**) T2.

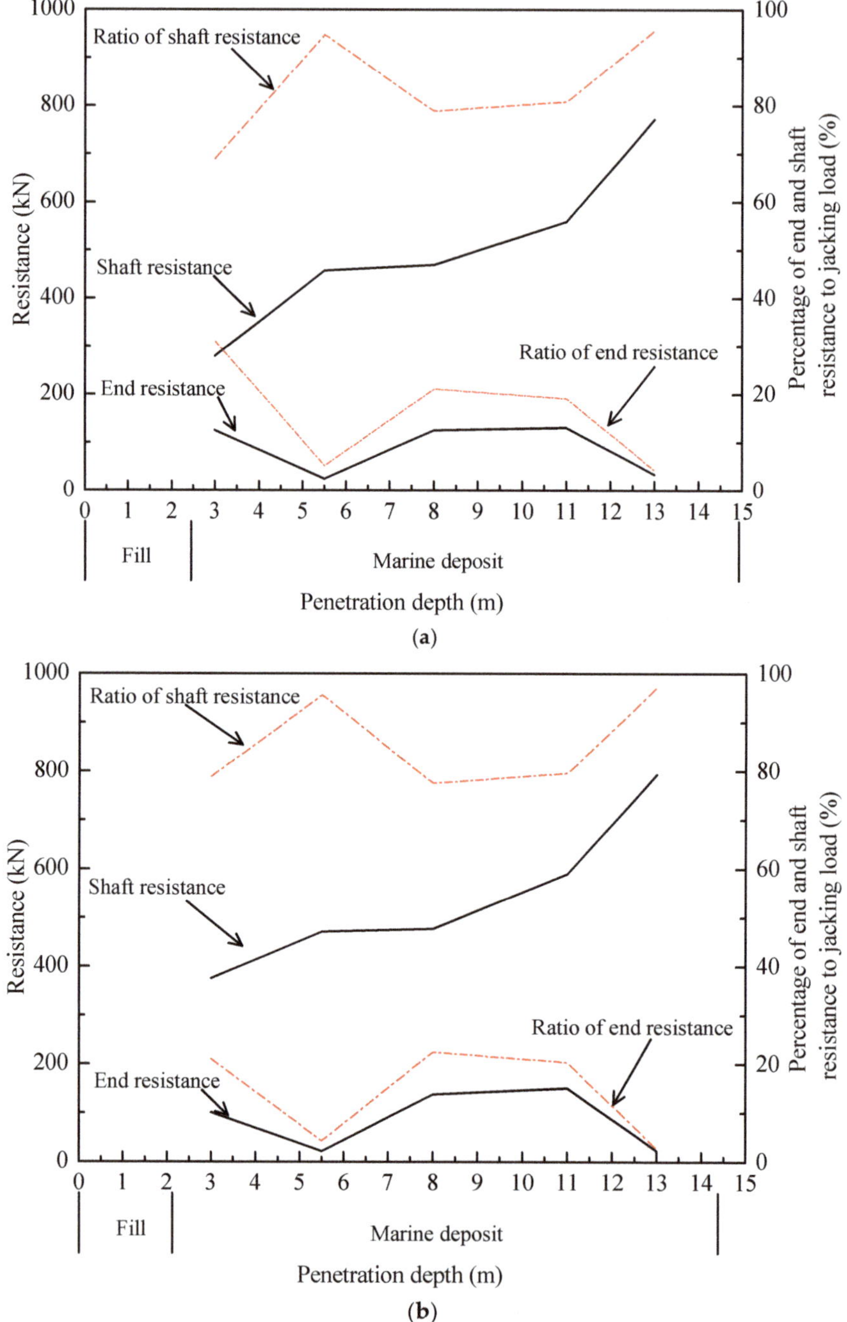

Figure 9. Variation of end and shaft resistance during jacking: (**a**) T1; and (**b**) T2.

The q_b is the force per unit cross section area, which the cross-section area is calculated as a closed-ended pipe piles. In order to compare with the cone resistance q_c from CPT, the q_c values at corresponding depths were also marked in the figure. It is apparent that these q_b values are around 0.8 q_c, which is in good agreement with those reported in previous database [35,36].

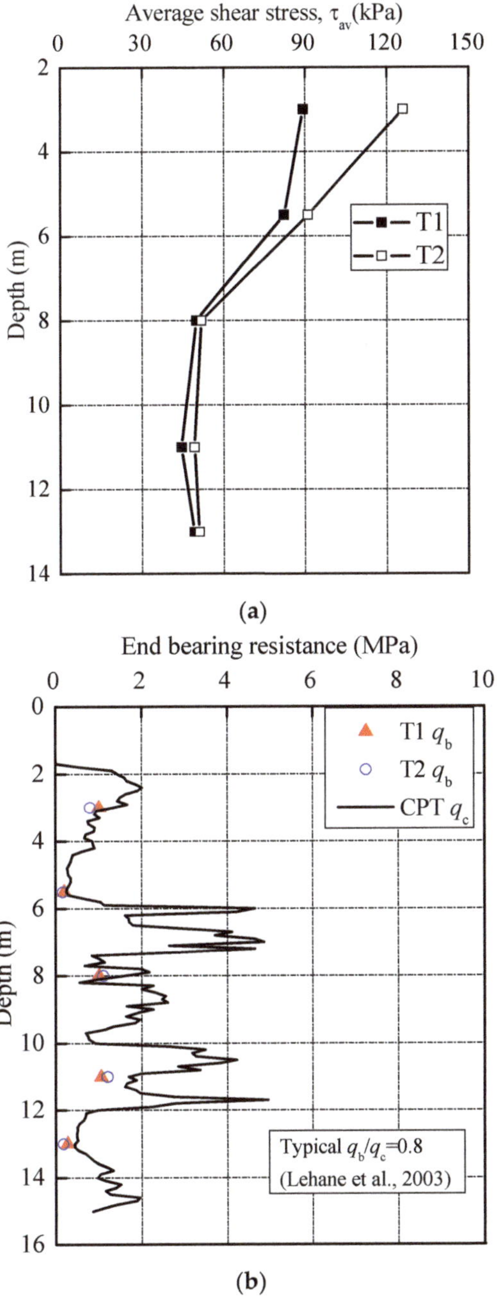

Figure 10. Pile installation stresses with depth. (a) Average shear stresses profile; (b) Average end bearing stress.

3.2. Soil Plug Behavior

Soil plugging plays an important role in controlling the behavior of open-ended pipe piles during installation and loading tests. The degree of soil plugging can be represented by plug length ratio (PLR) and incremental filling ratio (IFR), defined as:

$$\text{PLR} = H/L \tag{2}$$

$$\text{IFR} = dH/dL \times 100(\%) \tag{3}$$

where H/L is the length of soil plug H with reference to an installation length of piles L; dH/dL expresses the increase of soil plug length H per unit increase of installation depth L.

During installation, the soil plug condition can be divided into three categories: fully coring (IFR = 100%), fully plugged (IFR = 0) and partially plugged (0 < IFR < 100%). The variations of average soil plug length and IFR with installation are illustrated in Figure 11. The test piles were partially plugged from the outset of pile installation. The IFR decreased sharply from 23.2% to 5.3% at the depth of 6.0 m and then decreased to near zero at penetration depth of 11.0 m. The abrupt change of IFR is due to the existence of a relatively stiff interlayer that is verified by the CPT-q_c trace in Figure 2a. The PLR values of T1 and T2 recorded at the end of installation were 0.13 and 0.14, respectively. After the loading test, there was no change of the measured soil plug length. This reinforces the fact that soil plug behaviour of open-ended pipe piles is very different during installation and static loading test.

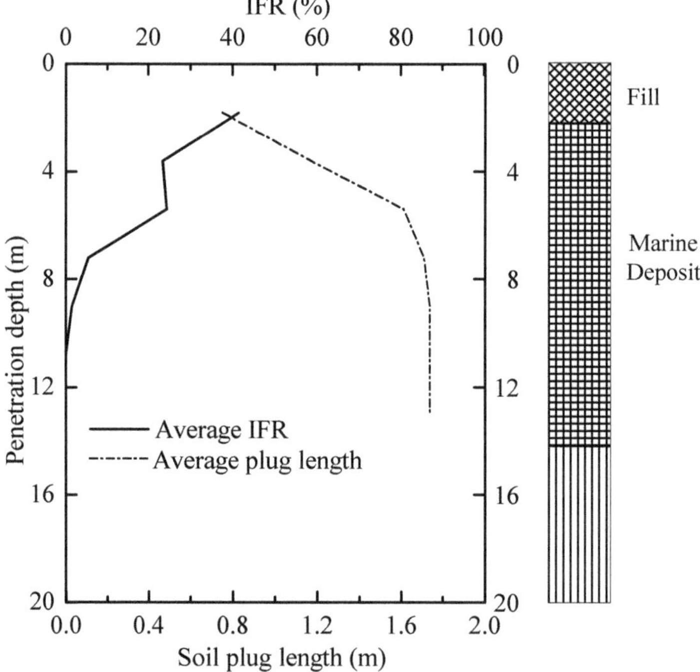

Figure 11. IFR and soil plug length versus penetration depth for test piles.

3.3. Residual Forces after Installation

At the end of each jacking stroke, in particular, after the last jacking stroke, the pile-soil interaction reached a static equilibrium with the recovery of elastic compression caused by compression/tension pulses during jacking. There are still some residual forces locked in the piles and these are always compressive at the pile toe [37]. The magnitude and distribution of locked residual forces in the piles after installation have significant effects on the interpretation of loading transfer, and then end and shaft resistances [38,39]. Through the comparison of FBG sensor readings of pre-installation and post-installation, the residual forces after installation can be determined, as shown in Figure 12. The FBG sensor readings after installation can be recorded several hours after jacking in order to let the pile-soil interaction reached the equilibrium. If the reading was recorded immediately after installation, the residual forces would be overestimated as the piles did not complete the recovery of elastic compression.

Figure 12 illustrates that the residual end forces of T1 and T2 were 9.5 and 8.2 kN, which were 25.2 and 23.8% of the end resistances at 13.0 m, respectively. The residual forces along the inner shaft of the test piles were not measured in this program due to the installation difficulties of FBG sensors on the inner shaft. Therefore, the recorded residual end forces were due to the residual soil plug and residual annulus forces. The magnitude and distribution of the residual forces along piles after installation are significantly affected by pile material, length, cross-sectional area and soil properties [40]. This can explain the similar trend of residual forces in the test piles after installation.

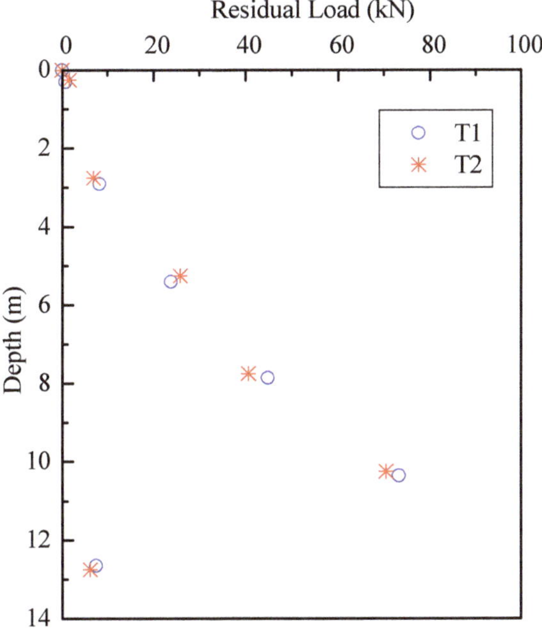

Figure 12. Residual loads distribution after installation.

3.4. Pile Behavior during Static Loading Test

The load versus settlement curve of T2 measured in the static loading test is drawn in Figure 13. At the loading of 600 kN, the settlement of piles head was 9.21 mm, about 2.3% of pile diameter. While at the maximum loading of 1200 kN, the settlement recorded was about 42.15 mm, which was 10.5% of pile diameter. Using the Chinese technical code for testing of building foundation piles [34], the ultimate bearing capacity of the pile T2 was determined to be 1180 kN, which was the load at which the piles settled by 10% of the pile diameter (40 mm). The ultimate end and shaft resistance at the displacement of 40 mm were 541 and 639 kN, respectively.

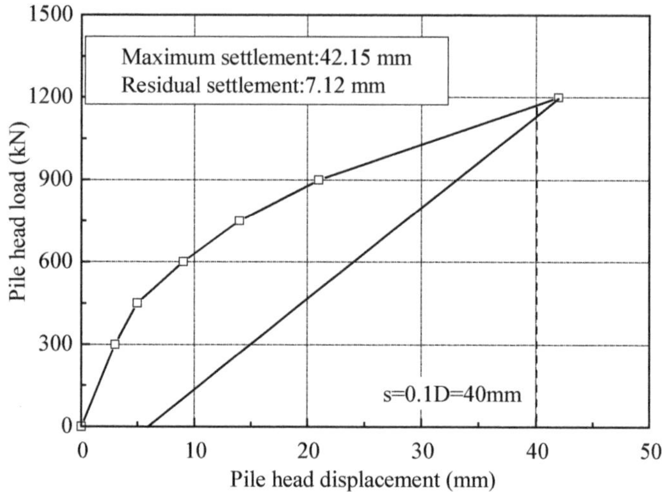

Figure 13. Load-settlement curve for static loading test of T2.

The axial forces distribution of the test piles T2 in the static loading test is shown in Figure 14. It can be seen that at each load level, a linear reduction of axial forces was observed. It implies that fairly uniform shaft resistance was mobilized in soil layers. It also indicates that the loads applied to the test piles was supported by shaft resistances in initial loading. The applied loads were then gradually transferred to the pile end.

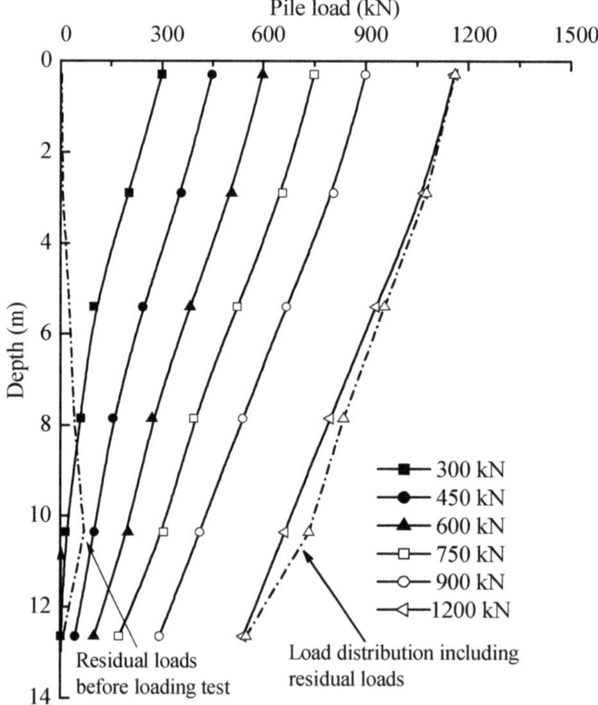

Figure 14. Load distribution curves for test pile.

The residual forces before loading test and the final load distribution including residual forces are plotted as dotted lines in Figure 15. The influence of residual forces appeared not to be negligible on the loading transfer in the static loading test. The end resistances of test pile T2 will be underestimated by 4.2% if not considering the residual forces in static loading test. For equilibrium to be established, the upward residual end forces must equal the downward resultant of the residual shaft forces. As such, the shaft resistances will be overestimated in the interpretation of pile bearing capacity if the residual forces are not taken into account. The test results are consistent with the findings of other researchers, e.g., [41,42]. As the aim of a loading test is to estimate the total bearing capacity, residual forces should not be considered as the summation of residual shaft and end forces for the pile must equal zero. However, we should account for residual forces if the goal of the loading test is to assess the end and shaft resistances for design.

To enable a better assessment, Figure 15 includes the end and shaft resistances (without residual forces) at each loading level, together with the values of the ratio of the end resistances to applied loads. The ratio of end resistances to applied loads of 300 kN was 0.3%; it was to 45.1% with the applied load of 1200 kN. The results indicated that most of the applied loads were undertaken by shaft resistances rather than end resistances in loading test. The distribution of unit shaft resistance can be deduced from Figure 15, as shown in Figure 16. It can be seen that the unit shaft resistance was gradually mobilized as the loads increased. The relative displacement δi between soil layer i and pile during static loading test can be expressed by:

$$\delta_i = S - \sum_{j=1}^{i} \frac{L_i}{2}\left(\varepsilon_j + \varepsilon_{j+1}\right) \tag{4}$$

where S is the pile head settlement at different loading; ε_j is the pile strain at j level; and L_i is the length of the pile located at i level.

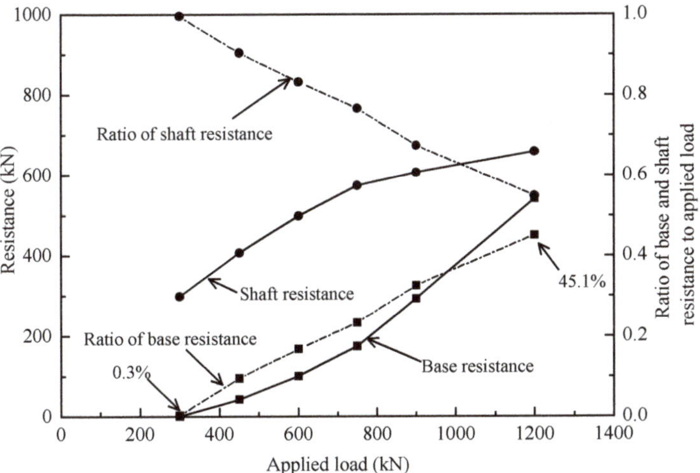

Figure 15. Variation of base and shaft resistance during loading test.

Figure 17 shows the relationship between local shaft resistance and local displacement at various depths for T2. The unit shaft resistance in Figures 16 and 17 does not include the residual forces. According to the study of [13,14], the local displacement could be defined as the differences between pile head settlement and the elastic compression of the pile above the depth under consideration, which can be induced from Equation (4). The test results indicated that the local unit shaft resistance had a good correlation with pile-soil relative displacement. At the same layer, the unit shaft resistance increased approximately hyperbolic to the peak value with pile-soil relative displacement. In the silty clay layer, the shaft resistance could be fully mobilized at a relative displacement of 15 mm, about 3.75% of pile diameter. The threshold of slip displacement for full mobilization of the shaft resistance was found to be 13 and 20 mm in sandy clay and muddy-silty clay layers respectively, about 3.25% and 5.0% of pile diameter. Note that the ultimate value of the shaft resistance in silty clay layer was up to 45 kPa. The values in sandy clay and muddy-silty clay layers were about 40 and 35 kPa, respectively. At the end of static loading test, the shaft resistance along the pile has been fully developed.

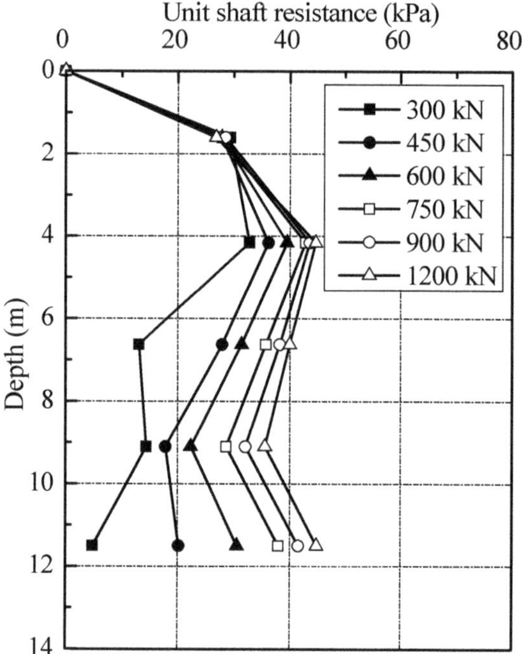

Figure 16. Distribution of unit shaft resistance of test pile.

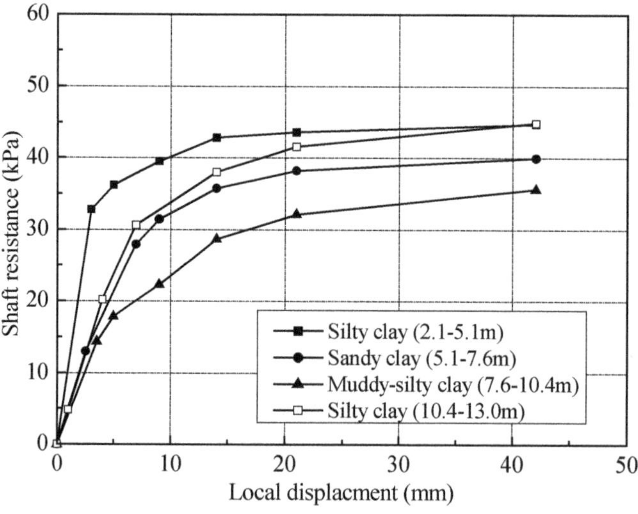

Figure 17. Local shaft resistance versus local displacement of test pile.

4. Conclusions

This paper described the results of a full-scale test on open-ended PHC pipe piles instrumented with FBG sensors in clay. The behavior of test piles was discussed during jacking and static loading test. The main conclusions could be obtained as follows:

(1) The FBG sensoring technology was proved be feasible to measure the axial forces of jacked open-ended PHC pipe piles in clay. It is revealed that the axial forces along PHC pipe piles can be obtained through FBG sensor multiplexing technology.

(2) The behaviour of open-ended PHC pipe piles is more complicated once the effect of soil plugging is considered. The open-ended PHC pipe piles were jacked into clay in a partially plugged mode while the behaved as fully plugged piles in loading tests. This implies that the soil plugging was very different under installation and static loading conditions.

(3) The residual forces in open-ended PHC pipe piles after installation were large and always were compressive at pile toe. The ratios of residual end forces to end resistances after installation were 25.2 and 23.8%, respectively. The residual forces also significantly affect the interpretation of the load distribution in static loading test. The end resistances of the test piles will be underestimated by 4.2% if the residual forces are not considered. However, the residual forces do not affect the total bearing capacity as the sum of residual shaft and end resistances must equal zero.

(4) Loading test results indicated that the shaft resistance has a good correlation with pile-soil relative displacement. The threshold of slip displacement for fully mobilizing the shaft resistance was found to be 15, 13 and 20 mm for silty clay, sandy clay and muddy-silty clay layers, respectively.

Author Contributions: Conceptualization, H.-L.K. and D.-W.Z.; Methodology, T.L.; Data curation, D.-L.Y.; Writing—review and editing, S.H.

Funding: This research was funded by the National Natural Science Fund of China (No. 51408439) and the Young Talent Program of Ocean University of China (No. 841712014). The last author is grateful to the Thailand Research Fund under the grant No. RTA5980005.

Acknowledgments: The authors gratefully acknowledge the financial support provided by the National Natural Science Fund of China (Nos. 51408439, 51879246) and the Young Talent Program of Ocean University of China (No. 841712014). The last author is grateful to the Thailand Research Fund under the grant No. RTA5980005.

Conflicts of Interest: The authors declare no conflict of interest.

References

1. Lehane, B.M.; Chow, F.C.; McCabe, B.A.; Jardine, R.J. Relationships between shaft capacity of driven piles and CPT end resistance. *Proc. Inst. Civ. Eng.-Geotech. Eng.* **2000**, *143*, 93–102. [CrossRef]

2. Zhang, L.M.; Ng, C.W.W.; Chan, F.; Pang, H.W. Termination criteria for jacked pile construction and load transfer in weathered soils. *J. Geotech. Geoenviron. Eng.* **2006**, *132*, 819–829. [CrossRef]

3. Yang, J.; Tham, L.G.; Lee, P.K.K.; Chan, S.T.; Yu, F. Behaviour of jacked and driven piles in sandy soil. *Geotechnique* **2006**, *56*, 245–259. [CrossRef]

4. Yang, J.; Tham, L.G.; Lee, P.K.K.; Yu, F. Observed performance of long steel H-piles jacked into sandy soils. *J. Geotech. Geoenviron. Eng.* **2006**, *132*, 24–35. [CrossRef]

5. Yang, J.; Mu, F. Use of state-dependent strength in estimating end bearing capacity of piles in sand. *J. Geotech. Geoenviron. Eng.* **2008**, *134*, 1010–1014. [CrossRef]

6. Yang, J.; Mu, F. Relating the maximum radial stress on pile shaft to pile base. *Geotechnique* **2011**, *61*, 1087–1092. [CrossRef]

7. Igoe, D.J.P.; Gavin, K.G.; O'Kelly, B.C. Shaft capacity of open-ended piles in sand. *J. Geotech. Geoenviron. Eng.* **2011**, *137*, 903–913. [CrossRef]

8. Randolph, M.F. Science and empiricism in pile foundation design. *Geotechnique* **2003**, *53*, 847–875. [CrossRef]

9. Brucy, F.; Meunier, J.; Nauroy, J.F. Behavior of pile plugs in sandy soils during and after driving. In Proceedings of the 23rd Annual Offshore Technology Conference, OTC 6514, Houston, TX, USA, 6–9 May 1991; Volume 1, pp. 145–154.

10. Smith, I.M.; To, P.; Wilson, S.M. Plugging of pipe piles. In Proceedings of the 3rd International Conference on Numerical Method in Offshore Piling, Nantes, France, 21–22 May 1986; pp. 53–73.

11. Paik, K.H.; Salgado, R. Determination of the bearing capacity of open-ended piles in sand. *J. Geotech. Geoenviron. Eng.* **2003**, *129*, 46–57. [CrossRef]

12. Kou, H.L.; Guo, W.; Zhang, M. Pullout performance of GFRP anti-floating anchor in weathered soil. *Tunn. Undergr. Space Technol.* **2015**, *21*, 437–442. [CrossRef]

13. Kou, H.L.; Chu, J.; Guo, W.; Zhang, M.Y. Field study of residual forces developed in pre-stressed high-strength concrete (PHC) pipe piles. *Can. Geotech. J.* **2015**, *53*, 696–707. [CrossRef]

14. Kou, H.L.; Guo, W.; Zhang, M.Y. Field study of set-up effect in open-ended PHC pipe piles. *Mar. Georesour. Geotechnol.* **2017**, *35*, 208–215. [CrossRef]

15. Yu, F.; Yang, J. Base capacity of open-ended steel pipe piles in sand. *J. Geotech. Geoenviron. Eng.* **2011**, *138*, 1116–1128. [CrossRef]

16. Lehane, B.M.; Schneider, J.A.; Xu, X. The UWA-05 method for prediction of axial capacity of driven piles in sand. In *Frontiers in Offshore Geotechnics*; CRC Press: Boca Raton, FL, USA, 2005; pp. 683–689.

17. Lehane, B.M.; Schneider, J.A.; Xu, X. Development of the UWA-05 design method for open and closed ended driven piles in siliceous sand. In *Contemporary Issues in Deep Foundations, ASCE 158*; American Society of Civil Engineers: Reston, VI, USA, 2007; pp. 1–10.

18. Moormann, C.; Labenski, J.; Aschrafi, J. Simulation of soil plug effects in open steel pipe piles considering the complex soil-structure-interaction during installation. In Proceedings of the 40th Annual Conference on Deep Foundations, Oakland, CA, USA, 12–15 October 2015; pp. 533–546.

19. Miller, G.A.; Lutenegger, A.J. Influence of pile plugging on skin friction in overconsolidated clay. *J. Geotech. Geoenviron. Eng.* **1997**, *123*, 525–533. [CrossRef]

20. Stevens, R.F. The effect of a soil plug on pile drivability in clay. In Proceedings of the 3rd International Conference on the Application of Stress Wave Theory to Piles, Ottawa, ON, Canada, 25–27 May 1988; pp. 861–868.

21. Paikowsky, S.G.; Whitman, R.V.; Baligh, M.M. A new look at the phenomenon of offshore pile plugging. *Mar. Georesour. Geotechnol.* **1989**, *8*, 213–230. [CrossRef]

22. Liu, J.W.; Zhang, Z.M.; Yu, F.; Xie, Z.Z. Case history of installing instrumented jacked open-ended piles. *J. Geotech. Geoenviron. Eng.* **2011**, *138*, 810–820. [CrossRef]

23. Igoe, D.; Doherty, P.; Gavin, K. The development and testing of an instrumented open-ended model pile. *Geotech. Test. J.* **2010**, *33*, 72–82.

24. Schmidt-Hattenberger, C.; Borm, G.; Amberg, F. Bragg grating seismic monitoring system. *Proc. SPIE* **1999**, *3860*, 417–424.

25. Schmidt-Hattenberger, C.; Straub, T.; Naumann, M.; Borm, G.; Lauerer, R.; Beck, C.; Schwarz, W. Strain measurements by fiber Bragg grating sensors for in-situ pile loading tests. In *Smart Structures and Materials 2003: Smart Sensor Technology and Measurement Systems*; The International Society for Optical Engineering: Bellingham, WA, USA, 2003; Volume 5050, pp. 289–294.

26. Klar, A.; Bennett, P.J.; Soga, K.; Mair, R.J.; Tester, P.; Fernie, R.; St John, H.D.; Torp-Peterson, G. Distributed strain measurement for pile foundations. *Proc. Inst. Civ. Eng. Geotech. Eng.* **2006**, *159*, 135–144. [CrossRef]

27. Doherty, P.; Igoe, D.; Murphy, G.; Gavin, K.; Preston, J.; McAvoy, C.; Byrne, B.W.; Mcadam, R.; Burd, H.J.; Houlsby, G.T.; et al. Field validation of fibre Bragg grating sensors for measuring strain on driven steel piles. *Geotech. Lett.* **2015**, *5*, 74–79. [CrossRef]

28. Schmidt-Hattenberger, C.; Naumann, M.; Borm, G. Dynamic strain detection using a fiber Bragg grating sensor array for geotechnical applications. In Proceedings of the European Workshop on Smart Structures in Engineering and Technology, Presquile de Giens, France, 21–23 May 2002; pp. 227–232.

29. Schmertmann, J.H. *Guidelines for Cone Penetration Test Performance and Design*; Report No. FHWA-TS-78-209; US Department of Transportation, Federal Highway Administration, Offices of Research and Development: Washington, DC, USA, 1978.

30. Paikowsky, S.G.; Whitman, R.V. The effect of plugging on pile performance and design. *Can. Geotech. J.* **1990**, *27*, 429–440. [CrossRef]

31. Choi, Y.; O'Nell, M.W. Soil plugging and relaxation in pipe pile during earthquake motion. *J. Geotech. Geoenviron. Eng.* **1997**, *123*, 975–982. [CrossRef]

32. Paik, K.; Salgado, R.; Lee, J.; Kim, B. Behaviour of open-and closed-ended piles driven into sands. *J. Geotech. Geoenviron. Eng.* **2003**, *129*, 296–306. [CrossRef]

33. Hong, C.Y.; Zhang, Y.F.; Zhang, M.X.; Leung, L.M.G.; Liu, L.Q. Application of FBG sensors for geotechnical health monitoring, a review of sensor design, implementation methods and packaging techniques. *Sens. Actuators A Phys.* **2016**, *244*, 184–197. [CrossRef]

34. MCC (Ministry of Construction of the People's Republic of China). *JGJ106-2014: Technical Code for Building Pile Foundations*; China Building Industry Press: Beijing, China, 2014.

35. Lehane, B.M.; Randolph, M.F. Evaluation of a minimum base resistance for driven pipe piles in siliceous sand. *J. Geotech. Geoenviron. Eng.* **2002**, *128*, 198–205. [CrossRef]

36. Gavin, K.; Lehane, B.M. Base load-displacement response of piles in sand. *Can. Geotech. J.* **2007**, *44*, 1053–1063. [CrossRef]

37. Fellenius, B.H.; Altaee, A.A. Critical depth: How it came into being and why it does not exist. *Proc. ICE-Geotech. Eng.* **1995**, *113*, 107–111. [CrossRef]

38. Cooke, R.W. Influence of residual installation forces on the stress transfer and settlement under working loads of jacked and bores piles in cohesive soils. In *Behavior of Deep Foundations*; Special Technical Publication STP; ASTM: West Conshohocken, PA, USA, 1979; pp. 231–249.

39. Poulos, H.B. Analysis of residual stress effects in piles. *J. Geotech. Eng. ASCE* **1987**, *113*, 216–229. [CrossRef]

40. Darrag, A.A. Capacity of Driven Piles in Cohesionless Soils Including Residual Stress. Ph.D. Thesis, Purdue University, West Lafayette, IN, USA, 1987.

41. Rieke, R.D.; Crowser, J.C. Interpretation of a pile load test considering residual stresses. *J. Geotech. Geoenviron. Eng.* **1987**, *113*, 320–334. [CrossRef]

42. Kraft, L.M. Performance of axially loaded pipe piles in sand. *J. Geotech. Eng. ASCE* **1991**, *117*, 272–296. [CrossRef]

Article

Fiber Bragg Grating Displacement Sensor with High Abrasion Resistance for a Steel Spring Floating Slab Damping Track

Yongxing Guo [1,2,*], Wenlong Liu [1,2], Li Xiong [1,2], Yi Kuang [1,2], Heng Wu [1,2] and Honghai Liu [3]

[1]   Key Laboratory of Metallurgical Equipment and Control Technology, Ministry of Education, Wuhan University of Science and Technology, Wuhan 430081, China; lwlwjc@126.com (W.L.); xiongli166@163.com (L.X.); kuangyi1993@163.com (Y.K.); wuheng6@126.com (H.W.)
[2]   Hubei Key Laboratory of Mechanical Transmission and Manufacturing Engineering, Wuhan University of Science and Technology, Wuhan 430081, China
[3]   State Key Laboratory of Mechanical System and Vibration, School of Mechanical Engineering, Shanghai Jiao Tong University, Shanghai 200240, China; honghai.liu@port.ac.uk
*   Correspondence: yongxing_guo@wust.edu.cn

Received: 22 May 2018; Accepted: 8 June 2018; Published: 11 June 2018

Abstract: This paper presents a fiber Bragg grating (FBG) displacement sensor with high abrasion resistance for displacement monitoring of a steel spring floating slab damping track. A wedge-shaped sliding block and an equal-strength beam form a conversion mechanism to transfer displacement to the deflection of the beam, and the deflection-induced strain is exerted on two FBGs. A special linear guide rail-slider and a precision rolling bearing have been adopted onto the conversion mechanism, which turned sliding friction into rolling friction and thus significantly reduced the friction during frequent alternating displacement measuring. Sensing principle and the corresponding theoretical derivation have been demonstrated. Experiment results show that the sensor has a sensitivity of 34.32 pm/mm and a high resolution of 0.0029 mm within a measurement range of 0~90 mm. Besides, the sensor has also a good measurement capability for micro-displacement within a range of 0~3 mm. The repeatability error and hysteresis error are 1.416% and 0.323%, respectively. Good creep resistance and high abrasion resistance for alternating displacement measurement have also been presented by a performance test. These excellent performances satisfy the requirements of high precision and long-term stability in structural health monitoring for machinery equipment and civil engineering, especially in the displacement monitoring of a floating slab damping track.

Keywords: Fiber Bragg grating (FBG); displacement sensor; floating slab track; abrasion resistance; structural health monitoring (SHM)

1. Introduction

Displacement measurement is a basic and important subject in structural health monitoring (SHM). As far as we know, the research of displacement monitoring of floating slab damping track based on FBG sensors has not been reported yet. It is difficult to measure the steel spring floating slab damping track's displacement of the subway or to monitor it in the long-term using the traditional electrical sensor, because of zero temperature drift, the mounted environment lacks sealing schedules, and difficulties of working in high voltage. Therefore, it is necessary to design a displacement sensor that can realize the long-term and real-time monitoring of the floating slab damping track under the effect of frequent alternating displacement. Various displacement sensors based on electronic or fiber-optic techniques have been developed and used in SHM. It's well known that electromagnetic principle-based displacement sensors need to work with a power supply and are susceptible to electromagnetic interference. Besides, they are also prone to tedious wires jointing and zero temperature drift. In recent years, as a particular class of optical fiber sensors, fiber Bragg grating (FBG) have widely attracted attention because of their inherent advantages, such as immunity to electromagnetic interference and optical power fluctuations, small profile, light weight, no zero-temperature drift and the fact that multiple FBGs can be arrayed along a single fiber [1,2]. During the past few decades, FBG sensors have been widely used in mechanical equipment, civil engineering, bridge scour monitoring, aircraft, and robot [3–8], and have successfully obtained the measurement of many physical quantities such as displacement, vibration, force and strain [9–13].

For the FBG displacement sensor, in order to measure the external displacement, a conversion mechanism must be designed to convert the displacement information into an axial strain exerted on FBG. Traditionally, most of the conversion mechanisms are designed as spring [14], ring-type [15,16], cantilever beam and its evolution type [17–19], and other composite structures [20–22]. However, these structure designs have the weaknesses of low measurement accuracy due to friction losses and poor durability, especially when the sensor works under frequent alternating displacement brought by the measured object.

In recent years, some different packing and structure designs of FBG displacement sensor have been reported. For instance, Zou et al. connected a spring and FBG in series and achieved high sensitivity measuring [23]. However, in this design, the elasticity coefficient of spring is liable to change, which could affect the measurement accuracy in the long-term monitoring. What's more, the practical application of the sensor is also limited by a small range. Tao et al. proposed an FBG displacement sensor based on a thin-wall ring [24,25]. Although the designed structure is compact and the temperature cross-sensitivity is also solved, it suffered from defects, such as low repeatability caused by inconsistency from the manually pasting process, and low resolution based on reflection spectrum bandwidth demodulation due to the use of an optical spectrum analyzer (OSA). Some different packaging designs have been also reported which can be designed into a temperature-insensitive displacement sensor based on cantilever beam structure and its evolution type [26–28], whose common characteristic is to detect the reflection spectrum broadening of FBG. As far as we know, the profile of the broadened spectrum of FBG is usually irregular due to the nonuniform strain on the grating, which may cause inaccurate experimental results. Although these designs can solve the problem of temperature cross-sensitivity, a professional spectrometer is usually needed for the monitoring of bandwidth. Furthermore, the advantage of the spectrometer is inferior to that of an FBG wavelength interrogator in terms of scanning frequency and cost, which is not conducive to long-term and real-time monitoring in practical engineering. Dong et al. [29] obliquely glued a single FBG onto the lateral side of the cantilever beam. However, it is difficult to guarantee that the FBG's midpoint coincides exactly with the zero-strain layer of beam, which will result in the FBG reflected spectrum fluctuations, thus it is difficult to obtain precisely reflected spectrum bandwidth variations. Furthermore, there is also the defect that a photo-detector is needed for compensating the optical power fluctuation and inconvenient in multiplexing due to the utilize of the power demodulation method.

To solve these issues and limitations, various structure designs and demodulation methods have been proposed by researchers to improve the performance of FBG sensors to meet the needs of practical engineering. Compared with the packing and structure designs of FBG as mentioned above, researchers provided some different structure designs based on the reflected wavelength demodulation. For instance, Li et al. designed two high-sensitivity FBG displacement sensors [30,31], a single FBG is pre-tensioned and the two side points are fixed, efficiently avoiding the unwanted chirp effect of grating. However, the wedge-shaped sliding block and T-shaped cantilever beam convert the horizontal displacement into vertical displacement by sliding rather than rolling, which may result in the poor abrasion performance in reciprocating displacement measurement in long-term. As mentioned previously, although the problems of low sensitivity and cross-sensitivity have been solved, most of those displacement sensors are limited by small measuring ranges and poor durability. Furthermore, the measurement accuracy can't be guaranteed due to friction losses when sensors are used for long-term monitoring under the effect of frequent alternating displacement and a harsh environment in the field of SHM.

In this paper, an improved FBG displacement sensor is proposed to improve the accuracy and abrasion resistance of the displacement measurement in the application of long-term monitoring of floating slab damping track. This improved structure design overcomes the defects such as the changeability of the elasticity coefficient of force-transmitting medium, and directional movement of sliding block which could bring negative influences on the measuring accuracy in the exiting design of FBG displacement sensors. This knowledge has been validated in our previous work [32]. Furthermore, compared with the displacement sensor in literature [32], a linear guide rail has been used instead of the sliding surface of the wedge-shaped sliding block, and the free end of the beam adopts a precision rolling bearing contacted with the top surface of the slider. Moreover, the packing method of these two gratings are all pasted instead of a grating being pasted while the other is bare grating. This configuration improves problems such as irregular errors due to the different packaging methods of gratings, and poor durability due to frictional losses under the effect of frequent alternating displacement. The sensing principle of the designed mechanical structure have been derived, and then the sensor prototype has been manufactured, installed and fully tested.

2. Principle and Structure Design

2.1. Mechanical Structure Design

In a subway line of Beijing, the dynamic impact by a passing high-speed train causes high stress in the rails and floating slab, thereby causing the alternating and creeping displacement. However, the allowable settlement range of the steel spring floating slab damping track is 0~3 mm. In order to guarantee the safe operation of the high-speed train, the quantitative value of rail alternating and creeping displacement must be monitored. All the sensors must be installed symmetrically in series along continuous floating slab damping track to form FBG sensing network for long-term and real-time displacement monitoring. All wavelength signal can be transmitted to an FBG interrogator by fiber optical cables and transmitted by the wavelength division multiplexing (WDM) technology. As shown in Figure 1, the FBG displacement sensor that meets design requirements is installed in between the outer barrel and inter barrel of vibration isolator.

In view of the requirements for high accuracy and high abrasion resistance of long-term monitoring of the floating slab damping track under the effect of frequent alternating displacement, this paper introduces our recent work on the design and investigation of a new FBG displacement sensor with high abrasion resistance. The mechanical structure design of the proposed sensor is shown in Figure 2. This sensor mainly consists of two FBGs with different wavelengths, a variable section cantilever beam, a wedge-shaped sliding block, a linear guide rail-slider, a pull rod and a restoring spring. Two FBGs in a single mode optical fiber are adhered on the upper and lower surfaces of the cantilever beam. The packing method of FBG can achieve differential measuring and solve the problem of temperature cross-sensitivity. Besides, in order to overcome the shortcoming of inaccurate measurement due to the frictional losses of mechanical conversion structures, the design of sensing structure has been improved. The detailed 3D diagram of cantilever beam and linear guide rail components are shown in Figure 3. A precision bearing has been adopted on the free end of the cantilever to contact with the surface of wedge-shaped slider. A special linear guide rail-slider has been used as a medium to fix the wedge-shaped slider to the base. These designs make all the sliding friction in our previous work [24] turn into rolling friction, which considerably reduces the friction. A special connector is designed to fix one end of the spring and the pull rod on the wedge-shaped slider. The pull rod receives the external displacement inputs and then drives the wedge-shaped slider. Meanwhile, the spring can effectively ensure the pull rod has a good reciprocating ability to move with the measured object. Therefore, the external displacement can be converted into the flexural strain of the cantilever beam, and the displacement-induced strain can be exerted on the two FBGs.

So the displacement is determined by the structural parameters of the conversion mechanism. The corresponding theoretical derivation will be presented as follows.

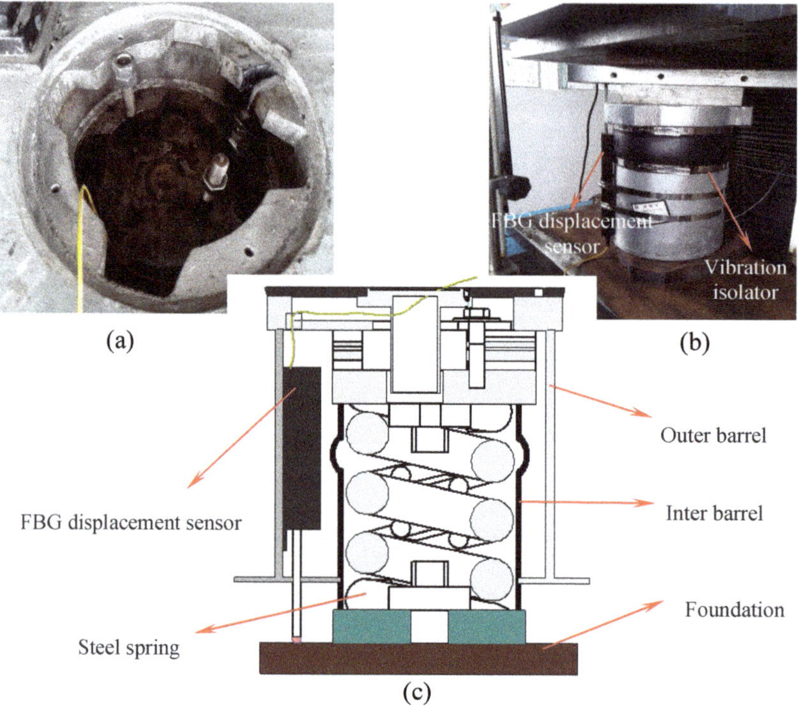

Figure 1. Installation situation of FBG displacement sensor: (**a**) Photo of field installation; (**b**) Installation testing; (**c**) Schematic diagram of installation.

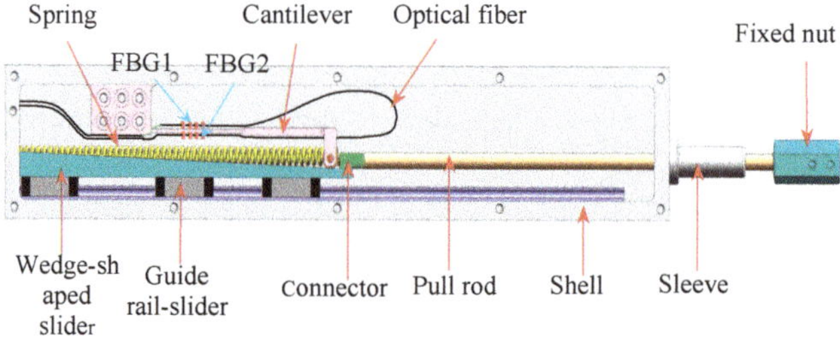

Figure 2. Structure of the proposed sensor.

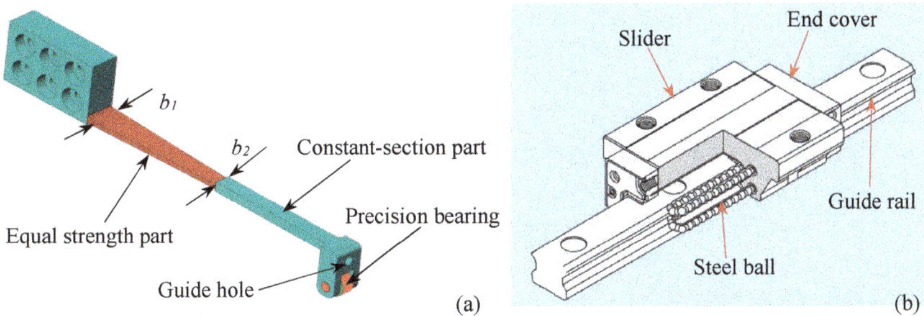

Figure 3. The detailed 3D diagram of designed structure: (**a**) Cantilever beam with a precision bearing; (**b**) Linear guide rail components.

2.2. Measuring Principle

The reflected wavelength shift of FBG is influenced by the changes of axial strain and environment temperature. The relation between center reflective wavelength shifts and strain/temperature can be described by:

$$\frac{\Delta\lambda}{\lambda} = (1 - P_e)\Delta\varepsilon + \left(\alpha_f + \xi\right)\Delta T \tag{1}$$

where $\Delta\lambda$ is the center wavelength shift of FBG, λ is the initial wavelength of FBG, α_f is the thermal expansion coefficient, ξ is the thermal-optic coefficient and P_e(≈ 0.22 at room temperature) is the effective photo-elastic coefficient.

As shown in Figure 2a, the cantilever beam includes an equal-strength part and a constant-section part. The thickness of the constant-section part is larger than the equal-strength part where the FBGs are glued, which could bring higher strain detection sensitivity for FBG because the bending strain concentrates on the thinner equal-strength part. Figure 4 shows the schematic diagram of the conversion mechanism. FBG1 and FBG2 are glued on the upper and lower surfaces of equal-strength part, respectively. Using the difference of the wavelength shifts brought by the two opposite surface strains as the sensing signal, the temperature cross-sensitivity could be avoided, and the sensitivity can be improved. The relationship between the measured displacement S and the output of wavelength shifts difference $\Delta\lambda_{2-1}$ is deduced below.

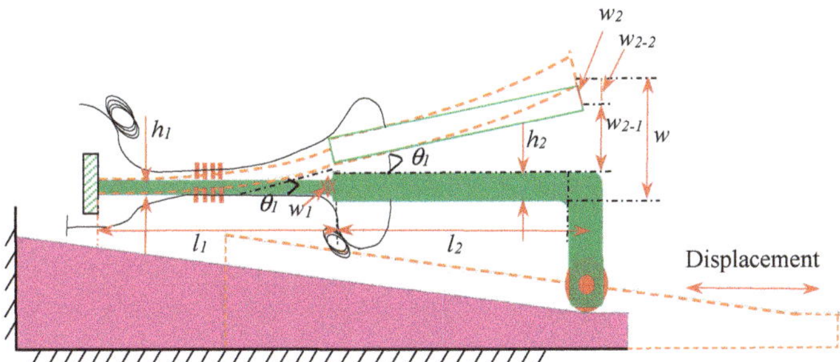

Figure 4. Schematic diagram for the sensing principle of the proposed sensor.

As shown in Figure 4, it can be seen that the height difference of the wedge-shaped slider is the total deflection W. What's more, the cantilever beam includes two parts, whereas these two FBGs are glued on the equal-strength part. Therefore, it is necessary to obtain the theoretical relationship between the deflection w_1 of the equal-strength part and the total deflection W. Meanwhile, we can see that vertical displacement of the constant-section part contains two parts: the displacement (w_{2-1})

because of the rotation angle θ_1 and the component of w_2 along the vertical direction (w_{2-2}). w_{2-1} equals $L_2 sin\theta_1$, and w_{2-2} equals $w_2 cos\theta_1$. So $W = w_1 + w_{2-1} + w_{2-2}$.

According to the transfer principle of the deflection for the cantilever, one can obtain the following expression:

$$w_1 = W - L_2 sin\theta_1 - w_2 cos\theta_1 \tag{2}$$

where L_2 is the length of the constant-section part.

According to material mechanics, the equal force F is exerted on the equal-strength part and constant-section part, the deflection w_1 of the equal-strength cantilever can be expressed as:

$$w_1 = \frac{FL_1^3}{EI} = \frac{6FL_1}{Eb_1 h_1^3}, \tag{3}$$

where b_1, h_1, L_1 is the width, thickness, and length of the equal-strength part, respectively; E, I is Young's modulus and the section moment of inertia, respectively.

According to material mechanics, the equations of rotation angle and moment of inertia can be expressed as:

$$\theta_x = \frac{dw}{dx} = \int \frac{F(L_1 - x)}{EI_x} dx + C \tag{4}$$

$$I_x = \frac{h^3}{12}\left(\frac{L_1 - x}{L_1}\right) b_1. \tag{5}$$

where x is the distance between equal-strength cantilever and the fixed end; I_x is the corresponding moment of inertia at x.

Substituting Equation (5) in Equation (4) we can get:

$$\theta_x = \frac{12FL_1}{Eb_1 h^3} x + C. \tag{6}$$

Then, the rotation angle θ_1 can be expressed as:

$$\theta_1 = \frac{12FL_1^2}{Eb_1 h_1^3}. \tag{7}$$

According to material mechanics, the deflection w_2 of the equal-section cantilever can be expressed as:

$$w_2 = \frac{4L_2^3}{Eb_2 h_2^3} \times F \tag{8}$$

where b_2, h_2 is the width and thickness of the equal-section cantilever, respectively; E is Young's modulus of the cantilever beam.

Combining Equations (2), (3), (7) and (8), the total deflection W can be written as:

$$W = \frac{6FL_1^3}{Eb_1 h_1^3} + sin\frac{12FL_1^2}{Eb_1 h_1^3} \times L_2 + cos\frac{12FL_1^2}{Eb_1 h_1^3} \times \frac{4FL_2^3}{Eb_2 h_2^3}. \tag{9}$$

For the design of cantilever beam in this paper, $sin\theta_1$ is small, which can be approximated to θ_1, and $cos\theta_1$ can be approximated to 1, so Equation (9) can be transformed to

$$W = \frac{6FL_1^3}{Eb_1 h_1^3} + \frac{12FL_1^2 L_2}{Eb_1 h_1^3} + \frac{4FL_2^3}{Eb_2 h_2^3}. \tag{10}$$

Substituting the relevant parameters of designed cantilever such as $b_1 = 6mm$, $b_2 = 3mm$, $L_1 = L_2 = 30mm$, $h_1 = 1mm$ and $h_2 = 2mm$ into Equation (10), the relationship between the deflection w_1 and the total deflection W can be regarded as: $w_1 = (6/19)W$. Then, combining the deflection w_1 and the strain ε of the equal-strength beam, we can get the equation: $\varepsilon = h_1 w_1 / L_1^2$. Moreover, according to the sensing principle of this proposed sensor, the relationship of displacement S, total deflection W and dip rotation angle θ can be written as $W = S \times tan\theta$.

So we can get the theoretical relationship between the measured displacement S and the strain ε can be expressed as:

$$S = \varepsilon \times \frac{19L_1^2}{6h_1 \tan \theta}. \tag{11}$$

Based on the differential measurement method and the cantilever structure, the Bragg wavelength shifts of two FBGs are in opposite directions with the same absolute values. Moreover, these two FBGs are very close to each other, their temperature-induced shifts are considered to be identical. So the difference $\Delta\lambda_{2-1}$ of the two shifted Bragg wavelengths can be expressed as:

$$\Delta\lambda_{2-1} = \Delta\lambda_2 - \Delta\lambda_1 = 2\lambda(1 - P_e)\varepsilon \tag{12}$$

Combining Equations (11) and (12), the relationship of the wavelength shift difference and the measured displacement can be expressed as:

$$\Delta\lambda_{2-1} = \lambda(1 - P_e)\frac{12Sh_1 \tan \theta}{19L_1^2} \tag{13}$$

Equation (13) illustrates that the wavelength shift difference $\Delta\lambda_{2-1}$ of two FBGs is linearly related to the displacement S, and the external displacement can be effectively obtained by the center wavelength shift of FBG.

3. Sensor Prototype Manufacturing and Experiments

3.1. Sensor Prototype Manufacturing

In addition to the parameters mentioned above, other parameters, such as the chute length in base, the height difference and length of wedge-shaped sliding block are regarded as 200 mm, 4 mm and 110 mm, respectively. So the measuring range of the proposed sensor can reach 90mm. Two FBGs with reflectivity of 90% and bandwidth of 0.18 nm have been glued on the cantilever beam by using a commercial adhesive (353ND, made by Epoxy Technology, Inc., Billerica, MA, USA). After assembling, a displacement sensor prototype with central wavelengths of 1530.1361 nm and 1533.2539 nm for FBG1and FBG2 has been manufactured as shown in Figure 5.

Figure 5. Photo of the FBG displacement sensor.

3.2. Experiments for Calibration and Test

Figure 6 shows the photo of the experiment setup for displacement calibration test. A vernier caliper (accuracy: 0.02 mm), a micrometer caliper (accuracy: 0.01 mm) and the sensor prototype have been fixed on the experimental platform. As shown in Figure 6a,b, the vernier scale and micrometer caliper have been fixed together with the pull rod, respectively, and the pull rod can freely slide along the horizontal direction. A homemade FBG interrogator (sampling rate: 100 Hz, accuracy:

5 pm, resolution: 0.1 pm) is used to record the wavelength changes of the two FBGs under the external displacement.

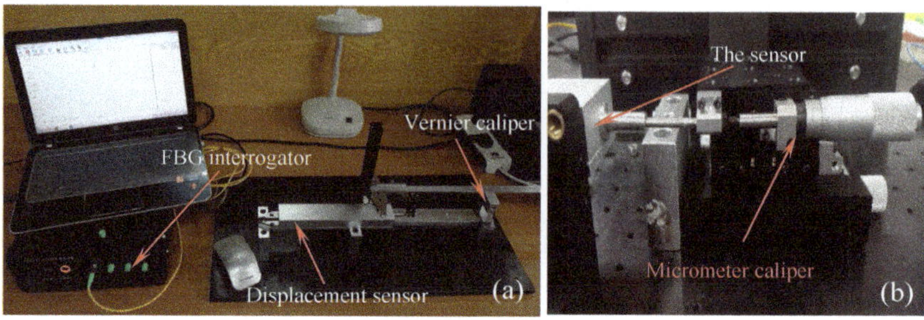

Figure 6. Photo of the experimental setup for displacement testing: (**a**) Large displacement measurement; (**b**) Micro-displacement measurement.

The calibration experiment begins by stretching the pull rod from 0 mm to 10 mm, then stretching it to full-scale (90 mm) with a step of 20 mm and keeping 3–5 s for each displacement point, then unloading the displacement to 0 mm with the same step. As mentioned above, because the allowable settlement range of steel spring floating slab damping track is 0~3 mm, the capability of sensor to measure micro-displacement must be tested. The micro-displacement experiment begins by stretching it to 3 mm with a step of 0.5 mm and keeping 3–5 s for each displacement point. This loading and unloading test process has been repeated three times at a stable room temperature. The time-history wavelength changes of the three cycling tests of full-scale and micro-displacement have been plotted in Figure 7.

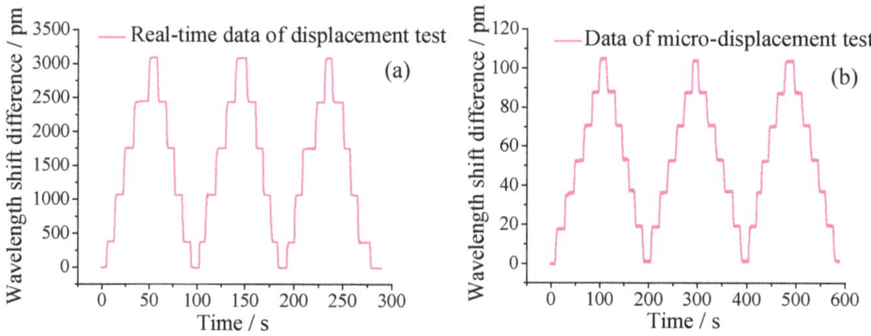

Figure 7. Time-history curve of three cycling tests: (**a**) Range of 0~90 mm; (**b**) Range of 0~3 mm.

Figure 8a shows the six curves of the relationship between different displacements and wavelength shift difference obtained from the time-history data. The results demonstrate that the variation patterns are linear and the repeatability error and hysteresis error for three cycling tests of the displacement sensor are 1.416% and 0.323%, respectively. Figure 8b shows variation of the average data of three repeated tests and the linear fitting curves which indicate good fits. From the fitted curve, the sensitivity of the sensor is observed to be 34.32 pm/mm with a linearity of 0.9999, and the measurement range is 0~90 mm. Besides, the thumbnails of Figure 8b is the fitted curve of the micro-displacement test, there is a little difference of the sensitivity between large-displacement and micro-displacement, it can be seen that the sensor has also a good measurement capability for micro-displacement. The wavelength resolution and accuracy of the used interrogator is 0.1 pm and 5 pm, consequently, this sensor can achieve a high resolution of 0.0029 mm and an accuracy of 0.15 mm.

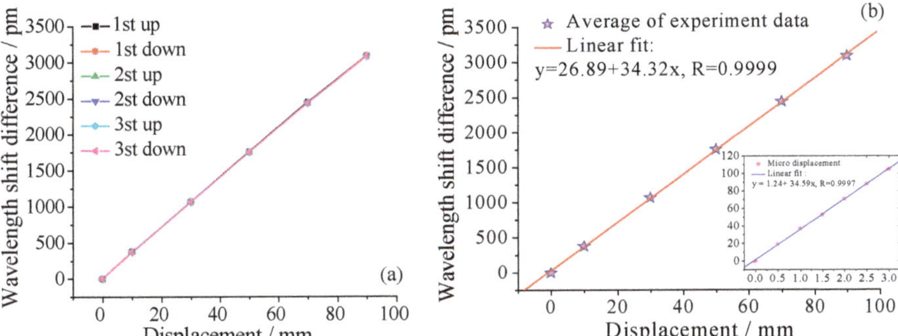

Figure 8. (**a**) The relationship of wavelength shift difference versus displacement; (**b**) Linear fit curve of average of experiment data and result of micro-displacement test.

Because the wavelength shift of FBG is influenced by both axial strain and temperature, knowledge of the temperature compensation ability of the displacement sensor is needed. As mentioned above, it can be seen that the temperature compensation method of the sensor is dual-grating difference output method with packaging full pasted FBGs. Compared with the reference grating method, the dual-grating difference output method reduces the big difference of the temperature sensitivity coefficient due to different packaging method of FBGs. As shown in Figure 9, a thermostat (OTF-1200X, HEFEI KE JING Materials Technology Co., Ltd. Hefei, China; accuracy: 1 °C, resolution: 0.1 °C) is used to change the surrounding temperature of the displacement sensor from 30 °C to 60 °C. Time-history curve of the two FBGs is shown in Figure 10. The maximum value of the wavelength shift difference in right Y axis is only 10 pm which indicates the sensor has good temperature compensation ability. The cause of the error is that the strain transfer ratio from FBG to substrate is affected by some random factors such as the length and thickness of adhesive area, consequently, which will affect the measurement accuracy of FBG.

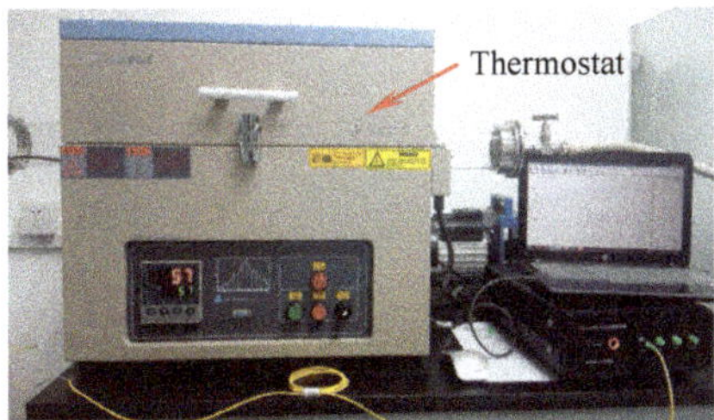

Figure 9. Photo of the experimental setup for temperature testing.

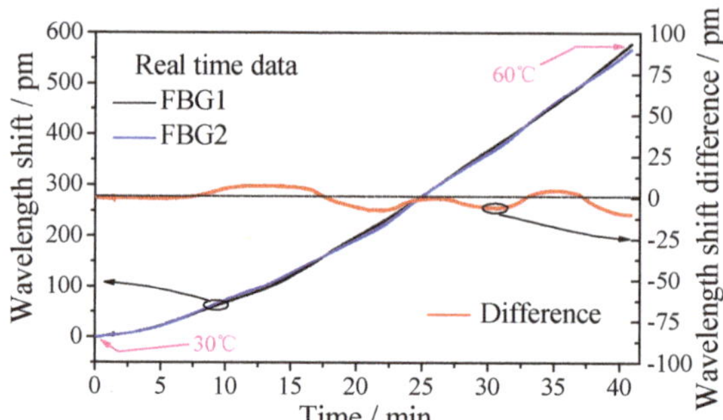

Figure 10. Time-history curve of temperature test.

As mentioned above, in the application of the floating slab damping track, the frequent alternating displacement brought about by external load often acts on the sensor in the long-term, which would affect the measurement accuracy due to the frictional losses of the mechanical structure in the sensor. What's more, if the material performance of the elastic cantilever beam and the robustness of the bonding of FBG are affected by the frequent alternating displacement, the sensitivity of sensor also will change. Therefore, in order to make sensor have an applicability for this working condition of frequent alternating displacements, an abrasion resistance test for a sensor is indispensable.

As shown in Figure 11, special experimental equipment with the function of reciprocating motion has been designed using a crank-slider mechanism. The crank is powered by a DC motor and the pull rod of sensor is connected and fixed on the slider. Continuous alternating displacement from 20 to 52 mm has been applied on the sensor and the sensitivity and the measuring range of the sensor have been tested at different cycle numbers. Figure 12 shows part of the experimental data recorded by FBG interrogator with a frequency of 200 Hz, from which we can see that the amplitude and frequency of the alternating displacement are 30 mm and 8 Hz, respectively. Table 1 shows the changes of sensitivity and range at different cycle numbers. The sensitivity values have no abnormal change, which demonstrate that the alternating displacement does not degrade the measurement capability.

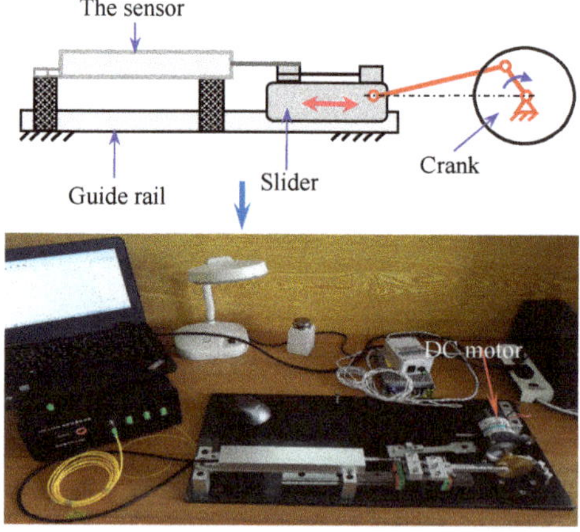

Figure 11. Principle and photo of the experimental equipment for alternating displacement test.

Table 1. Changes of sensitivity under alternating displacement test.

Number of Cycles	Sensitivity (pm/mm)
0	34.32
10^2	34.61
10^3	34.21
10^4	34.66
10^5	34.71

Creep performance has a significant influence on the measuring accuracy of sensor for long-term monitoring of floating slab damping track. Therefore, it is necessary to investigate the creep performance of the proposed sensor. During this test, the pull rod was pulled out to a certain position and fixed for more than 80 minutes. The FBG interrogator recorded the wavelength shifts change of the two FBGs. The time-history curve is shown in Figure 13, and the data of 36~41 min is selected and amplified. As we can see, the fluctuating value is within 2 pm, which demonstrates that the sensor has good creep resistance.

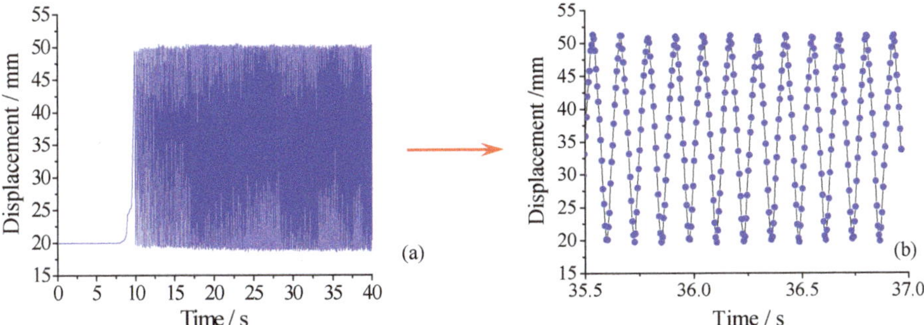

Figure 12. Part of the experimental data for alternating displacement test: (**a**) Real-time data of alternating displacement test; (**b**) The amplified data of 35.5 s~37 s.

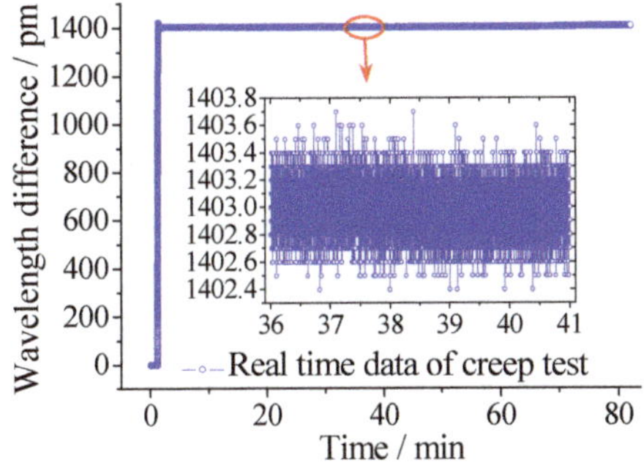

Figure 13. Result of creep performance test for displacement measurement.

4. Conclusions

This work proposed and demonstrated the design and performance test of an FBG displacement sensor with high abrasion resistance for a steel spring floating slab damping track. The use of a guide rail-slider system and a precision rolling bearing has been validated to effectively enhance the measurement accuracy and abrasion resistance of sensor. The measurement principle and theoretical model have been derived and experimentally verified. Full performance tests have been carried out in displacement calibration, temperature compensation, creep, and abrasion resistance. Experimental results show that the sensor has a good sensitivity of 34.32 pm/mm, a high resolution of 0.0029 mm over a measuring range of 0~90 mm, a good measurement capability of micro-displacement within a range of 0~3 mm, a repeatability error of 1.416%, and a hysteresis error of 0.323%. The creeping and alternating displacement tests illustrate good creep resistance and excellent abrasion resistance. Compared with the reported FBG displacement sensors, the proposed displacement sensor promises many potential applications in SHM, especially in the displacement monitoring of floating slab damping track that demands for high measurement accuracy and frequent alternating displacement-induced friction.

Author Contributions: In this paper, Y.G. conceived the idea, acquired the funding and supervised the work; W.L. organized the paper, finished the experiment and analyzed the data; L.X., Y.K., H.W., and H.L. contributed to data handling. All authors revised the paper for intellectual content.

Funding: This work was supported in part by the National Natural Science Foundation of China under Grant (No. 51605348, 51575407, 51575338, 51475427), in part by the Open fund of Hubei Digital Manufacturing Key Laboratory under Grant No. SZ1801.

Acknowledgments: This work was supported in part by the National Natural Science Foundation of China under Grant (No. 51605348, 51575407, 51575338, 51475427), in part by the Open fund of Hubei Digital Manufacturing Key Laboratory under Grant No. SZ1801.

Conflicts of Interest: The authors declare no conflict of interest.

References

1. Majumder, M.; Gangopadhyay, T.K.; Chakraborty, A.K.; Dasgupta, K.; Bhattacharya, D.K. Fibre Bragg gratings in structural health monitoring—Present status and applications. *Sens. Actuators A Phys.* **2008**, *147*, 150–164. [CrossRef]
2. Jia, D.D.; Zhang, Y.L.; Chen, Z.T.; Zhang, H.X.; Liu, T.G.; Zhang, Y.M. A self-healing passive fiber Bragg grating sensor network. *J. Lightwave Technol.* **2015**, *33*, 2062–2067. [CrossRef]
3. Zhao, Y.; Zhu, Y.N.; Yuan, M.D.; Wang, J.F.; Zhu, S.Y. A Laser-Based Fiber Bragg Grating Ultrasonic Sensing System for Structural Health Monitoring. *IEEE Photonics Technol. Lett.* **2016**, *28*, 2573–2576. [CrossRef]
4. Yazdizadeh, Z.; Marzouk, H.; Hadianfard, M.A. Monitoring of concrete shrinkage and creep using Fiber Bragg Grating sensors. *Constr. Build. Mater.* **2017**, *137*, 505–512. [CrossRef]
5. Kong, X.; Ho, S.C.M.; Song, G.B.; Cai, C.S. Scour Monitoring System Using Fiber Bragg Grating Sensors and Water-Swellable Polymers. *J. Bridge Eng.* **2017**, *22*, 04017029. [CrossRef]
6. Guo, H.L.; Xiao, G.Z.; Mrad, N.; Yao, J.P. Fiber optic sensors for structural health monitoring of air platforms. *Sensors* **2011**, *11*, 3687–3705. [CrossRef] [PubMed]
7. Guo, Y.X.; Kong, J.Y.; Liu, H.H.; Xiong, H.G.; Li, G.F.; Qin, L. A three-axis force fingertip sensor based on fiber Bragg grating. *Sens. Actuators A Phys.* **2016**, *249*, 141–148. [CrossRef]
8. Xiong, L.; Jiang, G.Z.; Guo, Y.X.; Liu, H.H. A Three-dimensional Fiber Bragg Grating Force Sensor for Robot. *IEEE Sens. J.* **2018**, *18*, 3632–3639. [CrossRef]
9. Li, D.S.; Li, H.N.; Ren, L.; Song, G.B. Strain transferring analysis of fiber Bragg grating sensors. *Opt. Eng.* **2006**, *45*, 024402. [CrossRef]
10. Zhao, X.F.; Gou, J.H.; Song, G.B.; Ou, J.P. Strain monitoring in glass fiber reinforced composites embedded with carbon nanopaper sheet using Fiber Bragg Grating (FBG) sensors. *Compos. Part B Eng.* **2009**, *40*, 134–140. [CrossRef]
11. Guo, Y.X.; Kong, J.Y.; Liu, H.H.; Hu, D.T.; Qin, L. Design and Investigation of a Reusable Surface-Mounted Optical Fiber Bragg Grating Strain Sensor. *IEEE Sens. J.* **2016**, *16*, 8456–8462. [CrossRef]
12. Li, R.Y.; Tan, Y.G.; Bing, J.Y.; Li, T.L.; Hong, L.; Yan, J.W.; Hu, J.M.; Zhou, Z.D. A Diaphragm-type Highly Sensitive Fiber Bragg Grating Force Transducer with Temperature Compensation. *IEEE Sens. J.* **2018**, *18*, 1073–1083. [CrossRef]
13. Li, R.Y.; Chen, Y.Y.; Tan, Y.G.; Zhou, Z.D.; Li, T.L. Sensitivity Enhancement of FBG-Based Strain Sensor. *Sensors* **2018**, *18*, 1607. [CrossRef] [PubMed]

14. Li, J.C.; Neumann, H.; Ramalingam, R. Design, fabrication, and testing of fiber Bragg grating sensors for cryogenic long-range displacement measurement. *Cryogenics* **2015**, *68*, 36–43. [CrossRef]

15. Li, W.L.; Zhang, Y.X.; Wang, Q.; Pan, J.J.; Liu, J.; Zhou, C.M. Displacement monitor with FBG deforming ring and its application in high speed railway. In Proceedings of the OFS2012 22nd International Conference on Optical Fiber Sensors, Beijing, China, 4 October 2012.

16. Ciminello, M.; Ameduri, S.; Flauto, D. Design of an FBG based-on sensor device for large displacement deformation. In Proceedings of the 2013 SBMO/IEEE MTT-S International Microwave & Optoelectronics Conference (IMOC), Rio de Janeiro, Brazil, 4–7 August 2013.

17. Zhao, Y.; Yu, C.B.; Liao, Y.B. Differential FBG sensor for temperature-compensated high-pressure (or displacement) measurement. *Opt. Laser Technol.* **2004**, *36*, 39–42. [CrossRef]

18. Zhao, Y.; Zhao, H.W.; Zhang, X.Y.; Meng, Q.Y.; Yuan, B. A novel double-arched-beam-based fiber Bragg grating sensor for displacement measurement. *IEEE Photonics Technol. Lett.* **2008**, *20*, 1296–1298. [CrossRef]

19. Zhang, Y.N.; Zhao, Y.; Wang, Q. Improved design of slow light interferometer and its application in FBG displacement sensor. *Sens. Actuators A Phys.* **2014**, *214*, 168–174. [CrossRef]

20. Chang, Y.T.; Yen, C.T.; Wu, Y.S.; Cheng, H.C. Using a fiber loop and fiber Bragg grating as a fiber optic sensor to simultaneously measure temperature and displacement. *Sensors* **2013**, *13*, 6542–6551. [CrossRef] [PubMed]

21. Chen, S.M.; Liu, Y.; Liu, X.X.; Zhang, Y.; Peng, W. Self-Compensating Displacement Sensor Based on Hydramatic Structured Transducer and Fiber Bragg Grating. *Photonic Sens.* **2015**, *5*, 351–356. [CrossRef]

22. Jiang, S.C.; Wang, J.; Sui, Q.M.; Cao, Y.Q. A novel wide measuring range FBG displacement sensor with variable measurement precision based on helical bevel gear. *Optoelectron. Lett.* **2015**, *11*, 81–83. [CrossRef]

23. Zou, Y.; Dong, X.P.; Lin, G.B.; Adhami, R. Wide range FBG displacement sensor based on twin-core fiber filter. *J. Lightwave Technol.* **2012**, *30*, 337–343. [CrossRef]

24. Tao, S.C.; Dong, X.P.; Lai, B.W. Temperature-insensitive fiber Bragg grating displacement sensor based on a thin-wall ring. *Opt. Commun.* **2016**, *372*, 44–48. [CrossRef]

25. Tao, S.C.; Dong, X.P.; Lai, B.W. A sensor for Simultaneous Measurement of Displacement and Temperature Based on the Fabry-Pérot Effect of a Fiber Bragg Grating. *IEEE Sens. J.* **2017**, *17*, 261–266. [CrossRef]

26. Wei, T.; Qiao, X.G.; Jia, Z.A. Simultaneous sensing of displacement and temperature with a single FBG. *Optoelectron. Lett.* **2011**, *7*, 26–29. [CrossRef]

27. Shen, C.Y.; Zhong, C. Novel temperature-insensitive fiber Bragg grating sensor for displacement measurement. *Sens. Actuators A Phys.* **2011**, *107*, 51–54. [CrossRef]

28. Zhong, C.; Shen, C.Y.; Chu, J.L.; Zou, X.; Li, K.; Jin, Y.X.; Wang, J.F. A displacement sensor based on a temperature-insensitive double trapezoidal structure with fiber Bragg grating. *IEEE Sens. J.* **2012**, *12*, 1280–1283. [CrossRef]

29. Dong, X.Y.; Yang, X.F.; Zhao, C.L.; Ding, L.; Shum, P.; Ngo, N.Q. A novel temperature-insensitive fiber Bragg grating sensor for displacement measurement. *Smart Mater. Struct.* **2005**, *14*, N7–N10. [CrossRef]

30. Li, T.L.; Shi, C.Y.; Ren, H.L. A Novel Fiber Bragg Grating Displacement Sensor with a Sub-micrometer Resolution. *IEEE Photonics Technol. Lett.* **2017**, *29*, 1199–1202. [CrossRef]

31. Li, T.L.; Tan, Y.G.; Shi, C.Y.; Guo, Y.X.; Najdovski, Z.; Ren, H.L.; Zhou, Z.D. A High-Sensitivity Fiber Bragg Grating Displacement Sensor Based on Transverse Property of a Tensioned Optical Fiber Configuration and Its Dynamic Performance Improvement. *IEEE Sens. J.* **2017**, *17*, 5840–5848. [CrossRef]

32. Guo, Y.X.; Xiong, L.; Kong, J.Y.; Zhang, Z.Y.; Qin, L. Sliding type fiber Bragg grating displacement sensor. *Opt. Precis. Eng.* **2017**, *25*, 50–58.

Article

Detection of Crack Initiation and Growth Using Fiber Bragg Grating Sensors Embedded into Metal Structures through Ultrasonic Additive Manufacturing

Sean K. Chilelli, John J. Schomer and Marcelo J. Dapino *

Department of Mechanical and Aerospace Engineering, The Ohio State University, Columbus, OH 43210, USA; chilelli.1@osu.edu (S.K.C.); schomer.11@osu.edu (J.J.S.)
* Correspondence: dapino.1@osu.edu

Received: 19 August 2019; Accepted: 8 November 2019; Published: 12 November 2019

Abstract: Structural health monitoring (SHM) is a rapidly growing field focused on detecting damage in complex systems before catastrophic failure occurs. Advanced sensor technologies are necessary to fully harness SHM in applications involving harsh or remote environments, life-critical systems, mass-production vehicles, robotic systems, and others. Fiber Bragg Grating (FBG) sensors are attractive for in-situ health monitoring due to their resistance to electromagnetic noise, ability to be multiplexed, and accurate real-time operation. Ultrasonic additive manufacturing (UAM) has been demonstrated for solid-state fabrication of 3D structures with embedded FBG sensors. In this paper, UAM-embedded FBG sensors are investigated with a focus on SHM applications. FBG sensors embedded in an aluminum matrix 3 mm from the initiation site are shown to resolve a minimum crack length of 0.286 $\pm$ 0.033 mm and track crack growth until near failure. Accurate crack detection is also demonstrated from FBGs placed 6 mm and 9 mm from the crack initiation site. Regular acrylate-coated FBG sensors are shown to repeatably work at temperatures up to 300 °C once embedded with the UAM process.

Keywords: ultrasonic additive manufacturing; UAM; fiber bragg grating; FBG; structural health monitoring; SHM; crack detection

---

1. Introduction

Structural health monitoring (SHM) provides improved safety and decreased costs through real-time sensor monitoring of engineering systems. Although SHM has been incorporated in a wide range of applications including civil, aerospace, and industrial systems, there is a need to develop more capable system monitoring and diagnostic tools. SHM has the potential to reduce the need for non-destructive inspection (NDI) which typically requires the system to be out of commission while costly, labor intensive testing is performed [1]. When NDI is relied upon, unsafe conditions due to defects arising during normal operation may go undetected until the next inspection. SHM systems could allow continuous and safer operations where defects are detected in real time. Furthermore, research is being conducted on the use of artificial intelligence, including neural networks, to improve the ability of sensors to detect defects in SHM applications [2].

SHM systems should include a minimally-invasive, long-term sensor network that can accurately detect damage in real time [3]. Fiber Bragg Grating (FBG) strain sensors are currently used in structural applications including buildings foundations [4], wind turbines [5], composite cure monitoring [6], bridges [7], concrete infrastructure [1], and full-scale composite structures [8]. Tosi [9] reviewed chirped FBG sensors that can provide local measurement of strain and temperature along the length of the fiber, typically 15–20 mm, with millimeter-level spatial resolution. Mao et al. [10] demonstrated the use of FBGs for identification of corrosion cracking and expansion in reinforced concrete based on Brillouin optical time domain analysis. FBG sensors have also been used for crack size prediction in conjunction with statistical models [11]. FBG sensors consist of a grating etched into an optical fiber that acts as a light transmission filter, which leads to a specific wavelength being reflected towards the signal source. As the fiber is subjected to either thermal or mechanical loading, strain changes within the sensor induce a change in the reflected wavelength [3]. There are several benefits to FBG sensors that make them effective for SHM. FBG sensors are immune to electromagnetic interference, can be multiplexed, are low cost, require no additional wiring, have high strain resolution, resist corrosion, and are lightweight with a minimally invasive geometry [4]. FBG sensors can provide structural loading data by ensuring strong coupling between the structural matrix and the sensor [12]. Although coupling has been achieved in many polymeric applications through direct embedment, metal systems have historically required sensors to be externally attached to the structure since internal embedment tends to involve process temperatures that are damaging to the sensors [13]. Attempts have been made to use metal-based additive manufacturing (AM) to embed FBG sensors. Selective laser melting (SLM) has been used to embed FBG sensors; however, the residual stresses left by this process tend to induce sensor damage along with poor coupling between the sensor and matrix [14,15]. Some investigations have demonstrated better AM embedment through the use of fibers with specialized coatings [16–18].

Unlike most metal forming techniques, UAM is a low-temperature process which allows for sensor embedment without producing mechanical or thermal damage to delicate components. In UAM, successive layers of thin foils are ultrasonically welded on top of each other to build up a metal part [19]. Combined with an incorporated CNC machine, subtractive operations allow for internal features and near-net-shape final parts. UAM builds are accomplished by creating solid-state bonds between layers of metal that are successively welded on top of a metallic baseplate. Metallurgical welding is achieved through direct metal-to-metal contact produced by the simultaneous application of lateral ultrasonic vibrations and mechanical pressure. The combination of high shear strain, normal pressure, and localized temperature increase has the effect of collapsing asperities, scrubbing away the oxide layer, and promoting atomic diffusion from one material to another [20]. The UAM system used in this study is the Fabrisonic SonicLayer 4000, whose welding assembly is shown in Figure 1. UAM can be used to join a variety of similar and dissimilar materials including aluminum alloys [20], steels [21], titanium [22], and carbon fiber composites [23]. Commercial FBG sensors embedded into metal through UAM have been shown to accurately track internal strain, demonstrating the potential of this approach for SHM applications [24–27]. Plastic flow around UAM-embedded fibers has been previously demonstrated, which leads to strong coupling between the fibers and metal matrix [28]. The main process parameters in UAM are vibration amplitude, down force, and weld speed [20].

In this study, we investigate the use of UAM-embedded FBG sensors in SHM prognostic applications. First, FBG sensors are embedded into compact tension (CT) specimens to determine their effectiveness in detecting crack initiation and growth. Second, embedded FBG sensors are thermally tested to investigate the upper temperature limit of the system. In Section 2, the experimental methods for manufacturing samples with embedded FBG sensors are presented along with the testing procedures. The results for both crack detection and elevated-temperature testing are presented in Section 3. Conclusions are discussed in Section 4.

Figure 1. Welding assembly of the 9 kW ultrasonic additive manufacturing welder employed in this study [24].

2. Experimental Methods

2.1. Sample Fabrication

Test specimens were built using aluminum 6061 due to the well-documented machine parameters available for this alloy [29], though the approach is applicable to other metals. The baseplate was of the T6 condition and the 0.154 mm thick foil layers were of the H18 condition. In this investigation, UAM parameters used for sample fabrication are a downward force of 5000 N, vibration amplitude of 32 μm, and weld speed of 84.6 mm/s.

The general coupon fabrication process is as follows. First, one layer of tape was welded onto the baseplate using UAM. Next, a 0.254 mm by 0.254 mm deep channel was cut using a ball end mill where the fiber would later be embedded. The channels help to avoid cross-sectional loading and sensor deformation during welding [25]. Standard acrylate-coated FBG strain sensors were used as supplied by Moog Inc. along with the wavelength interrogator (Insensys OEM 1030) and analysis software. The FBGs used have a wavelength operating range of 1545 nm to 1555 nm, which corresponds to ±4000 με within the fiber when the nominal wavelength is 1550 nm. The sensors were placed into the channels with the remaining fiber exiting the sample and additional layers were UAM-welded on top to fully encapsulate the FBG sensors at the center of the coupon geometry. After the sensor encapsulation, CNC milling operations were used to build the final sample geometry with the FBG sensor embedded halfway through the sample. The specific geometries created are outlined in the sections below as they vary for different tests. The embedded FBG sensors were examined for changes in their birefringence and polarization response, as it had been documented that the existence of birefringence-induced noise is indicative of an undersized channel and possibly poor bond quality [25]. When time synchronization with other data inputs was necessary, a custom-built analog-to-digital converter was used for converting the serial wavelength output of the interrogator into a voltage that was readable by the data acquisition system.

2.2. Crack Propagation

Tests were conducted to investigate the ability of UAM-embedded FBG sensors to detect cracks in samples. The geometry of the coupons and testing procedures are based on ASTM Standard E647:

Standard Test Method for Measurement of Fatigue Crack Growth Rates [30]. This test involves the cyclic tensile testing of notched compact tension (CT) specimens to induce crack growth. A clevis pin system was designed to be held in the load frame and attached to the CT specimen. A drawing of the CT specimen used in this testing is shown in Figure 2.

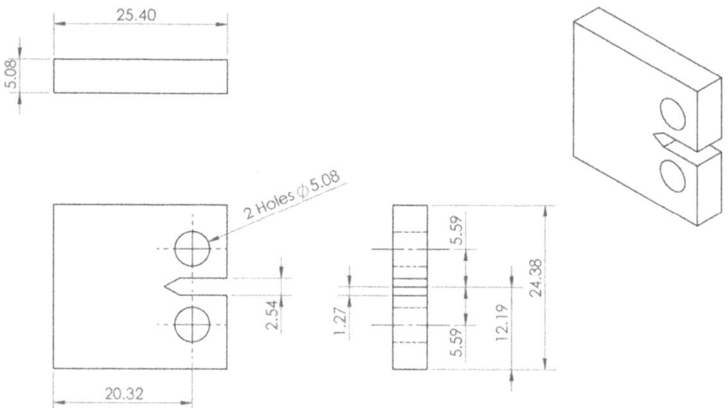

Figure 2. Compact tension (CT) specimen geometry used for the crack detection study. All dimensions are in mm. The FBG sensors were embedded perpendicular to the notch tip.

Although the purpose of ASTM E647 is for material characterization, we are interested in the strain signal throughout the lifetime of the coupon. For comparison to the strain measurements made by embedded FBG sensors, strain was also measured through digital image correlation (DIC) using a 5-megapixel camera with a 100 mm lens controlled by Vic-Snap 9 Image Acquisition software from Correlated Solutions. DIC is an image processing technique where a speckle pattern is placed on a sample and images are taken throughout testing. DIC analysis tracks the movement of individual speckles and determines the strain field at each image. An example DIC strain field is shown in Figure 3 which shows a strain field overlaid on a close-up image of a CT specimen just before failure. After a set number of cycles, the load frame triggers the DIC to capture an image by having a minimum tensile load slightly smaller than the other cycles. The vertical strain field reported by the DIC system was averaged over the same region as the Bragg grating and used as a comparison to the FBG signal. Crack growth was measured optically using images taken by the DIC system. The length of each pixel corresponded to 0.0054 mm. The loading profile was created using MTS TW Elite software for use with the MTS Criterion Model 43 load frame and a National Instruments 9215 data acquisition module. The experimental setup is shown in Figure 4. There were two main investigations into crack propagation with CT specimens: crack initiation and strain tracking, and prognostic analysis.

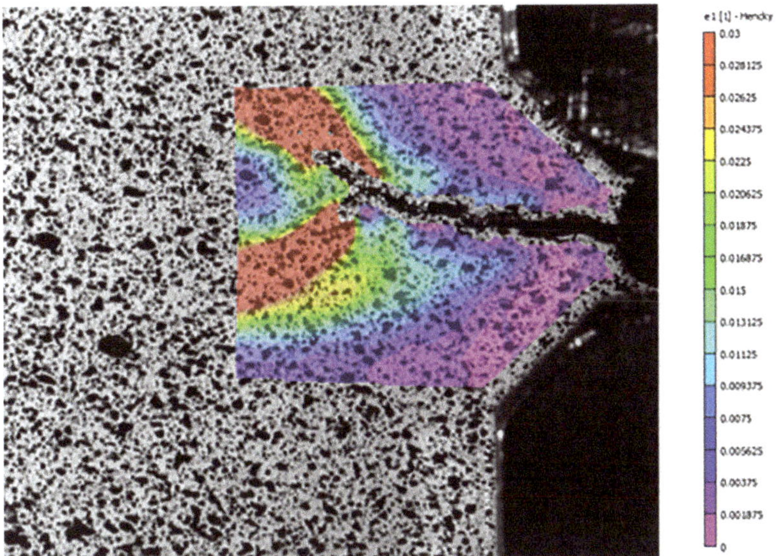

Figure 3. Strain field reported by the DIC system during cyclic loading of a CT specimen.

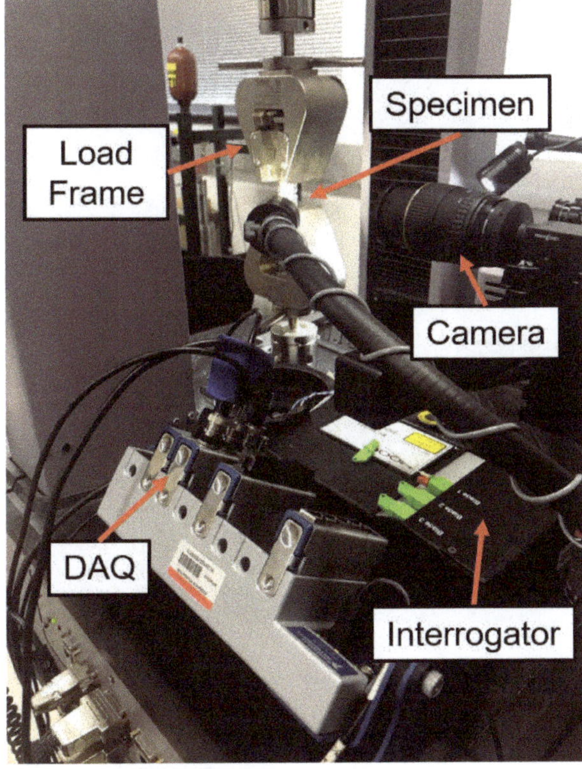

Figure 4. Close-up view of the crack-detection experiment showing a CT coupon installed in the load frame.

2.2.1. Crack Initiation and Strain Tracking

This experiment closely followed ASTM Standard E647. CT specimens with one FBG sensor embedded at a distance of 3 mm from the notch, as in Figure 5, were installed into the load frame and loaded cyclically until the crack reached 1 mm. These pre-cracking cycles consisted of a 0.25 Hz sine wave that alternated from 200 N to 2700 N. After the crack was measured to be greater than 1 mm, the main testing phase began, where the crack was grown until sample failure with the maximum load reduced to 800 N. The DIC system was used to take an image at peak load every 8 cycles during the pre-cracking phase and every 20 cycles during the remainder of the test. The FBG sensor collected data throughout the entire testing procedure. Data from the pre-cracking phase was used to determine the minimum crack size resolved by the FBG sensor. The mean was found for each set of peaks between DIC images. A distribution of the expected mean peak values for each set was found for the region before crack initiation occurred. Assuming this distribution is normal, we define the first set whose mean exceeds three standard deviations of the distribution mean to be the earliest that the FBG can detect the initiation of a crack. Data from the main testing phase was used to determine the ability of the embedded FBG sensors to track strain throughout the growth of a crack.

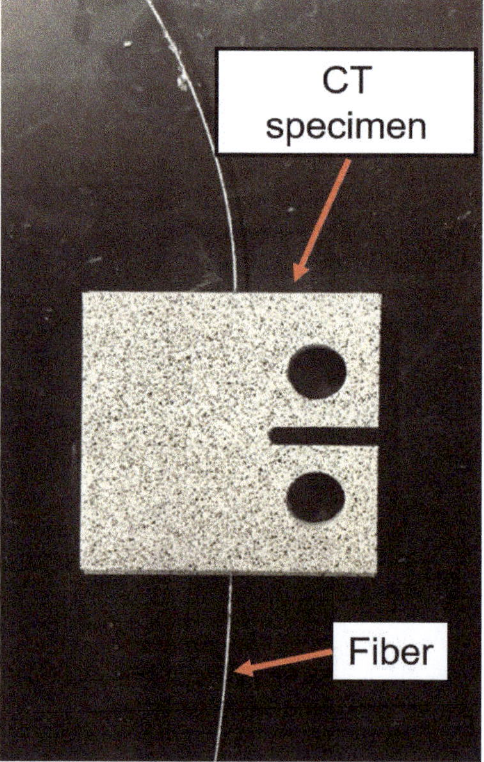

Figure 5. Compact tension (CT) specimen with one embedded FBG sensor built using UAM used in crack initiation and strain tracking. A fine speckle has been applied for DIC tracking. The FBG sensor is located 3 mm from the tip of the notch.

2.2.2. Prognostic Analysis

In this experiment, a CT specimen was built with three embedded FBG sensors spaced 3 mm, 6 mm, and 9 mm from the notch as pictured in Figure 6. Next, cyclic tensile loading was applied with a 0.25 Hz sine wave. The load profiles alternated between high-amplitude regions to induce crack growth for 200 cycles and low-amplitude regions to test FBG sensing at smaller loads for 30 cycles. In the high-load regions, the tensile loads alternated between 200 N and 2400 N; the tensile loads alternated between 200 N and 800 N in the low-load regions. Once the crack had formed and began propagating, the high-amplitude regions were reduced to 30 cycles. Similar to the previous crack-initiation testing, the wavelength at the peak loads was found before crack initiation had occurred. This was used to estimate the cycle where the expected peak signal exceeds normal bounds in both the high-amplitude and low-amplitude regions for all three embedded FBG sensors.

Figure 6. Compact tension (CT) specimen with three embedded FBG sensors built using UAM used in prognostic testing. The sensors are located 3 mm, 6 mm, and 9 mm from the tip of the notch.

2.3. Elevated-Temperature Testing

After pilot trials revealed accurate strain measurements at elevated temperatures, two tests were conducted to understand the operation of UAM-embedded FBGs at elevated temperatures. In the first test, the set point of the oven temperature increased every 30 min until the FBG failed to produce a signal. Set points used were 50, 75, 100, 150, 200, 250, 300, 350, 400, and 450 °C. In the second test, a coupon was thermally cycled at increasing temperatures. The sample was placed in a cool oven and heated to a set temperature, then the oven was turned off until the temperature returned to room temperature. This was repeated at increasing temperature set points until the FBG failed to produce a signal. Set points used were 50, 75, 100, 150, 200, 250, 300, and 350 °C. At the conclusion of the test, the sample was again heated to 350 °C to evaluate for permanent damage. Samples were then cross-sectioned and micrographs were taken for optical evaluation.

Coupons were based on ASTM Standard E8: Standard Test Methods for Tension Testing of Metallic Materials, with FBG sensors embedded in the middle of the sample [31]. A drawing of the coupons is shown in Figure 7. A Thermo Scientific Thermolyne furnace was used for testing. The temperature was verified using a K-type thermocouple and the strain was verified using an HPI Inc. HFK-12-125-6-ZCW high-temperature strain gauge that was fixed with epoxy to the coupon as pictured in Figure 8. Data was recorded using a National Instrument 9215 data acquisition module.

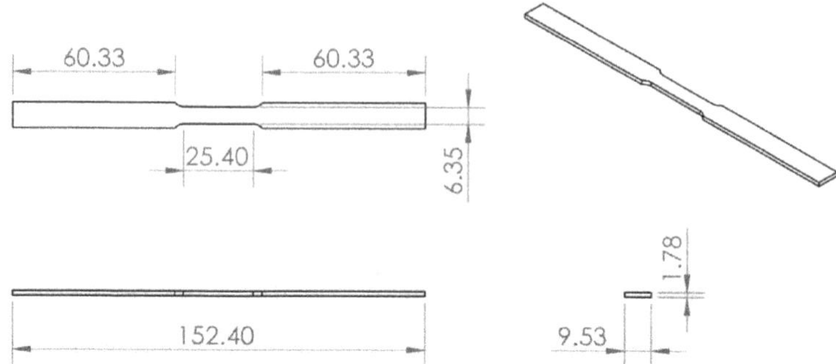

Figure 7. Drawing of coupon used in elevated-temperature testing. All dimensions shown are in mm.

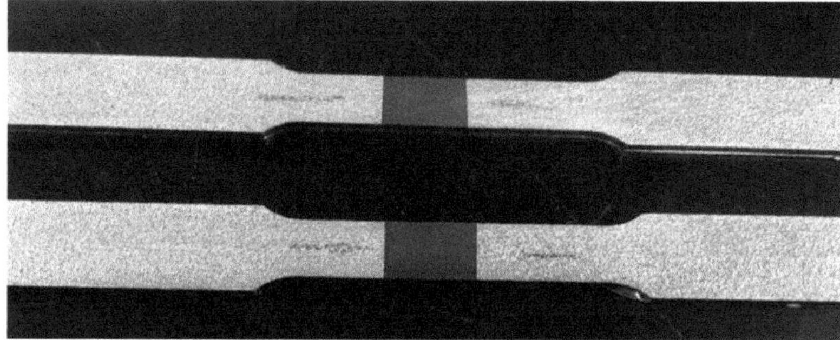

Figure 8. Elevated-temperature coupons built using UAM with a high-temperature strain gauge affixed to the outside of each coupon. FBG fibers are embedded down the length of each coupon.

3. Results and Discussion

3.1. Crack Propagation

3.1.1. Crack Initiation and Strain Tracking

Pre-cracking data was examined to determine the smallest crack size resolved by UAM-embedded FBGs. As illustrated in Figure 9, which shows the FBG signal peaks of each cycle in the pre-cracking phase, each time the minimum or maximum loads of the load frame were changed, there was a very small, but consistent decrease in the maximum and minimum points over the next few cycles. This effect was repeatable with different FBG sensors and builds, and was also observed in compression (albeit mirrored, with the peaks increasing). Examination of this phenomenon suggests that the decrease in signal peak consistently lasts 20–30 cycles and that the load frame does not exhibit this signal. Additional investigation is underway to understand this phenomenon, though its significance is minimal in practice as the actual strain drop is smaller than 1 microstrain.

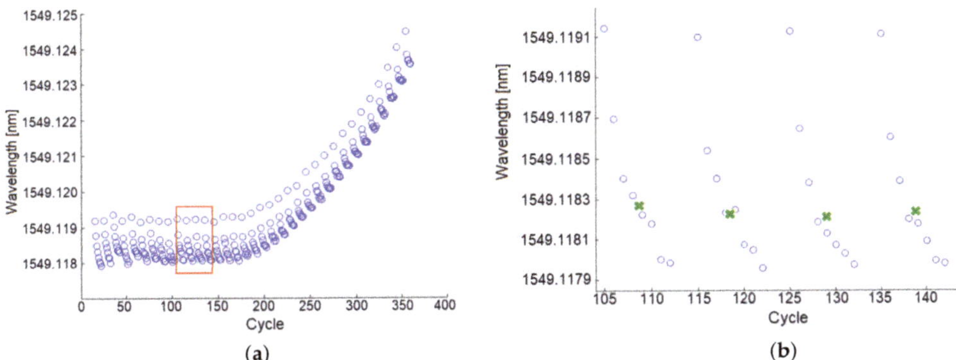

Figure 9. (**a**) Peaks of the FBG strain signal during the pre-cracking phase. The DIC was triggered with a low peak (not shown) every eight cycles. (**b**) Detailed view extracted from the red box. The green x markers indicate the mean of each set.

A consequence of this phenomenon is that the standard deviation of the normal peak distribution is increased, which reduces the effectiveness of initial crack detection. In the following discussion, the group of peaks after a DIC triggering cycle will be referred to as a set. Instead of using individual peaks before crack initiation, the mean is calculated for each set of 8 cycles, as illustrated in Figure 9b. The means of each set have a much smaller variation and allow for a more precise measurement of crack detection. The mean of the sets before the crack caused any deviation in the digital-to-analog converter signal are used to define a normal distribution with an overall mean of 1549.1181 nm and a standard deviation of 0.0097 nm. We can define the first set that exceeds three standard deviations from this normal mean to be the earliest detection of a crack. The first set whose mean exceeds this range is during cycles 195 to 205, corresponding to a crack length of 0.332 ± 0.046 mm. This analysis is repeated for two additional samples as presented Table 1.

Table 1. Length of crack in CT specimen at earliest detection using an UAM-embedded FBG sensor located 3 mm from the notch. Tolerances are due to crack entering a DIC speckle and consequently being unable to optically determine a precise endpoint.

Sample	Crack Size [mm]
1	0.332 ± 0.046
2	0.234 ± 0.054
3	0.291
Average	0.286 ± 0.033

This result illustrates the potential of embedded FBG sensors to detect early crack formation. Comparing this result to other Non-Destructive Investigation (NDI) techniques used in the aerospace industry, shown in Table 2, the UAM-embedded FBG sensors perform about an order of magnitude better than other methods.

Table 2. Comparison of minimum detectable crack size between UAM-embedded FBG sensors and common aerospace NDI techniques [32].

Technique	Crack Size [mm]
UAM FBG	0.286 ± 0.033
Eddy Current	2.54
Penetrant	3.81
Magnetic Particle	6.35

It is noted that literature values are the minimum crack detected 90% of the time using the NDI techniques on large parts with unknown cracks. For FBG sensors to be a viable alternative in industry applications, understanding of likely stress concentrations and crack initiation sites is needed in order to optimize FBG sensor location. The tests presented here were carried out with FBG sensors located 3 mm from the crack initiation site. Further investigation of the effects of distance are necessary for informing FBG placement into actual parts. There may be a trade-off between the area of influence of an FBG and the minimum crack size it can detect.

The results from the main phase of testing provide strong evidence for the ability of UAM-embedded FBG sensors to accurately track strain as a crack propagates toward the sensor. Figures 10–12 demonstrate accurate strain tracking for most of the sample's lifetime. At approximately 9500 cycles, the crack has grown to over 3 mm long and consequently has passed the fiber. At this point the FBG strain begins to deviate from the DIC-measured average strain as the fiber begins to slip within the channel. Prior to slipping, the FBG signal remains extremely accurate as illustrated in Figure 11. In the first 6000 cycles, a ten-cycle moving average of the peak strain of the FBG stays within 1% of the DIC strain, and remains within 5% after 8000 cycles. UAM-embedded FBG sensors are therefore a promising candidate for crack growth tracking in SHM applications.

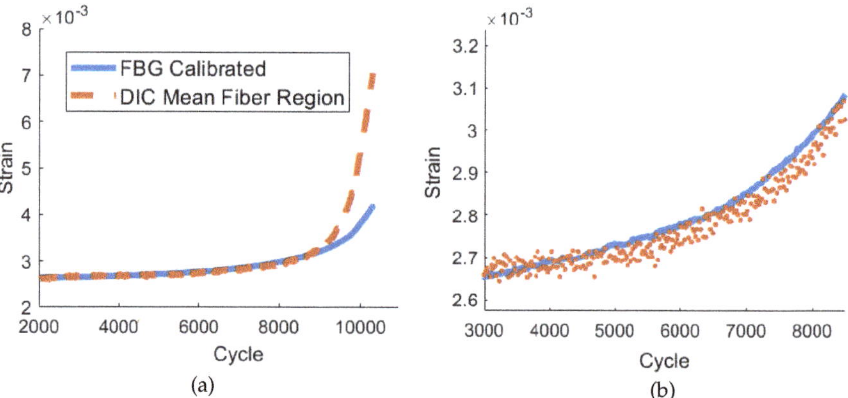

Figure 10. Vertical strain measurements from an FBG sensor UAM-embedded into a CT specimen and average strain from DIC in the region of the FBG; (**a**) from the end of the pre-cracking phase (∼cycle 2000) to sample failure (∼cycle 10500); (**b**) detailed view of (a) from approximately cycle 3000 to 8500.

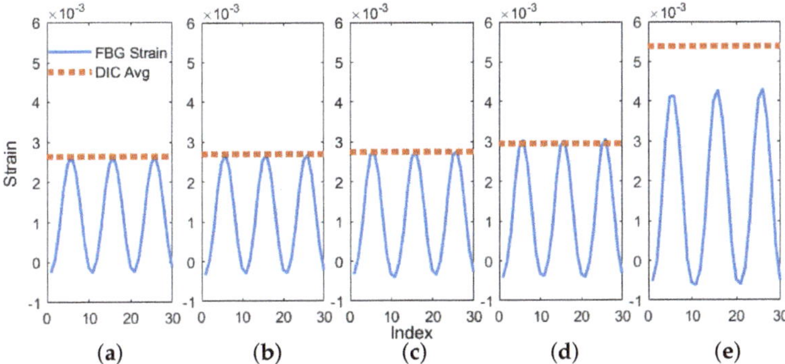

Figure 11. Comparison of the FBG strain signal and the average peak strain from DIC in the region of the FBG after (**a**) 2000, (**b**) 4000, (**c**) 6000, (**d**) 8000, and (**e**) 10,000 cycles.

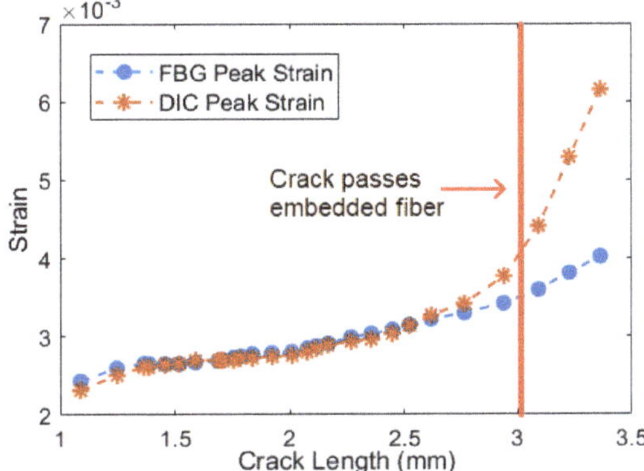

Figure 12. FBG strain signal and average peak strain from DIC in the region of the FBG as the crack length increases.

3.1.2. Prognostic Analysis

The prognostic analysis investigation used a CT specimen with three FBG sensors embedded at 3 mm, 6 mm, and 9 mm from the notch as shown in Figure 6. There were two main alternating phases of the load profile. In the high-amplitude, crack-growth phase, the investigation focused on determining the minimum detectable crack length. In the low-amplitude phase, the signal with no crack was compared to the case where a slowly-growing crack was present. The signals from all three embedded FBG sensors were analyzed to compare how distance from the initiation site affects the results. The test results are shown in Figures 13 and 14. The 3 mm and 6 mm FBG signals are out of phase with respect to the 9 mm FBG signal, as stress builds up unevenly in the CT specimen; this causes the side opposite the notch to be compressed when the load frame loads the coupon in tension [33].

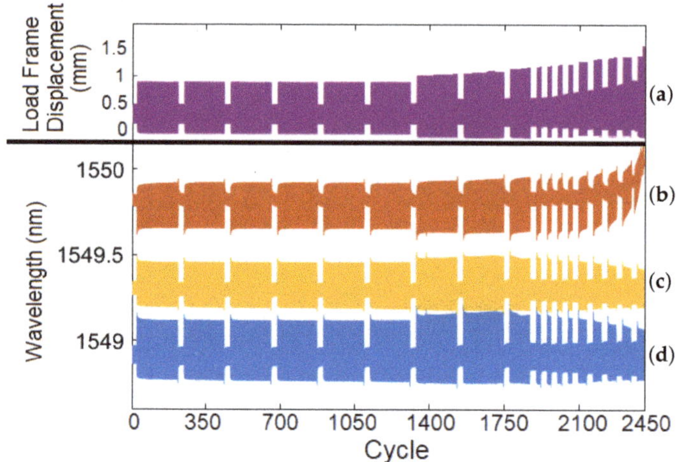

Figure 13. (**a**) Load frame displacement and FBG signals from sensors embedded (**b**) 9 mm, (**c**) 6 mm, and (**d**) 3 mm from the notch over the entire prognostic analysis test.

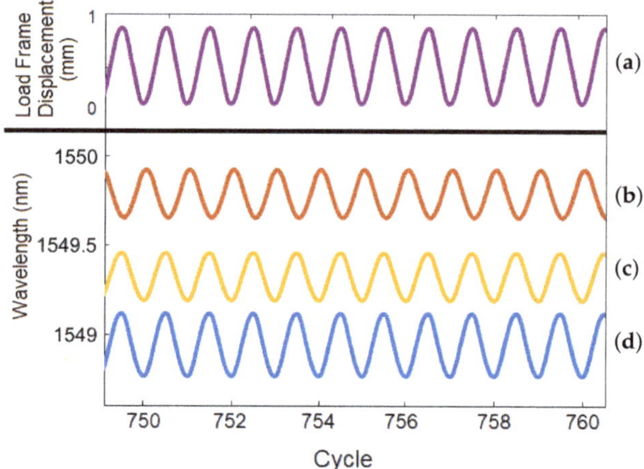

Figure 14. Detailed view extracted from Figure 13 between cycles 749 and 761. (**a**) Load frame displacement and FBG signals from sensors embedded (**b**) 9 mm, (**c**) 6 mm, and (**d**) 3 mm from the notch. Note the alternating crack-growth region (high wavelength amplitude) and low-amplitude region (low wavelength amplitude). The number of cycles of the crack-growth phase was reduced toward the end of the test to ensure that enough data was obtained as the crack grew.

During the test, the crack initiated and grew in a single cycle to a length of 0.350 mm. Using the analysis technique described in Section 3.1.1, this increased crack length was detected by all three fibers in both the crack-growth phase and the low-amplitude phase. The following figures show the FBG signal as a solid line and the calculated upper bound of set means with no crack as a dotted line. The red circles indicate the mean of each set. When the red circle crosses above the dotted line, the FBG detects a crack. As illustrated by Figures 15–20, all embedded FBGs successfully detect the crack after the seventh high-amplitude set when the crack reaches a length of 0.350 mm.

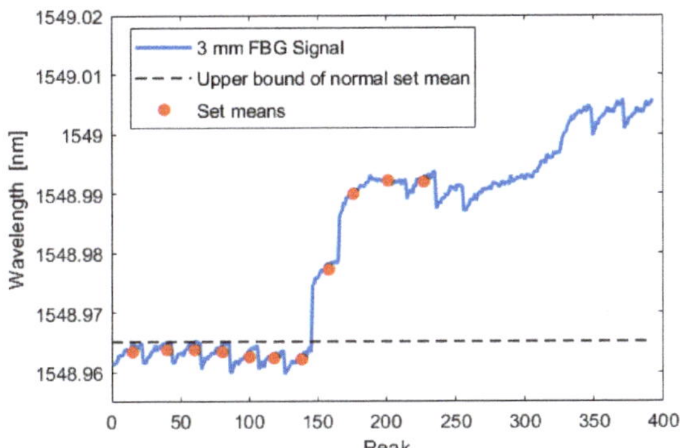

Figure 15. Crack detection analysis during low-amplitude cyclic loading for the FBG embedded 3 mm from the notch tip. For this and subsequent images, the dashed line represents three standard deviations above the normal set mean.

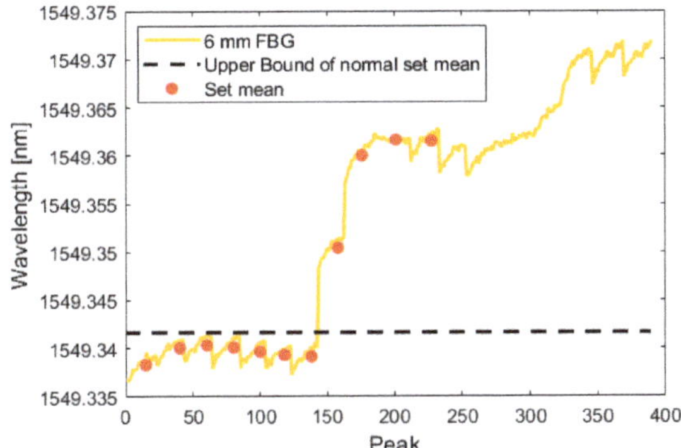

Figure 16. Crack detection analysis during low-amplitude cyclic loading for the FBG embedded 6 mm from the notch tip.

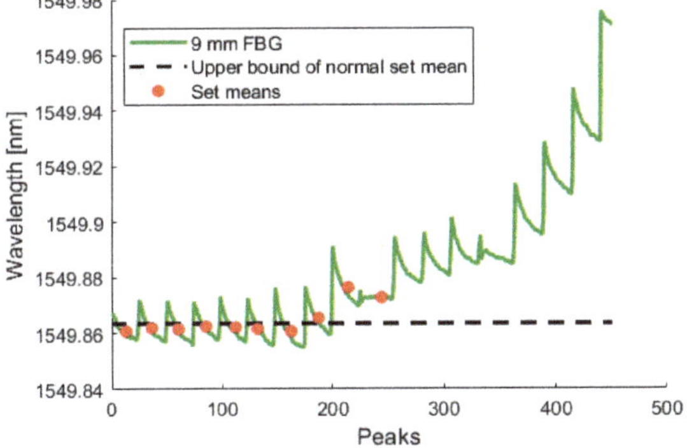

Figure 17. Crack detection analysis during low-amplitude cyclic loading for the FBG embedded 9 mm from the notch tip.

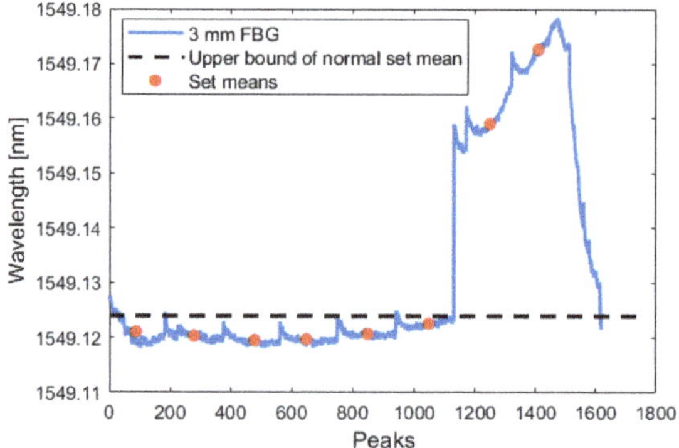

Figure 18. Crack detection analysis during high-amplitude cyclic loading for the FBG embedded 3 mm from the notch tip.

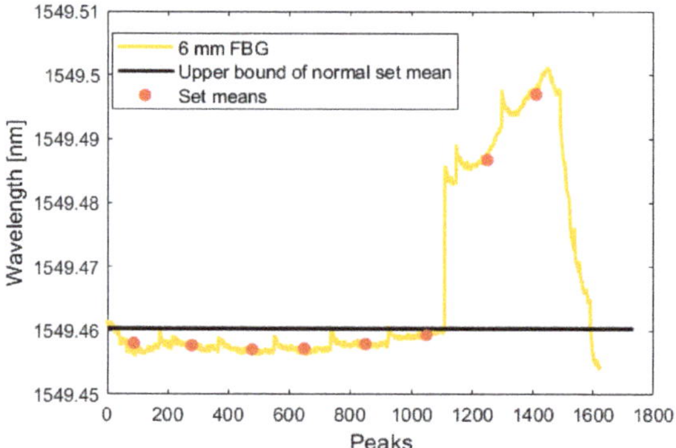

Figure 19. Crack detection analysis during high-amplitude cyclic loading for the FBG embedded 6 mm from the notch tip.

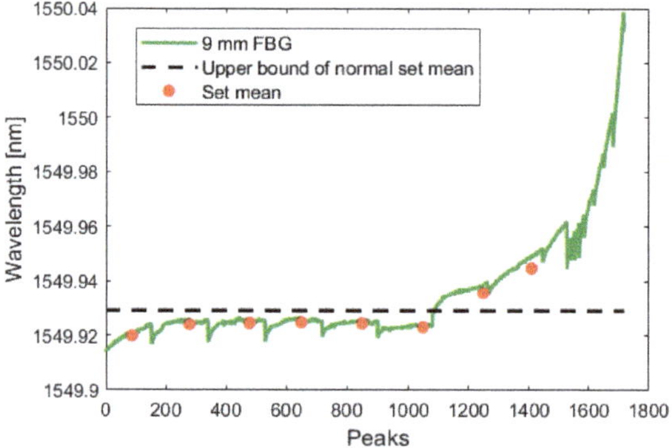

Figure 20. Crack detection analysis during high-amplitude cyclic loading for the FBG embedded 9 mm from the notch tip.

Although the signal change was less clear for FBG signals farther away from the notch, the fact that all three FBG sensors detect the crack growth illustrates the potential robustness of UAM-embedded FBG sensors in detecting crack initiation. Furthermore, the detection of the crack in the low-amplitude phase illustrates the ability of FBG sensors to detect minute defects even at normal operating conditions. Additional testing with more FBG sensors at more locations could provide insight into the relationship between crack detection and embedded sensor locations.

3.2. Elevated-Temperature Testing

Initial pilot investigations show that UAM-embedded FBG sensors may be able to accurately detect strain at temperatures higher than typical acrylate-coated FBG sensors. To investigate this further, the test shown in Figure 21 was conducted. This test illustrates that the FBG signal is able to accurately track the increasing temperature set points up to temperatures between 200 and 275 °C.

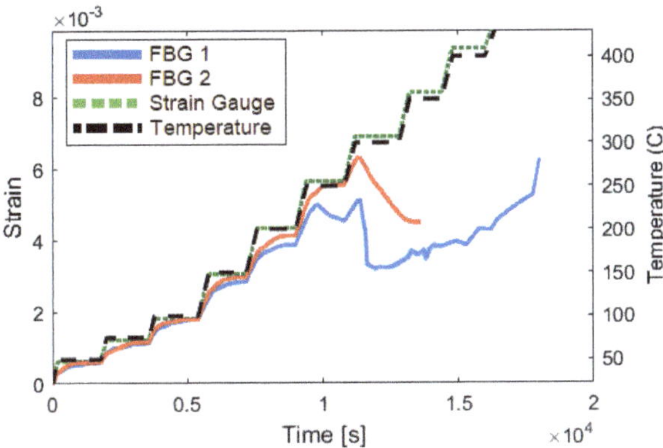

Figure 21. Elevated-temperature testing of two UAM-embedded FBG sensors. The strain measurements of two coupons with embedded FBG sensors and strain gauges is shown using the left scale. The temperature trace (black line) follows the scale on the right.

Figure 22 shows a second elevated-temperature test where a sample exhibits a repeatable signal up to 300 °C. At the 350 °C set point, there is a clear deviation from the previous signal. Additional testing confirmed that the fiber was permanently degraded.

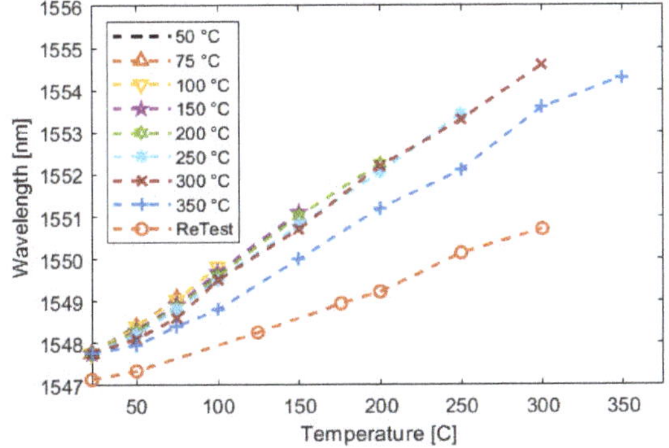

Figure 22. Elevated-temperature cyclic testing of an UAM-embedded FBG sensor. The FBG signal was measured during a cyclic temperature test with increasing oven set points. Once noticeable deviation occurred at 350 °C, the test was performed one more time to confirm that permanent damage had occurred.

Samples were then cross-sectioned and examined through optical microscopy to identify any visible changes in structure. As presented in Figure 23, the control sample shows the clear and intact structures of the fiber: outer coating, inner coating, and cladding (the core is not visible unless illuminated from behind). However, in the sample that was subjected to a temperature of 350 °C, the outer and inner coatings appear mostly gone and the cladding is no longer centered in the channel. This result was consistent across samples from both tests.

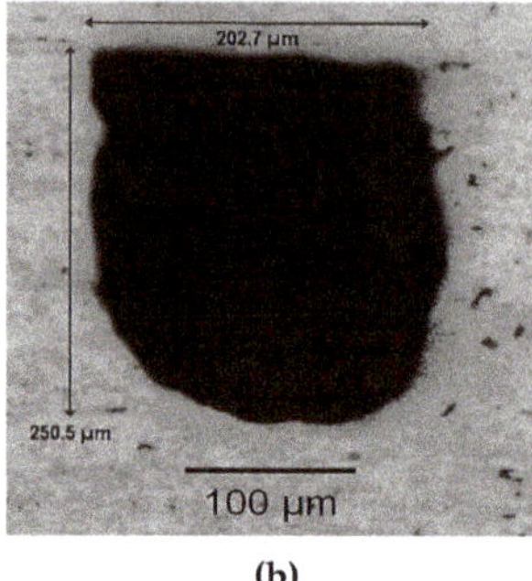

(a) (b)

Figure 23. (a) Cross-section of an UAM-embedded FBG that has not undergone thermal loading and (b) cross-section of the UAM-embedded FBG that underwent cyclic thermal testing.

4. Concluding Remarks

Structural Health Monitoring (SHM) techniques are underused in many industries due to sensor limitations. Fiber Bragg Grating strain sensors that have been embedded into metal through Ultrasonic Additive Manufacturing (UAM) are promising for SHM applications. UAM-embedded FBG sensors were shown to detect and track crack growth through the life of a CT specimen. Embedded FBG sensors were able to closely monitor crack growth until the crack passed the embedded fiber and the fiber began to slip. Embedded FBG sensors can enable early crack detection, with sensors 3 mm from the crack initiation point detecting cracks with a length of 0.286 ± 0.033 mm, an order of magnitude better than traditional NDI techniques. Even at distances of 6 mm and 9 mm, a crack size of 0.350 mm was resolved during both high loads (crack-growth phase) and low loads (normal operating phase). This result demonstrates the potential for FBG sensors to act as prognostic tools during the operation of components. Embedded FBG sensors were also shown to work at elevated temperatures, highlighting the possibility of higher temperature applications up to 300 °C. UAM-embedded FBG sensors have been shown to be an effective tool for structural health monitoring of complex systems.

Author Contributions: Conceptualization, S.K.C., M.J.D. and J.J.S.; methodology, S.K.C., M.J.D. and J.J.S.; validation, S.K.C. and M.J.D.; formal analysis, S.K.C. and M.J.D.; investigation, S.K.C.; resources, S.K.C. and M.J.D.; data curation, S.K.C. and M.J.D.; writing—original draft preparation, S.K.C. and M.J.D.; writing—review and editing, S.K.C. and M.J.D.; visualization, S.K.C. and M.J.D.; supervision, M.J.D.; project administration, S.K.C. and M.J.D.; funding acquisition, M.J.D.

Funding: This research was supported by the member organizations of the Smart Vehicle Concepts Center, a National Science Foundation Industry-University Cooperative Research Center established under grant IIP-1238286 (www.SmartVehicleCenter.org).

Acknowledgments: The authors wish to acknowledge the technical support of Thomas Greetham, Paul Guerrier and George Small of Moog, Inc. and Paul Deadman, Charles England, and Andy Strohacker of Moog, Inc. (Insensys, Ltd.)

Conflicts of Interest: The authors declare no conflict of interest.

Abbreviations

The following abbreviations are used in this manuscript:
UAM   Ultrasonic Additive Manufacturing
FBG   Fiber Bragg Grating
SHM   Structural Health Monitoring
CT   Compact Tension
DIC   Digital Image Correlation
NDI   Non-Destructive Inspection

References

1. Taheri, S. A review on five key sensors for monitoring of concrete structures. *Constr. Build. Mater.* **2019**, *204*, 492–509. [CrossRef]
2. Chang, C.M.; Lin, T.K.; Chang, C.W. Applications of neural network models for structural health monitoring based on derived modal properties. *Measurement* **2018**, *129*, 457–470. [CrossRef]
3. Das, S.; Saha, P. A review of some advanced sensors used for health diagnosis of civil engineering structures. *Meas. J. Int. Meas. Conf.* **2018**, *129*, 68–90. [CrossRef]
4. Majumder, M.; Gangopadhyay, T.K.; Chakraborty, A.K.; Dasgupta, K.; Bhattacharya, D.K. Fibre Bragg gratings in structural health monitoring-Present status and applications. *Sens. Actuators A Phys.* **2008**, *147*, 150–164. [CrossRef]
5. Arsenault, T.J.; Achuthan, A.; Marzocca, P.; Grappasonni, C.; Coppotelli, G. Development of a FBG based distributed strain sensor system for wind turbine structural health monitoring. *Smart Mater. Struct.* **2013**, *22*. [CrossRef]
6. Stolov, A.A.; Wrubel, J.A.; Simoff, D.A. Thermal stability of specialty optical fiber coatings. *J. Therm. Anal. Calorim.* **2016**, *124*, 1411–1423. [CrossRef]
7. Worden, K.; Cross, E.J. On switching response surface models, with applications to the structural health monitoring of bridges. *Mech. Syst. Signal Process.* **2018**, *98*, 139–156. [CrossRef]
8. Murayama, H.; Kageyama, K.; Naruse, H.; Shimada, A.; Uzawa, K. Application of fiber optic distributed sensors to health monitoring for full-scale composite structures. *J. Intell. Mater. Syst. Struct.* **2003**. [CrossRef]
9. Tosi, D. Review of Chirped Fiber Bragg Grating (CFBG) Fiber-Optic Sensors and Their Applications. *Sensors* **2018**, *18*, 2147. [CrossRef] [PubMed]
10. Mao, J.; Xu, F.; Gao, Q.G.; Liu, S.; Jin, W.; Xu, Y. A Monitoring Method Based on FBG for Concrete Corrosion Cracking. *Sensors* **2016**, *16*, 1093. [CrossRef] [PubMed]
11. He, J.; Yang, J.; Wang, Y.; Waisman, H.; Zhang, W. Probabilistic Model Updating for Sizing of Hole-Edge Crack Using Fiber Bragg Grating Sensors and the High-Order Extended Finite Element Method. *Sensors* **2016**, *16*, 1956. [CrossRef] [PubMed]
12. Zhou, J.; Zhou, Z.; Zhang, D. Study on strain transfer characteristics of fiber Bragg grating sensors. *J. Intell. Mater. Syst. Struct.* **2010**, *21*, 1117–1122. [CrossRef]
13. Sony, S.; Laventure, S.; Sadhu, A. A literature review of next-generation smart sensing technology in structural health monitoring. *Struct. Control. Health Monit.* **2019**, *26*, 1–22. [CrossRef]
14. Havermann, D.; Mathew, J.; Macpherson, W.N.; Maier, R.R.J.; Hand, D.P. In-situ strain sensing with fiber optic sensors embedded into stainless steel 316. *Sens. Smart Struct. Technol. Civ. Mech. Aerosp. Syst.* **2015**, *9435*, 94352W. [CrossRef]
15. Havermann, D.; Mathew, J.; MacPherson, W.N.; Hand, D.P.; Maier, R.R.J. Measuring residual stresses in metallic components manufactured with fibre Bragg gratings embedded by selective laser melting. In Proceedings of the 24th International Conference on Optical Fibre Sensors, Curitiba, Brazil, 28 September–2 October 2015; Volume 9634. [CrossRef]
16. Grandal, T.; Fraga, S.; Castro, G.; Vazquez, E.; Zornoza, A. Laser cladding-based metallic embedding technique for fiber optic sensors. *J. Light. Technol.* **2018**, *36*, 1018–1025. [CrossRef]
17. Mathew, J.; Hauser, C.; Stoll, P.; Kenel, C.; Polyzos, D.; Havermann, D.; Macpherson, W.N.; Hand, D.P.; Leinenbach, C.; Spierings, A.; et al. Integrating fiber fabry-perot cavity sensor into 3-D printed metal components for extreme high-temperature monitoring applications. *IEEE Sens. J.* **2017**, *17*, 4107–4114. [CrossRef]

18. Maier, R.R.J.; Havermann, D.; Schneller, O.; Mathew, J.; Polyzos, D.; MacPherson, W.N.; Hand, D.P. Optical fibre sensing in metals by embedment in 3D printed metallic structures. In Proceedings of the 23rd International Conference on Optical Fibre Sensors, Santander, Spain, 2–6 June 2014; Volume 9157. [CrossRef]

19. Graff, K. Ultrasonic additive manufacturing. In *Welding Fundamentals and Processes*; Lienert, T.; Siewert, T.; Babu, S.; Acoff, V., Eds.; ASM International: Cleveland, OH, USA, 2011; pp. 731–741. [CrossRef]

20. Wolcott, P.; Dapino, M. Ultrasonic additive manufacturing. In *3D Printing Handbook: Product Development for the Defense Industry*; Badiru, A., Ed.; CRC Press: Boca Raton, FL, USA, 2017; Chapter 17, pp. 275–313.

21. Levy, A.; Miriyev, A.; Sridharan, N.; Han, T.; Tuval, E.; Babu, S.S.; Dapino, M.J.; Frage, N. Ultrasonic additive manufacturing of steel: method, post-processing treatments, and properties. *J. Mater. Process. Technol.* **2018**, *256*, 183–189. [CrossRef]

22. Wolcott, P.J.; Sridharan, N.; Babu, S.S.; Miriyev, A.; Frage, N.; Dapino, M.J. Characterisation of Al–Ti dissimilar material joints fabricated using ultrasonic additive manufacturing. *Sci. Technol. Weld. Join.* **2016**, *21*, 114–123. [CrossRef]

23. Guo, H.; Gingerich, M.B.; Headings, L.M.; Hahnlen, R.; Dapino, M.J. Joining of carbon fiber and aluminum using ultrasonic additive manufacturing (UAM). *Compos. Struct.* **2019**, *208*, 180–188. [CrossRef]

24. Schomer, J.J.; Dapino, M.J. High temperature characterization of fiber bragg grating sensors embedded into metallic structures through ultrasonic additive manufacturing. In Proceedings of the Conference on Smart Materials, Adaptive Structures and Intelligent Systems, Snowbird, UT, USA, 18–20 September 2017; pp. 1–8.

25. Schomer, J.J.; Hehr, A.J.; Dapino, M.J. Characterization of embedded fiber optic strain sensors into metallic structures via ultrasonic additive manufacturing. *Proc. SPIE* **2016**, *9803*. [CrossRef]

26. Hehr, A.; Norfolk, M.; Wenning, J.; Sheridan, J.; Leser, P.; Leser, P.; Newman, J.A. Integrating fiber optic strain sensors into metal using ultrasonic additive manufacturing. *JOM* **2018**, *70*, 315–321. [CrossRef]

27. Mou, C.; Saffari, P.; Li, D.; Zhou, K.; Zhang, L.; Soar, R.; Bennion, I. Smart structure sensors based on embedded fibre Bragg grating arrays in aluminium alloy matrix by ultrasonic consolidation. *Meas. Sci. Technol.* **2009**, *20*, 1–6. [CrossRef]

28. Hehr, A.; Dapino, M.J. Interfacial shear strength estimates of NiTi-Al matrix composites fabricated via ultrasonic additive manufacturing. *Compos. Part B Eng.* **2015**, *77*, 199–208. [CrossRef]

29. Wolcott, P.J.; Hehr, A.; Dapino, M.J. Optimized welding parameters for Al 6061 ultrasonic additive manufactured structures. *J. Mater. Res.* **2014**, *29*, 2055–2065. [CrossRef]

30. *E647: Standard Test Method for Measurement of Fatigue Crack Growth Rates*; Technical Report; ASTM International: West Conshohocken, PA, USA, 2015. [CrossRef]

31. *E8: Standard Test Methods for Tension Testing of Metallic Materials*; Technical Report; ASTM International: West Conshohocken, PA, USA, 2013. [CrossRef]

32. *Non-Destructive Evaluation Requirements for Fracture Critical Metallic Components*; Technical Report NASA-STD 5009; NASA-STD-5009; National Aeronautics and Space Administration: Washington, DC, USA, 2008.

33. Kong, X.; Li, J.; Laflamme, S.; Bennett, C.; Matamoros, A. Characterization of a soft elastomeric capacitive strain sensor for fatigue crack monitoring. *Sens. Smart Struct. Technol. Civ. Mech. Aerosp. Syst.* **2015**, *9435*, 94353I. [CrossRef]

Article

Strain Monitoring of a Composite Drag Strut in Aircraft Landing Gear by Fiber Bragg Grating Sensors

Agostino Iadicicco [1,*]**, Daniele Natale** [2]**, Pasquale Di Palma** [1]**, Francesco Spinaci** [3]**,
Antonio Apicella** [3] **and Stefania Campopiano** [1]

[1]   Department of Engineering, University of Naples "Parthenope", Centro Direzionale Isola C4,
   80143 Napoli, Italy; pasquale.dipalma@uniparthenope.it (P.D.P.); campopiano@uniparthenope.it (S.C.)

[2]   Infratel Italia S.p.A, 00187 Roma, Italy; daniele.natale@uniparthenope.it

[3]   Magnaghi Aeronautica SpA, via Galileo Ferraris, 76, 80142 Napoli, Italy;
   fspinaci@magnaghiaeronautica.it (F.S.); aapicella@magnaghiaeronautica.it (A.A.)

*   Correspondence: iadicicco@uniparthenope.it; Tel.: +39-081-5476718

Received: 11 April 2019; Accepted: 12 May 2019; Published: 15 May 2019

Abstract: This work reports on the use of Fiber Bragg Grating (FBG) sensors integrated with innovative composite items of aircraft landing gear for strain/stress monitoring. Recently, the introduction of innovative structures in aeronautical applications is appealing with two main goals: (i) to decrease the weight and cost of current items; and (ii) to increase the mechanical resistance, if possible. However, the introduction of novel structures in the aeronautical field demands experimentation and certification regarding their mechanical resistance. In this work, we successfully investigate the possibility to use Fiber Bragg Grating sensors for the structural health monitoring of innovative composite items for the landing gear. Several FBG strain sensors have been integrated in different locations of the composite item including region with high bending radius. To optimize the localization of the FBG sensors, load condition was studied by Finite Element Method (FEM) numerical analysis. Several experimental tests have been done in range 0–70 kN by means of a hydraulic press. Obtained results are in very good agreement with the numerical ones and demonstrate the great potentialities of FBG sensor technology to be employed for remote and real-time load measurements on aircraft landing gears and to act as early warning systems.

Keywords: Fiber Bragg gratings; fiber optic sensors; aircraft landing gear; load monitoring system; composite device

---

1. Introduction

The aerospace sector is constantly searching for new solutions to optimize the component lifetime and the maintenance of aircraft. Research efforts on smart-martials, intelligent systems and/or innovative monitoring technologies are widely welcome. In this context, the aircraft manufacturers started to use composite materials in aircraft components since the early 1980s [1]. Composite materials are a key solution in aircraft structures due to the lightness of composites and the corrosion problem in aluminum or metallic items. However, the introduction of novel materials requires appropriate systems for monitoring of their health state before enabling a wide use [2].

The landing gear (LG) system is an important component in aircraft [3]. There are several types of LGs and arrangements of LGs in aircraft: the most common LG arrangement is the tricycle-type one. The LG aims to provide a suspension system during the take-off and landing phases and during the taxi operation. It is designed to absorb and dissipate the energy of landing impact. Moreover, the LG facilitates braking of the aircraft and provides directional control of the aircraft on ground using a wheel steering system [1,3]. From these considerations, the LG must be an extremely resistant structure. At same time, it should be as light as possible because it is not used for most of the flight time [3]. On this line of argument, LGs realized in composite material represent a unique solution if the mechanical robustness can be kept. To address this ambitious goal, appropriate structural health monitoring (SHM) systems both for the testing and for real operation phases are welcome.

The current SHM technology used in composite materials is done using sensors, especially optical sensors [4–8]. This is since the traditional strain gauge is sensitive to lightning, current leakage and corrosion, whereby inaccurate readings will be shown. On contrary, in last 20 years the potential of optical sensors has been widely demonstrated [9]. Optic sensors provide several advantages for aerospace applications that include insensitivity to electromagnetic interference, lightweight and flexible harness, multiplexing and multi-parameter sensing, high measurement accuracy, low power requirements per sensor, remote interrogation and operation, and the potential to embed in composite structures. Among the several optical fiber sensor configurations [10], Fiber Bragg Sensors (FBGs) are considered the favorite technology [4–8] for their capability to build a dense multiplexing SHM network, durability under extreme weather conditions and ease of application to the surfaces of different structures. As matter, FBG sensors represent the most mature and assessed sensing platform ready to be used in real industrial scenarios [4–8,11,12].

Several dozens of FBG sensors, including temperature and strain can be written on the same optical fiber, and can be simultaneously interrogated by a single interrogator unit. A very important point in FBG based SHM system is the method to couple the FBG sensors with the composite structure, and different ways to accomplish this have been experimented [10]. The most used and simple ways consist in the FBG surface bonding with the composite structure previously created. There is a big literature knowledge [13] that can help to identify the best glue and cure procedure for the specific application: some authors reported that the most important factors are the adhesive thickness and the length of bonding [14]. A second most integrated way to apply the sensors is to insert the FBG between two composite structures linked together, but this method is possible only in situations that provide for assembly of different parts. The last and more integrated method to apply the FBG to the structure is incorporation of FBG during the composite production, in order to place the FBG just some layers underneath the surface and ensure the sensors network from surface accidental contacts [1] and detect damage in different locations inside the material [15].

In this context, this manuscript deals with the activities carried-out within the framework of a National Projects (PON03PE 00135, "CAPRI - Carrello per Atterraggio con Attuazione Intelligente") aimed to identify novel composite items of a standard LG and appropriate fiber optic sensors to investigate its SHM during loading testing. In particular, in this paper we present the successful use of several FBG sensors for the real-time, continuous and remote monitoring of the strain profile of novel composite items of an aircraft LG during increasing loads test. We designed and subsequently tested several FBG strain sensors installed on a true composite item of a real LG, provided by the company Magnaghi Aeronautica Spa. To optimize the localization of the FBG sensors, load condition was numerically studied by finite element method (FEM) numerical analysis. Successively, several experimental tests have been done in range 0–70 kN by means of a hydraulic press, up to the breaking of the innovative item. Obtained results are in very good agreement with the numerical ones and demonstrate the great potentialities of FBG sensors technology to be employed for remote and real-time load measurements on aircraft landing gears. It is worth noting that in most of literature works FBG sensors were proposed to sensorize large components such as turbines, wings or hull elements that have a large dimension. Moreover, recent works successfully demonstrate the use of FBG sensors to detect defects or cracks in mechanical structure even with reduced size [16–18]. In our work, the FBG sensors are successfully applied on complex geometry characterized by a high surface curvature.

2. Materials and Method

2.1. Composite Item to Be Sensorized

The design of the landing gear, according to the Airworthiness Regulations, must take into account several requirements in terms of safety, strength, stability, etc., under all possible in-service loading conditions (weather conditions included). The landing gear is one of the main structural component characterizing an aircraft. It is aimed to support the aircraft during the landing, the tacking off and ground operations. Among such loading conditions, the landing phase defines the design specifications, since it is the most burdensome. As a result, this loading condition determines its structural size.

Most of LG components are currently steel made in order to provide a robust suspension system during the take-off and landing phases. The design and fabrication of LG items in composite materials represent open challenges offering significant lightening of the landing gear. Within the Italian Project CAPRI, the project leader Magnaghi Aeronautica Spa proposed and selected the drag brace lower arm of the nose landing gear of the Alenia C27J aircraft (plotted in Figure 1a) as a target sample to be realized in composite materials.

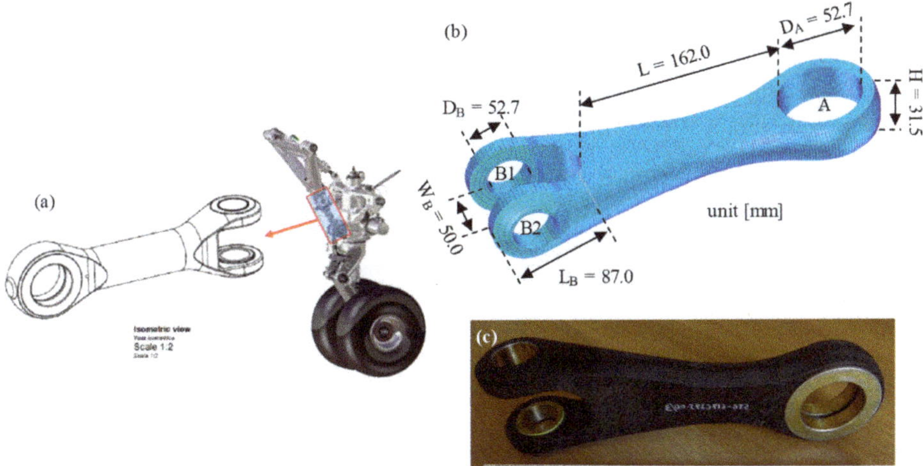

Figure 1. (**a**) Schematic view of the landing gear (LG) and standard drag brace; and (**b**) design and (**c**) picture of the composite drag brace.

The aim was to realize a composite material drag brace and demonstrate the capability to use optical fiber sensors as a real-time strain state monitoring system. The innovative item was proposed (see Figure 1b) and realized (see Figure 1c) by Hexply 914-40%-G803, or similarly Cytec 977-2A/HTA. The total weight was less than 1.0 Kg when compared with the steel version of about 3.0 Kg. Concerning the fixing points with the rest of the LG, it presented an upper fixing point (A) equipped with metallic ring with inner diameter $D_A = 52.7$ mm. The body length was L = 162.0 mm. The down fixing point consisted of two arms with lengths $L_B = 87.0$ mm and a separated width of $W_B = 50.0$ mm. The arms are equipped with metallic rings, in following named B1 and B2, with inner diameters $D_B = 32.7$ mm. The shape created a bent surface between B1 and B2 with a bending radius of 24.0 mm, which was critical for strain surface distribution.

2.2. FBG Sensors

A Fiber Bragg Grating sensor represents a robust and efficient sensor for precise determination of the deformations of the selected item during loading test. A FBG consists in a periodic modulation of the core refractive index along the core of a standard single-mode optical fiber with typical length of

1–20 mm. Consequently, FBG reflects a specific wavelength, called Bragg wavelength λ_B that depends on the period Λ of the perturbation according to the following formula [7,11]:

$$\lambda_B = 2n_{eff}\Lambda \tag{1}$$

where n_{eff} is the effective refractive index of the guided core mode.

The effective refractive index of the core and the spatial periodicity of the grating are both affected by changes in strain ($\Delta\varepsilon$) and temperature (ΔT). As a result, the Bragg wavelength changes its position, according to the following equation:

$$\frac{\Delta\lambda_B}{\lambda_B} = S_T\Delta T + S_\varepsilon\varepsilon \tag{2}$$

where S_T is the thermal sensitivity coefficient and S_ε is the strain sensitivity coefficient. According to Equation (2), strain and temperature may be deduced from the measurement of the Bragg wavelength shift and strain and temperature contributions can be separated through the implementation of specific sensing configurations. For our purposes, FBGs were employed for accurate measurement of the strain profile of the structure where they are embedded. Additionally, the thermal effect has been properly compensated for by using an additional free and unstrained FBG, so that it only recorded temperature changes [7].

Moreover, it is worth noting that in most SHM applications, bare FBGs are too fragile to be used as robust solutions, especially when large stresses or loads are applied. In such cases, FBGs integrated in protective packages, commercially available, are often preferred. However, most packages significantly increase the final size of the sensors and thus are undesired for SHM of small items with complex geometry. In our case, the most stringent requirement was keeping the sensing area as small as possible in order to be able to investigate the strain profile in a small region with fast changes of the strain profile. Thus, the attention was focused on unpackaged FBGs with lengths of 1.0–10.0 mm depending on sensors layout. All sensors were protected by a tight polyamide coating, 20 µm thick. During load test, the Bragg wavelengths of all sensors as function of the item load stress were detected by Bragg meter (from Fiber sensing) with eight channels and operating in range of 1500–1600 nm with a scanning rate of 1/2 Hz and wavelength resolution of 1.0 pm.

2.3. Sensors Bounding Procdure and Characterization

The use of FBG as optoelectronic strain gauge sensors is well known [4–8] However, it demands specialist solutions for each application field. The gluing of sensors with items needing to be sensorized represents the most critical issues to be carefully addressed to favor maximum strain transfer to sensors. This section presents and discusses preliminary experimental tests aimed to identify the more appropriate adhesive to fix the fiber optic sensors to the composite surface. We considered standard electric strain gauges (SGs), provided from Micro-Measurements with well-known gauge factors as references. We also investigated the effects of different adhesives on the response of electric SGs.

The preliminary tests were provided by a sensorized composite panel with dimensions of 40 cm × 20 cm with one end (20 cm long) fixed, as schematically illustrated in Figure 2a. On the top surface of the panel, several electric SGs and several FBGs were fixed at same distance from the wall by different adhesives while incremental loading force (five steps from zero to about 470 N) was applied on the opposite side acting as significant strain on the bounded sensors.

Concerning SGs, different kind of adhesives were preliminarily tested; the best results have been obtained by using cyanoacrylate-based adhesives, according with [19,20]. Figure 2b plots the strain profiles of two SGs fixed by different commercial cyanoacrylate adhesives, Micro-Measurements M-Bond 200 (more elastic feedback), and Loctite Super Attak (more rigid feedback), respectively, when the composite panel was subject to increasing and decreasing step-by-step force sequence. The Loctite Super Attak provided a better strain transfer, recording a higher strain measurement of about 15% as compared with M-Bond glue. Maximum strain surface of almost 2000 µε was measured by a load force of 471 N. Similar analysis and conclusions were achieved concerning FBG sensors. Consequently, in the following tests all electrical and optical sensors were fixed by Attack adhesive.

Moreover, Figure 3a,b plot the spectra of a 1.0 mm long FBG and of a 10.0 mm long one, respectively, fixed to the composite panel by means of Attack adhesive in the different load conditions. The maximum applied load inducing a superficial strain of about 2000 $\mu\varepsilon$ forced Bragg wavelength red shifts of 2.54 nm and 2.42 nm for the short and long FBG, respectively. It is important to highlight that despite the high strain values, the shape and maximum reflectivity of the FBGs were unchanged, confirming that the bonding process was able to transfer a uniform strain state to the sensors.

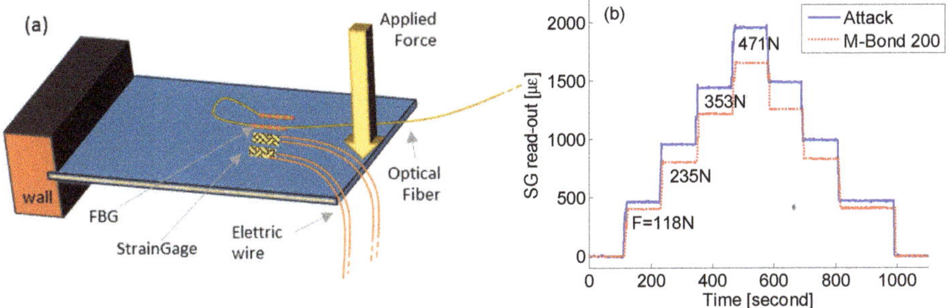

Figure 2. (**a**) Schematic view of the sensorized composite panel test; and (**b**) strain reading of two strain gauges (SGs) fixed by means of Attack and M-Bond 200 adhesive, respectively, for different load states.

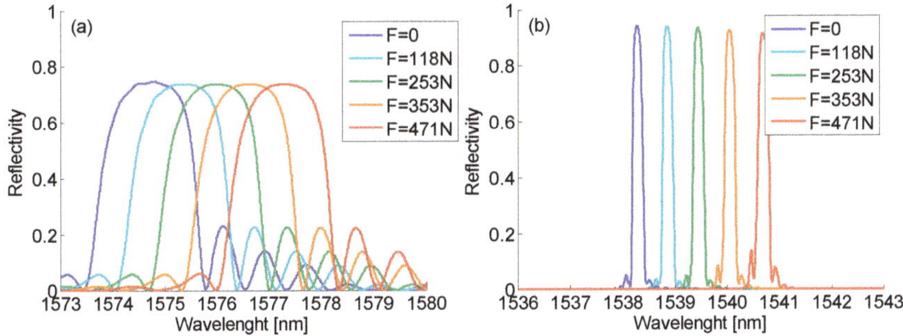

Figure 3. Fiber Bragg Grating (FBG) spectra in different strain states: (**a**) a 1.0 mm long FBG; and (**b**) a 10.0 mm long FBG.

Finally, the FBGs' strain sensitivity was estimated by the comparison of electrical and optical sensors. First, the deformation profile of the used panel has been displayed in Figure 4a where the strain values of the electrical sensor were plotted versus the applied force: it exhibited a linear behavior with slope of 4.15 $\mu\varepsilon$/N. Instead, Figure 4b plots the relative wavelength shifts of the short and long FBG versus the surface strain: the behavior is absolutely linear over the range of 0–2000 $\mu\varepsilon$. Moreover, as expected, the sensitivity values for the 1 mm and the 10 mm long FBGs were practically the same, with negligible differences probably inducted by the experimental measurements and the position alignment between the two FBG of different sizes on the panel surface: the experimental sensitivity was estimated to be $S_{strain} = (8.2 \pm 0.05) \cdot 10^{-7} \, [\mu\varepsilon^{-1}]$.

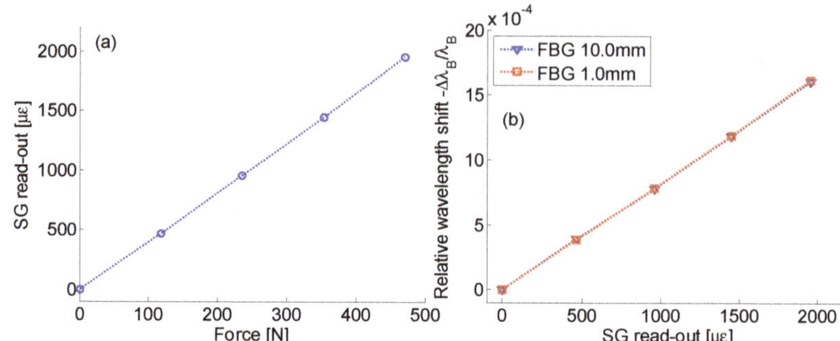

Figure 4. (**a**) Surface strain (from electrical SG) vs. applied force; and (**b**) relative wavelength shifts of FBGs vs. surface strain.

2.4. Identification of the Best Locations for Sensors Installation

This section deals with the identification/selection of the locations to be sensorized on the item surface. This analysis should take into consideration the most stressed regions or the geometric characteristics.

To this aim, Magnaghi Aeronautica provided a static stress numerical analysis of the composite item by means of the FEM. The model permits the estimation of the strain distribution along the main axis of the item structure when subjected to the load. Figure 5 plots the FEM result in terms of strain surface along *y*-axis when the item is loaded with 50 kN. It is worth highlighting that the numerical model does not take into consideration metallic ring in the fixing points and thus it is reasonable to believe that the numerical results close the fixing points can be different from the real ones. However, a more accurate FEM model is far from the aim of this work.

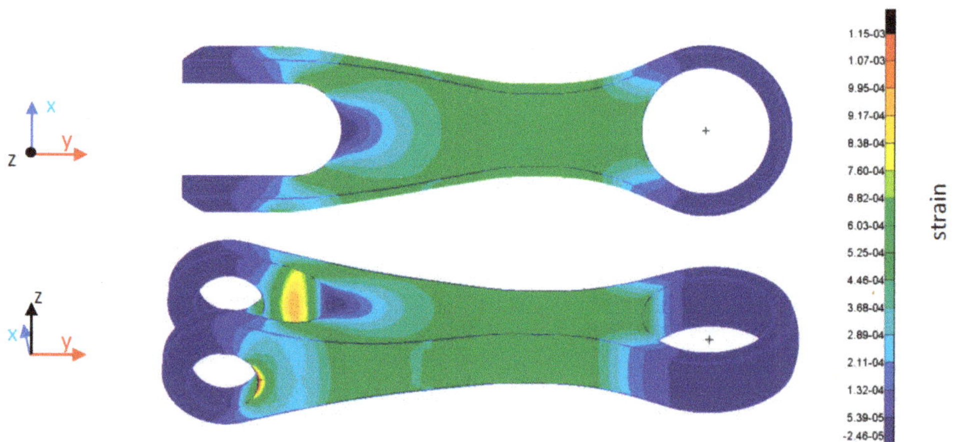

Figure 5. *y*-axis strain by finite element method (FEM) analysis of the composite item.

From data reported in Figure 5, we designed a custom FBG array, Y-array, aimed to measure the y-strain profile, as schematically plotted in Figure 6a. Y-array, including six FBG sensors, was arranged on the center of the lateral surface to measure the longitudinal (along *y*-axis) strain profile. The sensors were non-uniformly spatially distributed depending on numerical strain profile slope. Sensors Y_1, Y_2, Y_3, Y_4, Y_5 and Y_6 were thus positioned along *y*-axis at 5 mm, 10 mm, 22 mm, 38 mm, 70 mm and 135 mm, respectively.

Finally, a four FBG sensor array, C-array, was designed by the analysis of geometric characteristics. It took into consideration the bending region and thinning of the arms forming the double fixing point, B1 and B2. C-array was designed to measure the surface strain in the bent region. It was selected to demonstrate the goodness of the bonding procedure and the potentiality of the FBG technology in monitoring of surface strain on composite items with complex geometrical shapes. The sensors of C-array, C_1, C_2, C_3 and C_4 were equally distributed along the c-axis, starting from the center of the bent region and separated by 10 mm (as one can straighten the bent region along the c-axis). The sensor C_1 was in the center of the bent region whereas the C_4 sensor was the closest to the fixing point. In the Cartesian-system plotted in Figure 6a, the bent c-axis moved in the plane xy for z = −19.5 mm.

All sensors were glued by means of the procedure discussed in previous section and consisted of unpackaged FBGs with 20 μm thick polyamide coating. Moreover, most of the sensors were based on 1.0 mm long FBGs, in order to decrease the sensing area. Only the sensors Y5 and Y6 were selected to be 10.0 mm long because they are expected to be fixed in a region with low strain variation, as from numerical results. Figure 6b plots the spectra of all FBG sensors: Y-array and C-array. All sensors 1.0 mm long exhibited a FWHM (Full Width at Half Maximum) bandwidth of about 1.6 nm whereas the 10.0 mm long gratings showed a bandwidth of about 0.2 nm. Figure 6c–e shows some pictures during the gluing of sensors. Moreover, the fiber coming out from the composite item was then protected by a red silicone glue.

Finally, one more FBG sensor, not fixed to the item, was included for the compensation of the thermal changes during the experimentation. To this aim we supposed that the item as well as all the glued FBGs were in the same thermal state. During the temperature test variations lower than 3 °C were measured.

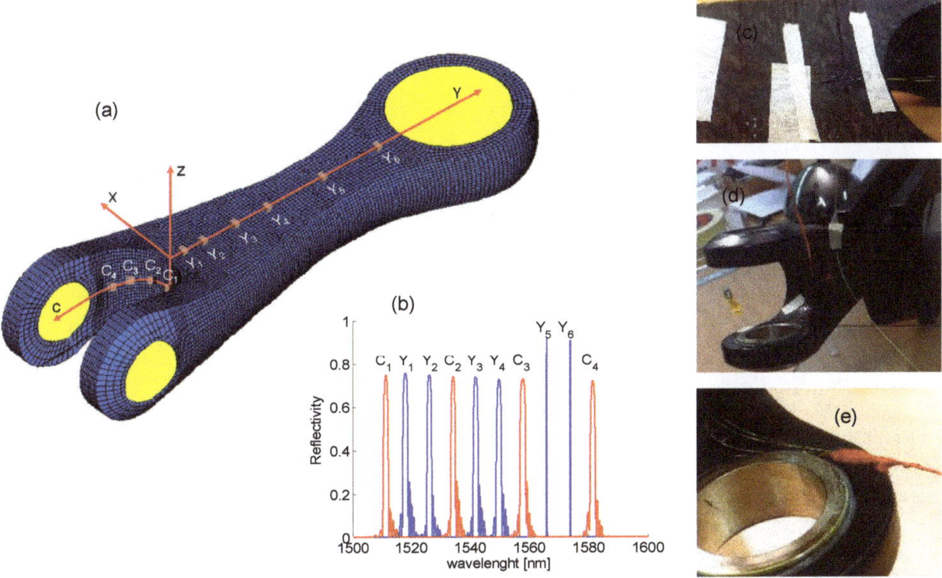

Figure 6. (**a**) Scheme of layout of FBG arrays Y-array and C-array; (**b**) Y-array and C-array spectra; and (**c**–**e**) pictures during the gluing of sensors.

2.5. Experiemntal Setup

The real-time structural monitoring of the composite item consisted of a continued monitoring of the FBG arrays response when the item was subjected to an incremental load up to 50 kN. This permitted the emulation of a true stress situation for the counterattack, replicating a typical traction during the landing phase of the vehicle.

The response of the FBG arrays was measured by means of FBG interrogation unit (Bragg meter) exhibiting eight-channel operating in a range 1500–1600 nm and with scanning rate of up to 1/2 Hz. The Bragg meter communicated with a standard notebook by a custom software that permitted the acquisition of full-spectra of the sensors or Bragg wavelengths of all sensors.

Finally, the loading test was carried out by means of an appropriate press machine for high load condition provided from a Magnaghi co-worker. The Figure 7a,b shows pictures of the sensorized item fixed in the load-machine by means of two proper metallic holders (upper and lower ones). The load-machine was programmed to apply an incremental traction with a rate of 250 N/s.

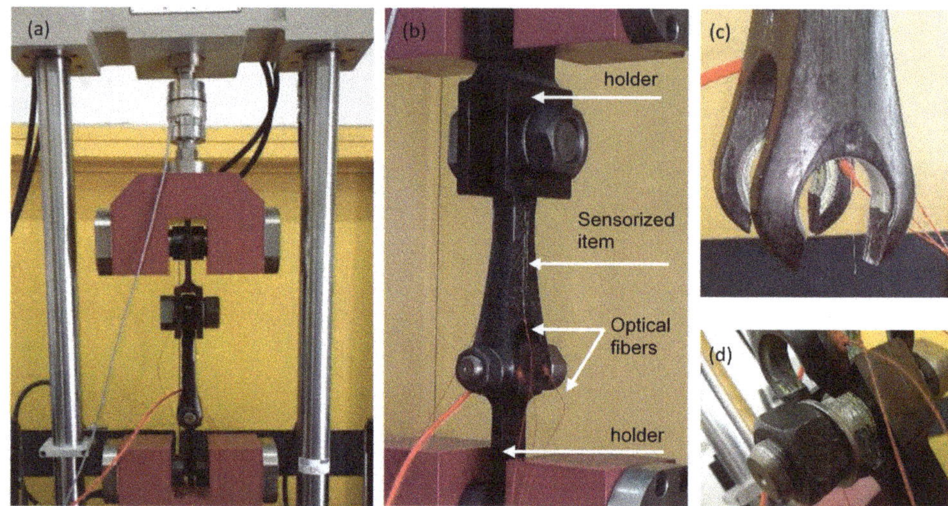

Figure 7. (**a**) and (**b**) Sensorized item fixed in the loading press; (**c**) and (**d**) view of the broken region.

3. Results and Discussion

This section presents and discusses the experimental results, in terms of responses of FBG strain sensors, when the drag brace item is loaded by a longitudinal force linearly incrementing with rate of 250 N/s. In particular, comparative analysis of the behavior of the Y- and C-array in different sections is take into consideration.

Figure 8a plots the strain values returned from the Y-array and C-array, versus acquisition time during the first test section, named test A. The applied force starts from zero (at 10 s) and reaches the maximum load of 25 kN at 110 s. During the incremental load ramp, as expected, all sensors of Y-array explore positive longitudinal strain/traction. In particular, the measured strain increases moving from Y_1 to Y_5, accordingly with numeric results and with the decreasing of the cross-section area of the composite item. On the contrary, the strain measured from Y_6 is significantly lower than Y_5 one. It can be attributed to the further increase of the item cross section (as moving from the Y_5 section to the Y_6 section) and to the proximity to the metallic ferule of the upper fixing point.

The C-array monitors a much more critical region of the item due to the rapid change of the surface strain state and its geometry shape. The sensor C_1, fixed in the center of the bent region between the ferules B1 and B2, returned a negative response (compressive strain). Due to its location, it measures strain along x-axis, which is compressed when the item is longitudinally stretched. Moving along the bent region, thousands of C_2, C_3 and C_4 sensors, the compressive effect of the x-axis decreases and the longitudinal elongation along y direction appears. The sensor C_4 measures the highest positive strain.

Finally, after 110 s the maximum load of 25 kN was kept unchanged for several minutes (up to 1020 s). During this interval, all sensors showed unchanged response as well. Furtherly, the item was suddenly unloaded and thus all sensors recovered their original unperturbed response. The sensor C_4 showed atypical additive noise, as compared with other sensors, that we were not able to understand.

The Figure 8b plots the response of all sensors in the second test section, test B. Here the applied force linearly increments up to 50 kN (at about 210 s) with same rate. All sensors measure strain increasing with the applied force and the relative behavior of all sensors is in good agreement with the previous test section.

However, at 170 s (approximately 40 kN) an anomalous behavior was observed for the sensors of C-array, especially for C_3 and C_4: they explore an improvised reduction of the strain of about 110 µε and 420 µε, respectively. Furtherly they remain almost unchanged up to the end of the test section. It is worth noting that, there was no apparent reason to understand such behavior (at 170 s) because the composite item does not show visible damages and sensor integration with composite surface were still good. However, it is obvious that the sudden response change acts as an early warning since it can be related to a partial breaking of bulk composite layers of the composite item (not visible on external surface). Note that this apparently unfounded hypothesis has been successively confirmed.

Concerning other sensors, the returned strain values continued to increase (decrease for C_1 and C_2 reading a compressive state) up to 210 s when the maximum load of 50 kN was reached. Successively, the input load as well as the measured strain were unchanged for several minutes (up to 1120 s). Finally, all sensors recovered to the original unperturbed value when the load state was removed. Concerning the sensors C_4 the behavior of the recovery curve is like the previous case.

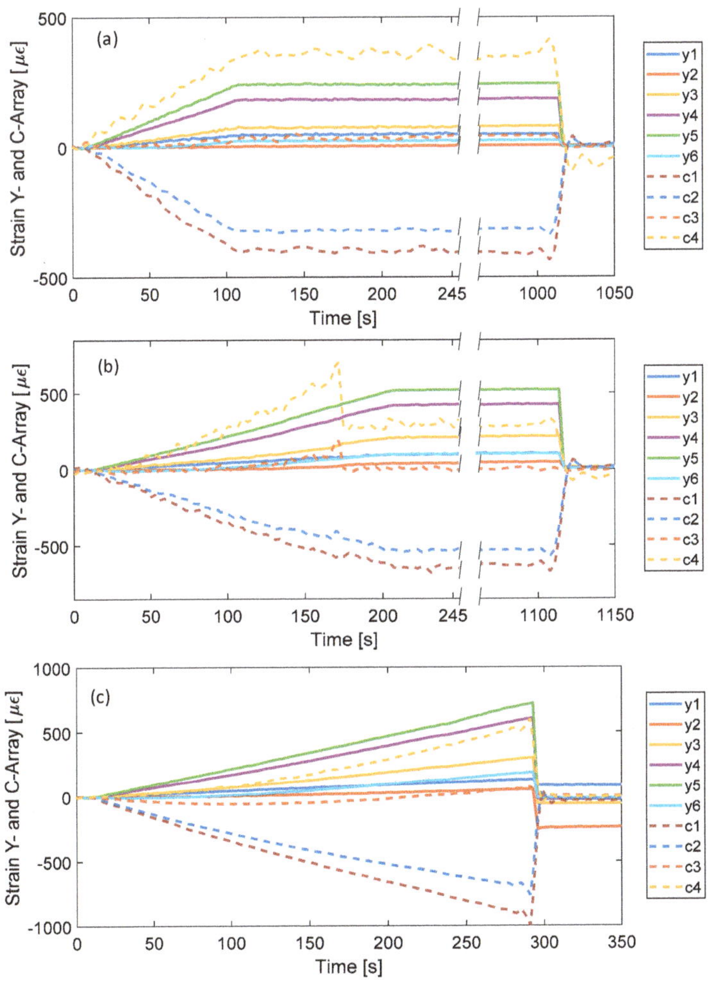

Figure 8. Time response of all FBG sensors in different test sections: (**a**) test A, (**b**) test B, and (**c**) test C.

Finally, Figure 8c shows the returned strain values during C section. It was planned to reach a maximum load of 75 kN, significantly higher than the expected working load. Strain measured from all sensors starts the linear shift in perfect agreement with previous experiments. However, after 282 s of the incremental load ramp (approximately applied load of 70 kN) the composite item suddenly breaks. In particular, the double fixing point breaks the anchoring of both B1 and B2 metallic rings. Figure 7c,d reports some pictures of the broken parts of the item. After the breaking point, the strain values of all sensors recovered to the original values. A not perfect recovery was measured for sensors Y_1 and Y_2 probably because they were the closest to the break region. From this consideration, we believe that the anomalous situation occurred during test B at 170 s is due to a partial breaking of the item.

A better comparison between different load tests was conducted in the Figure 9a,b for sensor Y_5 and for sensors C_1 and C_4, respectively. The response of sensors Y_5 and C_1 in different test ramps show a very good agreement. With reference to sensor Y_5, the maximum difference occurs between test B and test C curves at 210 s and is less than 5%. Differently, the sensor C_4 clearly demonstrated that its behavior during the load test is completely modified by the event during test B, confirming the hypothesis of partial breaking.

In order to demonstrate the capability of FBG sensors technology to be used as real-time tool for structural health monitoring of composite items in aeronautic applications, Figure 10a,b plots 3D real-time maps vs. *y*-axis and *c*-axis, respectively. The map images give a clear understanding of the complexity of the investigated surface in terms of the strain value and of the rate of changes along the spatial axis (*y*- and *c*-axis). Also, the Figure 10b clearly shows the capability to identify the anomalous situation occurs at 170 s and thus shows the capability to act as an early warning monitoring system.

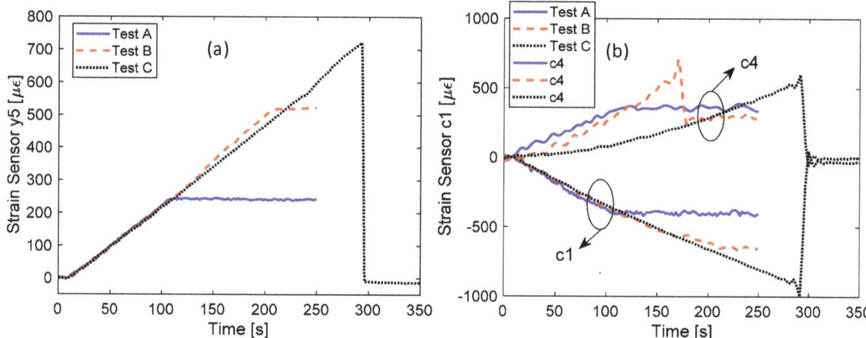

Figure 9. Comparison between the time responses in different test sections for the following sensors: (**a**) sensor Y_5; and (**b**) sensors C_1 and C_4.

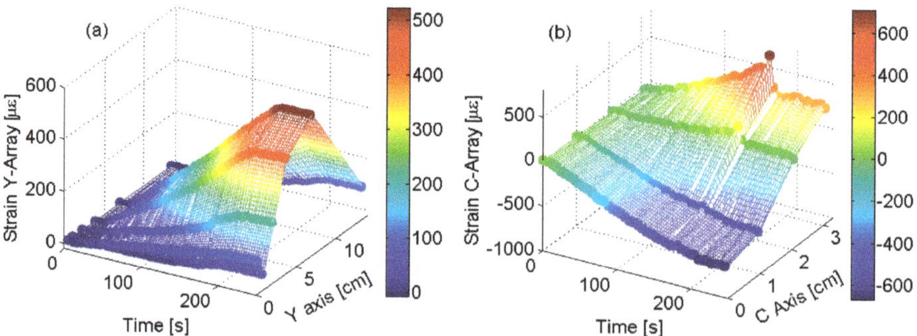

Figure 10. FBG Surface strain profile vs. time: (**a**) strain profile along *y*-axis; and (**b**) strain profile along *c*-axis.

Finally, Figure 11a,b shows 2D plots of the strain profile, versus *y*-axis and *c*-axis, respectively, in different load states during test ramp. This analysis is still focused on the test B section and for three load states, at 90 s, 130 s and 170 s, before the breaking event, and the last one at the end of the loading ramp (after the breaking) at 210 s. Concerning the y-strain profiles, for comparison purposes the Figure 11a also plots the strain profile from numerical results (as reported in Figure 5) on the sensors line and in the same load states. It shows impressive agreement between numerical and experimental results in the center on the composite item. Some differences are evident around the sensors Y_6 where the experiments return strain values much less that the FEM based expected one. We believe that such behavior is due to the presence of the metallic ring in the fixing point not considered in the numerical model.

The shape of the C-array response in Figure 11b highlights that the region between B1 and B2 exhibits a more critical strain behavior. With increasing applied tension, the dome of the curved region (position of C_1) is compressed whereas moving to the ends of the arms (to B_1 and B_2 fixing points) the positive strain significantly increases. Moreover, the anomaly in the shape after the partial breaking is clearly observable with red curve. In that case, with a load of F = 50 kN, one can expect significant increases of the positive strain of the sensors C_3 and C_4: on contrary, the real curve measures lower strain values and an increment of the compression in C_1.

Based on these results, the proposed FBG monitoring system demonstrated the capability to detect anomalous mechanical behavior as an early warning system.

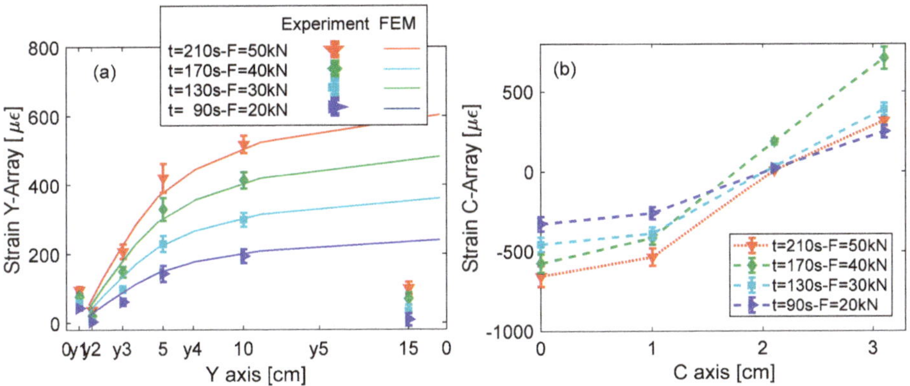

Figure 11. FBG Surface strain profile in different strain states: (**a**) Y-array; and (**b**) C-array.

4. Conclusions

In this work, we presented a detailed study devoted to assessing the capability of FBG based optical fiber sensing technology to act as in-situ, remote and real-time monitoring platforms of the load applied to a single component of aircraft landing gear. FBG based optical fiber sensor network involving two FBG arrays of six and four gratings was properly designed, implemented and integrated on a real drag brace. The design and integration of the sensors network were assisted by numerical studies of the composite item selected as target of the present project. Several experimental tests have been done where the composite item was subjected to stress in range 0–70 kN by means of a hydraulic press. Obtained results are in very good agreement with the numerical ones and demonstrate the great potentialities of FBG sensors technology to be employed for remote and real-time load measurements on aircraft landing gears. The proposed FBG monitoring system demonstrated the capability to detect strange behavior as an early warning system and demonstrated the capability to detect breaking of the item when the early warning is ignored.

Author Contributions: A.I., A.A. and S.C. conceived and designed the experiments; D.N. and P.D.P. performed preliminary characterization of gratings and bounding; F.S. and A.A. performed FEM numerical simulations; A.I., D.N. and S.C. designed the sensors layout; D.N. sensorized the item; A.I., D.N., A.A. and S.C. performed the load test; A.I. and D.N. analyzed the data and wrote the paper; all authors revised the paper.

Acknowledgments: The work is supported by the Italian Ministry of University and Research under the National Project PON03PE_00135_1, "CAPRI - Car-rello per Atterraggio con Attuazione Intelligente". The authors wish to thank Department of Industrial Engineering of University of Naples Federico II permitting us to use the facilities for load test.

Conflicts of Interest: The authors declare no conflict of interest.

References

1. Ramly, R.; Kuntjoro, W.; Rahman, M.K.A. Using Embedded Fiber Bragg Grating (FBG) Sensors in Smart Aircraft Structure Materials. *Procedia Eng.* **2012**, *41*, 600–606. [CrossRef]

2. Baker, A.; Scott, M. *Composite Materials for Aircraft Structures*, 3rd ed.; American Institute of Aeronautics and Astronautics, Inc: Washington, DC, USA, 2016; ISBN 978-1-62410-326-1.

3. Divakaran, V.N.; Kumar, G.R.; Rao, P.S. Aircraft Landing Gear Design and Development. 2015. Available online: https://pdfs.semanticscholar.org/23b1/2a2ba32d0020d49e28e7cb3f73216f1aa37c.pdf (accessed on 14 May 2019).

4. Allwood, G.; Wild, G.; Hinckley, S. Fiber Bragg Grating Sensors for Mainstream Industrial Processes. *Electronics* **2017**, *6*, 92. [CrossRef]

5. Qiao, X.; Shao, Z.; Bao, W.; Rong, Q.; Caucheteur, C.; Guo, T. Fiber Bragg Grating Sensors for the Oil Industry. *Sensors* **2017**, *17*, 429. [CrossRef] [PubMed]

6. Iadicicco, A.; Della Pietra, M.; Alviggi, M.; Canale, V.; Campopiano, S. Deflection Monitoring Method Using Fiber Bragg Gratings Applied to Tracking Particle Detectors. *IEEE Photonics J.* **2014**, *6*, 1–10. [CrossRef]

7. Palumbo, G.; Iadicicco, A.; Messina, F.; Ferone, C.; Campopiano, S.; Cioffi, R.; Colangelo, F. Characterization of Early Age Curing and Shrinkage of Metakaolin-Based Inorganic Binders with Different Rheological Behavior by Fiber Bragg Grating Sensors. *Materials* **2018**, *11*, 10. [CrossRef] [PubMed]

8. Iele, A.; Leone, M.; Consales, M.; Persiano, G.; Brindisi, A.; Ameduri, S.; Concilio, A.; Ciminello, M.; Apicella, A.; Bocchetto, F.; et al. Load monitoring of aircraft landing gears using fiber optic sensors. *Sens. Actuators A: Phys.* **2018**, *281*, 31–41. [CrossRef]

9. Hiche, C.; Liu, K.C.; Seaver, M.; Chattopadhyay, A. Characterization of impact damage in woven fiber composites using fiber Bragg grating sensing and NDE. In Proceedings of the SPIE Smart Structures and Materials + Nondestructive Evaluation and Health Monitoring, San Diego, CA, USA, 8 April 2009.

10. López-Higuera, J.M.; Cobo, L.R.; Incera, A.Q.; Cobo, A. Fiber optic sensors in structural health monitoring. *J. Light. Technol.* **2011**, *29*, 587–608. [CrossRef]

11. Meltz, G.; Hill, K. Fiber Bragg grating technology fundamentals and overview. *J. Light. Technol.* **1997**, *15*, 1263–1276.

12. Moccia, M.; Consales, M.; Iadicicco, A.; Pisco, M.; Cutolo, A.; Galdi, V.; Cusano, A. Resonant hydrophones based on coated fiber bragg gratings. *J. Light. Technol.* **2012**, *30*, 2472–2481. [CrossRef]

13. Lin, Y.B.; Chang, K.C.; Chern, J.C.; Wang, L.A. Packaging methods of fiber-Bragg grating sensors in civil structure applications. *IEEE Sens. J.* **2005**, *5*, 419–424.

14. Wan, K.T.; Leung, C.K.Y.; Olson, N.G.; Leung, C.K.Y. Investigation of the strain transfer for surface-attached optical fiber strain sensors. *Smart Mater. Struct.* **2008**, *17*, 35037. [CrossRef]

15. Luycks, G.; Voet, E.; Lammens, N.; Degrieck, J. Strain measurements of composite laminates with embedded fibre Bragg gratings: Criticism and opportunities for research. *Sensors* **2011**, *11*, 384–408. [CrossRef] [PubMed]

16. Cięszczyk, S.; Kisała, P. Inverse problem of determining periodic surface profile oscillation defects of steel materials with a fiber Bragg grating sensor. *Appl. Opt.* **2016**, *55*, 1412. [CrossRef] [PubMed]

17. Jin, B.; Zhang, W.; Zhang, M.; Ren, F.; Dai, W.; Wang, Y.; Yuan, S. Investigation on Characteristic Variation of the FBG Spectrum with Crack Propagation in Aluminum Plate Structures. *Materials* **2017**, *10*, 588. [CrossRef] [PubMed]

18. Zhou, J.; Cai, Z.; Zhao, P.; Tang, B. Efficient Sensor Placement Optimization for Shape Deformation Sensing of Antenna Structures with Fiber Bragg Grating Strain Sensors. *Sensors* **2018**, *18*, 2481. [CrossRef] [PubMed]

19. Leang, K.K.; Shan, Y.; Song, S.; Kim, K.J. Integrated Sensing for IPMC Actuators Using Strain Gages for Underwater Applications. *IEEE/ASME Trans. Mechatron.* **2012**, *17*, 345–355. [CrossRef]

20. Coelho, C.K.; Das, S.; Chattopadhyay, A.; Papandreou-Suppappola, A.; Peralta, P. Detection of fatigue cracks and torque loss in bolted joints. In Proceedings of the Health Monitoring of Structural and Biological Systems 2007, San Diego, CA, USA, 19–22 March 2007.

Article

Identification of Ground Intrusion in Underground Structures Based on Distributed Structural Vibration Detected by Ultra-Weak FBG Sensing Technology

Weibing Gan [1], Sheng Li [1,*], Zhengying Li [2] and Lizhi Sun [3]

[1] National Engineering Laboratory for Fiber Optic Sensing Technology, Wuhan University of Technology, Wuhan 430070, Hubei, China; ganweibing@whut.edu.cn

[2] School of Information Engineering, Wuhan University of Technology, Wuhan 430070, Hubei, China; zhyli@whut.edu.cn

[3] Department of Civil and Environmental Engineering, University of California, Irvine, CA 92697-2175, USA; lsun@uci.edu

* Correspondence: lisheng@whut.edu.cn

Received: 11 April 2019; Accepted: 6 May 2019; Published: 9 May 2019

Abstract: It is challenging for engineers to timely identify illegal ground intrusions in underground systems such as subways. In order to prevent the catastrophic collapse of subway tunnels from intrusion events, this paper investigated the capability of detecting the ground intrusion of underground structures based on dynamic measurement of distributed fiber optic sensing. For an actual subway tunnel monitored by the ultra-weak fiber optic Bragg grating (FBG) sensing fiber with a spatial resolution of five meters, a simulated experiment of the ground intrusion along the selected path was designed and implemented, in which a hydraulic excavator was chosen to exert intrusion perturbations with different strengths and modes at five selected intrusion sites. For each intrusion place, the distributed vibration responses of sensing fibers mounted on the tunnel wall and the track bed were detected to identify the occurrence and characteristics of the intrusion event simulated by the discrete and continuous pulses of the excavator under two loading postures. By checking the on-site records of critical moments in the intrusion process, the proposed detection approach based on distributed structural vibration responses for the ground intrusion can detect the occurrence of intrusion events, locate the intrusion ground area, and distinguish intrusion strength and typical perturbation modes.

Keywords: subway tunnel safety; ground intrusion detection; ultra-weak FBG; distributed vibration; dynamic measurement

1. Introduction

As an important carrier of the urban population, the subway system has greatly eased the pressure of ground transportation. In recent decades, research on early warning and treatment of various hazards that may affect the safety of subways has attracted widespread attention. Due to fewer indicators and concise detection principles, compared with structural safety monitoring of subway infrastructure, more commercial applications have emerged in the field of subway fire monitoring. Overviews and applications related in this area were reported in [1–3]. However, when long and large range needs to be considered, especially for subway tunnels, it is still a great challenge to find viable measures to meet the diverse needs of structural safety monitoring.

Since the tunnel lining structure generally uses concrete as the construction material, cracks on the lining surface are often used to reflect the safety status of a subway tunnel. One method based on digital images to detect cracks was reported by Zhang et al. [4]. For the indicators of temperature and strain, Ye et al. [5] described the study of tunnel safety monitoring during the construction stage based on quasi-distributed sensing technology of a limited number of fiber Bragg grating (FBG) sensors. Recently, references [6,7] reported some research advances in FBG-based sensors that combine the Internet of things or 3D printing techniques to detect damage or movement of underground structures. In addition to the conventional indicator, the study conducted in [8] indicates that some influencing factors, such as buried depth and operation age should also be collected when assessing the state of the tunnel. Reference [9] pointed out that particular emphasis should be paid on time and space-continuous monitoring for environmental and geotechnical underground structures. However, the above studies were primarily based on discrete response information with respect to time or space, which is difficult to meet the real-time requirements of structural safety monitoring for the entire line of the actual underground structure, such as a subway. Due to the advantages of large-scale monitoring, high sensitivity, and multiplexing capacity, distributed fiberoptic sensing technology is widely considered to be an ideal means for the safety monitoring of tunnel structures. Dewynter et al. [10] indicated the feasibility of continuous monitoring of soil movements while tunneling based on Brillouin optical time domain reflectometry (BOTDR) technology. Recent studies on using BOTDR to trace the distributed strain to secure tunnel safety can be found in [11–14].

Although distributed sensing technology provides a viable way for understanding the responses of tunnel structures in view of continuous time and space, existing studies primarily focus on the static measurement based on strain or temperature. In the past decades, research of distributed dynamic measurement [15] was mainly based on distributed acoustic sensing (DAS) techniques [16,17], which have been another research hotspot in the field of engineering monitoring. In the railway fields, DAS techniques were researched for railway perimeter security [18] and condition monitoring of the train and rail [19]. In geophysical engineering, the need and application concerning simultaneous vibration and temperature sensing technology based on DAS were reviewed in [20]. Moreover, Rao [21] reported the feasibility of using DAS technology to monitor the illegal or unauthorized third-party intrusion (TPI) in oil pipelines. However few studies pay attention to the impact of ground construction on the safety of underground structures covering a long-distance range through distributed dynamic measurements. He et al. [22] proposed that by processing images taken by unmanned aerial vehicle was a feasible way to detect ground drilling construction which may affect tunnel safety. However, this method is undoubtedly susceptible to climatic conditions and occlusion of ground buildings. Moreover, the method based on the analysis of UAV-images is still difficult to meet the timely warning needs of the entire subway line. Compared with oil pipelines, subway tunnels have deeper buried depths, and more complicated boundary conditions and load propagation paths. Therefore, different test accuracy, sensitivity, response speed, signal-to-noise ratio (SNR), and other parameters need to be considered when using DAS technology to tackle the similar need for these two different fields. This may be the reason for less reports on DAS-based tunnel TPI.

Comparing with DAS technology using ordinary optic fiber, ultra-weak FBG array based on the draw tower [23,24] using sensing optic fiber, integrates both advantages of fiber optic point sensors and distributed sensors. This technology is an alternative way to achieve high-precision, fast, and wide coverage distributed measurement. Previous research around this technology focused more on monitoring strain, temperature or strain-based deformation for the object of interest [25,26]. In addition, a multi-parameter measurement system based on ultra-weak FBG array with sensitive material was proposed in [27]. However, all the research is still limited to static indicators. Actually, ultra-weak FBG array is also adept to perform dynamic monitoring [28] in addition to the above positive characteristics usually witnessed in static measurement. From the reports in [15,29–32], the comparison results in Table 1 reveal that the ultra-weak FBG array can be not only used for both static and dynamic measurements, but also has higher SNR than that of DAS sensors. Moreover, higher SNR often leads to better sensing performances, such as higher measurement accuracy, faster response time, and simpler detection circuit, so ultra-weak FBG array is more suitable than DAS when dealing with distributed vibration and other scenarios requiring high-speed measurement. Therefore, based on such performance advantages in distributed dynamic measurement, this paper explores the detection capability of ultra-weak FBG sensing array fabrication by the draw tower for ground intrusion. The experimental results of detecting the ground intrusion event of an actual subway tunnel were reported. The detection approach of ground intrusion is the second part of this paper, followed by the details on the design and implementation of a field experiment. Finally, the effectiveness of the proposed method to detect and identify a simulated intrusion is discussed based on the on-spot experimental results from the ultra-weak FBG distributed sensing technology.

Table 1. Comparisons between common sensors for underground structure monitoring and ultra-weak fiber Bragg grating (FBG).

Sensors	Static/Dynamic Measurement	Multiplexing Capacity	Reflectivity index [1]	Transmission Medium
Electronic	Both	Weak	N.A. [2]	Electric cable
FBG	Both	Median	0.1–1	Sensing optic fiber
Rayleigh-based OTDR/OFDR	Static			Ordinary optic fiber
Brillouin-based BOTDR/BOTDA	Static	Strong	10^{-9}–10^{-7}	Ordinary optic fiber
Rayleigh-based DAS	Both			
Ultra-weak FBG	Both		10^{-5}–10^{-4}	Sensing optic fiber

[1] Higher reflectivity index usually means better signal-to-noise ratio (SNR); [2] Not applicable.

2. Detection Methodology of Ground Intrusion

Figure 1 illustrates the distributed vibration sensing principle used to detect ground intrusion. The phenomenon of light interference caused by the reflection signals of two adjacent ultra-weak FBGs is used to detect the vibration of the object of interest. Here, the ultra-weak FBG is regarded as a mirror, and L represents the distance that causes light interference. The spatial resolution of the distributed vibration along the sensing fiber is typically determined by the parameter L. The sensitivity and the frequency response of the vibration signal measured by the strain-induced phase variation between two ultra-weak FBGs are improved by the interferometer. Here, Faraday rotating mirrors are utilized in the demodulation process of ultra-weak FBG array to suppress the polarization effect. Moreover, the 3-by-3 coupler phase demodulation algorithm is used to reconstruct the time domain signal, and restore the phase information of the vibration signal, through which the interrogation of the vibration frequency and amplitude can be realized. Further, optical time domain reflectometry technique is utilized to achieve vibration localization.

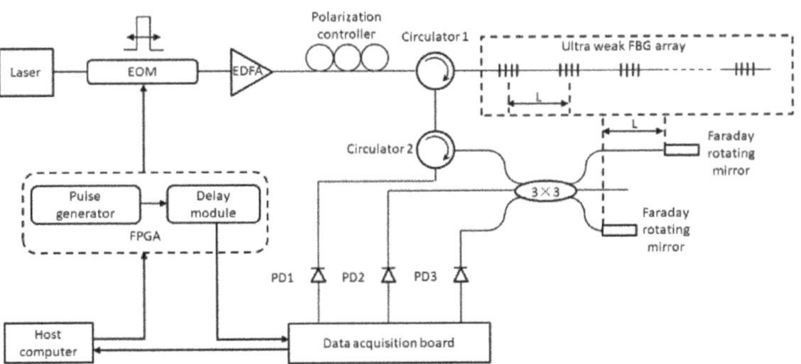

EOM: electro-optic modulator; EDFA: erbium-doped fiber amplifier; PD: photodetector; 3 × 3: symmetrical 3-by-3 coupler; FPGA: field programmable gate array.

Figure 1. Sensing principle of distributed vibration detection based on ultra-weak FBG array.

The high sensitivity of large-scale ultra-weak FBGs and the corresponding demodulation system of high speed [33] make the sensing fiber particularly suitable for locating abnormal perturbations occurring within a long-distance range. In addition, the previous study [34] revealed the repeatability of such a sensor is around 3.41 nε. When an illegal ground intrusion event occurs, the propagation path of the intrusion load from the ground to the tunnel wall and track bed can generally be demonstrated, as shown in Figure 2. Based on this assumption, the study used armored distributed sensing fibers to measure the distributed vibration of the tunnel wall and the track bed. Five-meter equidistance between adjoining FBGs along the sensing fiber determined the spatial resolution of the detection target, and this resolution almost can meet the positioning accuracy requirement for an actual subway tunnel. The approach used to quickly indicate whether a ground intrusion occurs was conducted by monitoring the distributed structural vibration responses along the tunnel and analyzing the difference in responses between the immediate state and the normal baseline state. Since the light interference region indicated by the address of ultra-weak FBG can be interrogated with the time- and wavelength-division multiplexing method [35,36] and has a corresponding relationship with the mileage information of the monitoring structure, locating the intrusion can be achieved by identifying the light interference region corresponding to the abnormal vibration responses.

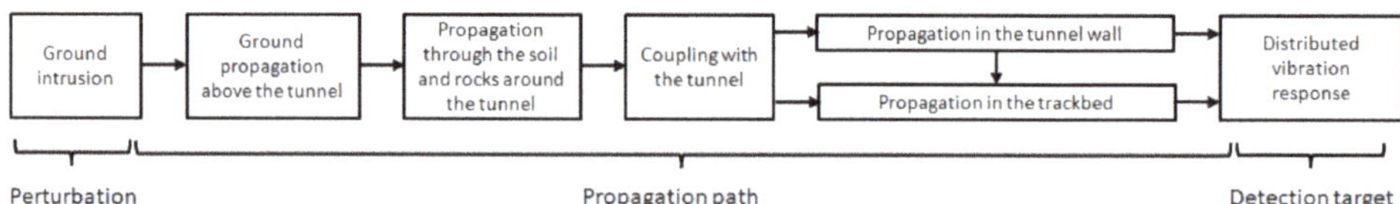

Figure 2. Propagation path of the intrusion load assumed by the detection principle.

3. Experimental Design and Implementation

3.1. Engineering Background of the Experimental Scheme

An actual tunnel structure (Wuhan Metro Line 7) was used in this study. Before the operation of the subway, the ultra-weak FBG sensing fibers were installed on the structural surfaces of the tunnel wall and the track bed. It covered a range of nearly three kilometers, aiming to detect the distributed structural vibration response of the monitoring zones. Figure 3 displays the actual layout of sensing fibers on the spot. The real-time vibration responses with 1 kHz sampling rate were fully transmitted back to the platform monitoring center and processed by the demodulator and servers. According to the spatial resolution of the sensing fiber and the on-spot layout of the tunnel structure, more than 500 vibration regions along the tunnel wall and the track bed can be distinguished based on the interrogated address of the light interference.

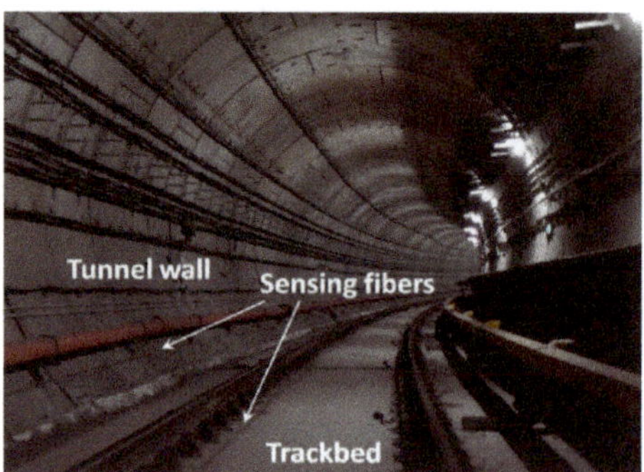

Figure 3. Distributed vibration sensing fibers mounted on the surfaces of the tunnel wall and the track bed.

3.2. Design of the Ground Intrusion

The intrusion perturbation was simulated by drilling ground through a small hydraulic excavator. According to the on-spot survey and structural design blueprints of the subway, both the available ground region for the simulated perturbation and the facade relationship between the ground and underground tunnel structure were taken into account to determine the intrusion sites and path, as illustrated in Figure 4. The plane distances between the underground tunnel and five intrusion sites correspond to the right part of the plot provided in Figure 4. Here, based on design and survey data, the average buried depth of the tunnel under the selected area was approximately 22.6 m. From the facade, the position P1 was placed just above the sensing fiber that monitored the tunnel wall.

Figure 4. Sites and path of the simulated ground intrusion.

3.3. Implementation of the Intrusion Perturbation

Two types of perturbation postures of the experiment excavator displayed in Figure 5 were set to simulate the different strengths of the ground intrusion. Discrete and continuous pulses were applied sequentially for each perturbation posture to simulate different intrusion modes. In order to reduce the damage of the pavement, when perturbations were applied, a steel plate was placed in advance at each position given in Figure 4.

(a)   (b)

Figure 5. Two perturbation postures of the excavator: (**a**) fully and (**b**) partially touching the ground.

All tests were scheduled to be carried out in the early hours to reduce interference with ground traffic. As depicted in Figure 4, field tests were sequentially performed at the positions of P1, P2, P3, P4, and P5. The purpose of considering varied distance was to explore the identifiability of distributed vibration responses under different strengths and modes. After the excavator reached the designated invasion site, discrete and continuous pulses were applied sequentially in the two postures shown in Figure 5. Perturbation in each place lasted for 1 to 1.5 min. In addition to the duration time for each specified intrusion point, other critical moments in each process were also recorded, including moments about excavator movement, loading posture adjustment, and pre-intrusion repositioning.

4. Result Analysis and Discussion

This section reports the characteristics of distributed vibration responses of underground tunnel structures under simulated intrusions. The ability to detect and distinguish the strength and mode of intrusion loads with different distances was investigated and discussed.

4.1. Responses of the Whole Intrusion Process

The #159 vibration zone of the tunnel structure just below the invasion site P1 was taken as an example to illustrate the detectability of the dynamic structure responses to the intrusion load. The vibration responses of the tunnel wall and the track bed under the perturbations of multiple intrusion sites are shown in Figures 6 and 7, in which dotted lines based on the field records mark the whole process of each intrusion site. It is shown that with the increase of the distance between the perturbation sites and the tunnel, the identification effect of intrusion based on the amplitude feature of structural vibration responses gradually decreased. Specifically, the results at position P5 showed that it became difficult to distinguish the vibration responses caused by the intrusion load from those of the excavator movement and pre-intrusion repositioning. Further, the response magnitude of the track bed in each loading process was significantly smaller than that of the tunnel wall. This phenomenon was consistent with the assumption described in Figure 2, indicating the dissipation of intrusion loads during propagation from the tunnel wall to the track bed. In addition, this indicated that the response of the tunnel wall was more suitable for inferring ground intrusion.

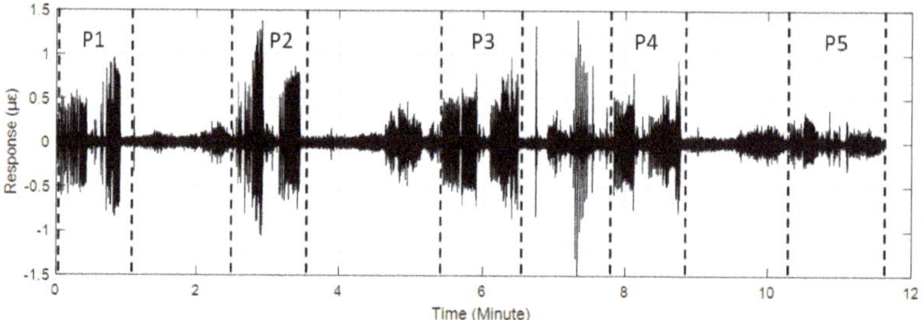

Figure 6. Distributed structural vibration of #159 zone of the tunnel wall under multiple position perturbations.

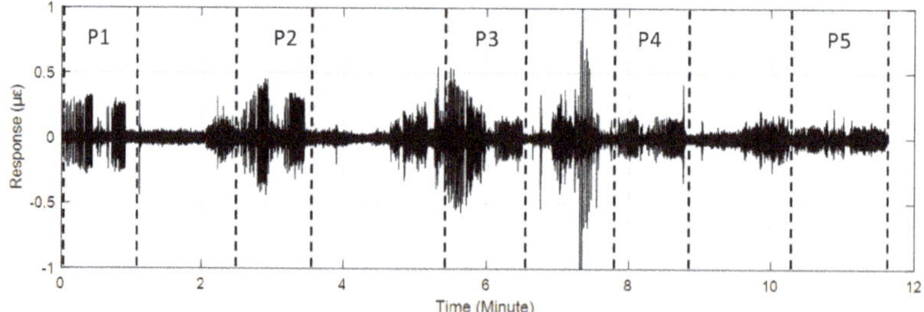

Figure 7. Distributed structural vibration of #159 zone of the track bed under multiple position perturbations.

4.2. Identifiability of Intrusion Characteristics

The above study discussed the relationship between the vibration response of#159 zone regarded as the most conducive to receiving perturbations and ground intrusions at different distances. By further focusing on a complete intrusion process of one intrusion site in Figure 6, the feasibility of identifying the characteristics of the simulated intrusion load was discussed. Figure 8 depicts the details of the intrusion process of the position P1 marked in Figure 6, which clearly distinguishes the detailed features of each stage of the simulated intrusion loading. During the test, the steel plate used for protecting the pavement in each intrusion site was only temporarily placed rather than fixed. This resulted in the phenomenon that the plate often bounced off and deviated from the initial position, especially in the first perturbation stage of the discrete pulse. In order to address this problem that occurred during pulse loading, the drill bit was typically used to reposition the plate, so that it was apparent from the first perturbation phase of Figure 8 that the amplitude uniformity of the vibration response was inconsistent with theoretical expectations. For the second perturbation stage, the duration of the discrete pulses was short because of the difficulty in maintaining the posture stability of the excavator. Due to this reason, there was no obvious increment in the magnitude of the response in this stage. The increase in load strength due to the change in the perturbation posture was apparent in the continuous pulse process of the second perturbation posture shown in Figure 8, although the response amplitude of the first half of the continuous pulse still deviated from the expectation. The deviation discrepancy was mainly attributed to disturbances caused by the unstable posture of the excavator and the process of re-adjusting the plate.

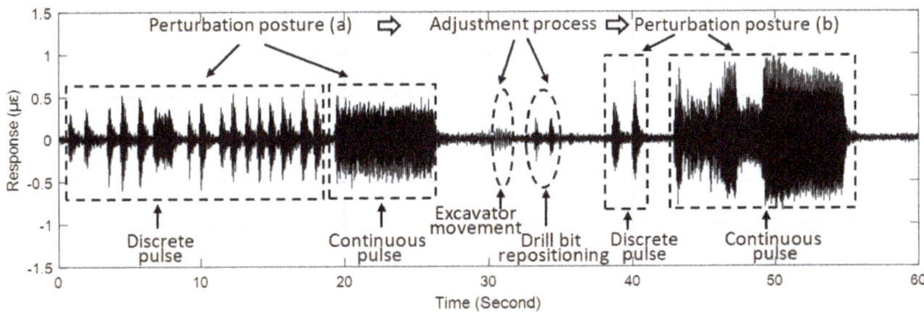

Figure 8. Distributed vibration response of #159 zone of the tunnel wall under the perturbation of position P1.

Moreover, both the excavator movement and the repositioning of the drill bit during the adjustment process can be observed in Figure 8, which was more obvious in the time–frequency spectrum shown in Figure 9, based on short-time Fourier transform. As can be seen from Figure 9a, different stages of the simulated intrusion corresponded to different frequency response ranges of the tunnel wall. For the excavator movement, the structure frequency was approximately 8 Hz. When the pulse loads acted on the ground, more structure frequencies were presented. Also, Figure 9a indicates the fact that the structure frequencies caused by each pulse mode were identical, in which the discrete pulses stimulated frequencies around 37 Hz, 47 Hz, and 115 Hz. Instead, continuous pulses made more frequencies with a maximum frequency around 267 Hz. Furthermore, as indicated in Figure 9b, the vibration energy caused by discrete or continuous pulses in the second perturbation posture was apparently greater than those in the first perturbation posture. These experimental results were in agreement with the expectations of the design and further demonstrated the ability of the proposed method of identifying typical load patterns.

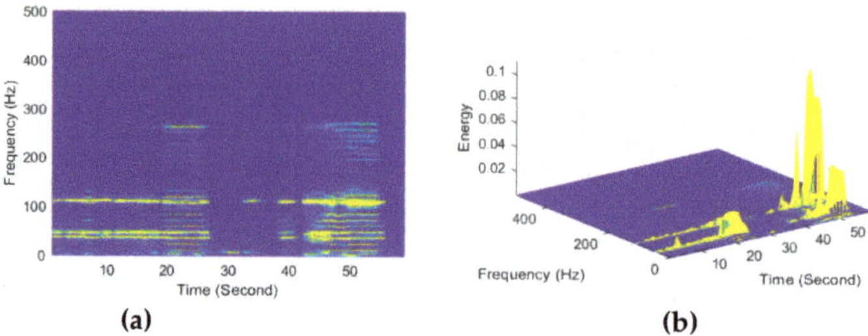

Figure 9. Time–frequency spectrum of #159 zone of the tunnel wall under the perturbation of position P1 viewed in (**a**) 2D and (**b**) 3D.

4.3. Detection Range of the Simulated Intrusion

The coordinate system depicted in Figure 10 was defined to analyze the influence range of the simulated intrusion. The XOY and XOY' represent the planes of the ground and the underground buried tunnel, respectively, where X', Y, and Z axes indicate the tunnel mileage, intrusion path, and depth from the ground, respectively. The origins of the planes XOY and $X'O'Y'$ were set to the intrusion position P1 and the tunnel #159 area, respectively.

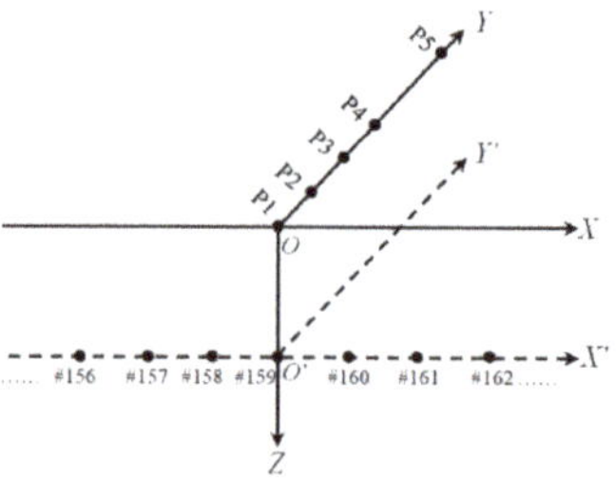

Figure 10. Coordinate system defined for the analysis of influence range of the simulated intrusion.

Figures 11–15 plot the waterfall diagrams of the distributed structural vibration responses of 13 consecutive monitoring zones of the tunnel wall with a length of 65 m along the X' axis, where the response of the #159 area in each waterfall diagram was set in the center and highlighted in a different color. Comparisons among Figure 11 show that at least 9 test zones with the #159 zone as the symmetry center had the ability to identify the simulated intrusion. Since the position P5 was farther away from the tunnel than other intrusion sites, it was not always easy to distinguish the simulated intrusion based on weak responses for most of the monitoring zones shown in Figure 15. Figures 11–15 also revealed that the amplitude of the distributed vibration response along the defined X' axis and the response symmetry of the monitoring zones centered on the #159 zone gradually became weak as the intrusion site moved away from the tunnel. Here, the asymmetry may be related to the fact reflected in Figure 4 that the designed intrusion path was not completely orthogonal to the tunnel.

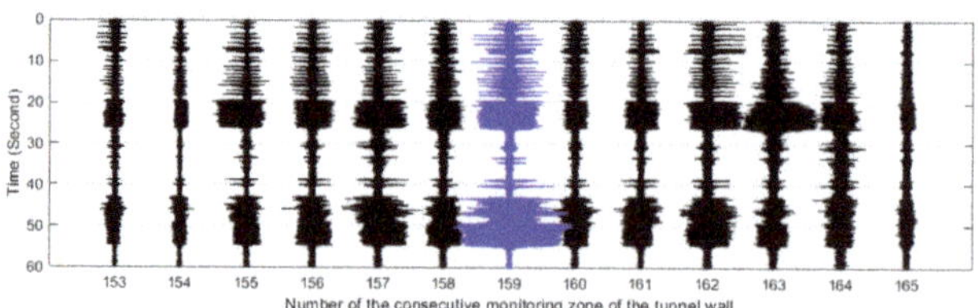

Figure 11. Distributed vibration responses of 13 consecutive monitoring zones under the perturbation of position P1.

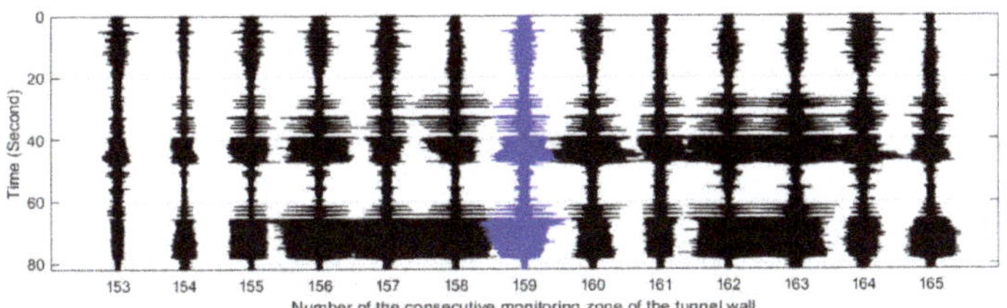

Figure 12. Distributed vibration responses of 13 consecutive monitoring zones under the perturbation of position P2.

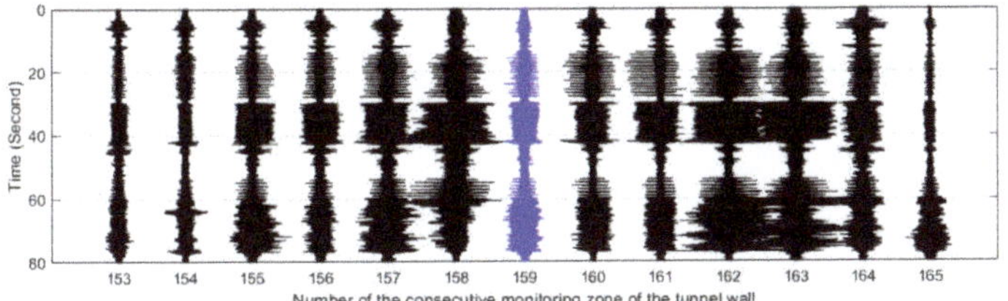

Figure 13. Distributed vibration responses of 13 consecutive monitoring zones under the perturbation of position P3.

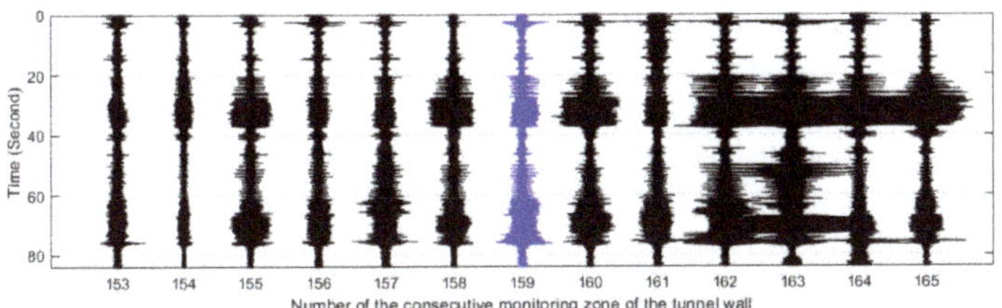

Figure 14. Distributed vibration responses of 13 consecutive monitoring zones under the perturbation of position P4.

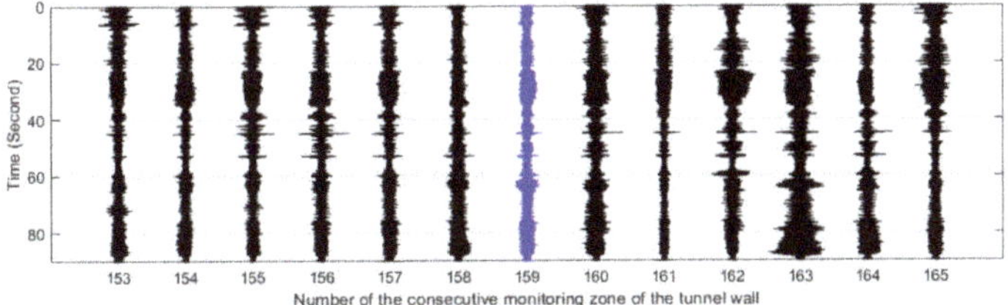

Figure 15. Distributed vibration responses of 13 consecutive monitoring zones under the perturbation of position P5.

To further analyze the characteristics of the distributed vibration responses along the Y and X' axes, the intrusion response sensitivity k was defined as,

$$k = \frac{r_{\text{intrusion}}}{r_{\text{stationary}}} \tag{1}$$

where $r_{\text{intrusion}}$ and $r_{\text{stationary}}$ represented the root mean square values of the vibration signals of the intrusion state and the stationary state, respectively. The duration used to determine $r_{\text{intrusion}}$ and $r_{\text{stationary}}$ came from the field records. In order to search for potential regularity, we computed the k

values of 13 consecutive regions in five simulated intrusion tests and obtained a k matrix with a shape of 5 by 13. Then, the medians of k were calculated based on the k matrix according to the intrusion site and the tunnel monitoring zone, respectively. After this statistical operation, the median-based distributions along the defined Y and X' axes are shown in Figure 16.

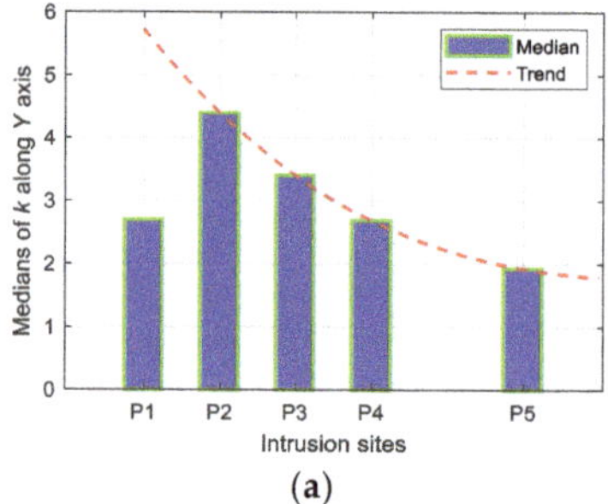
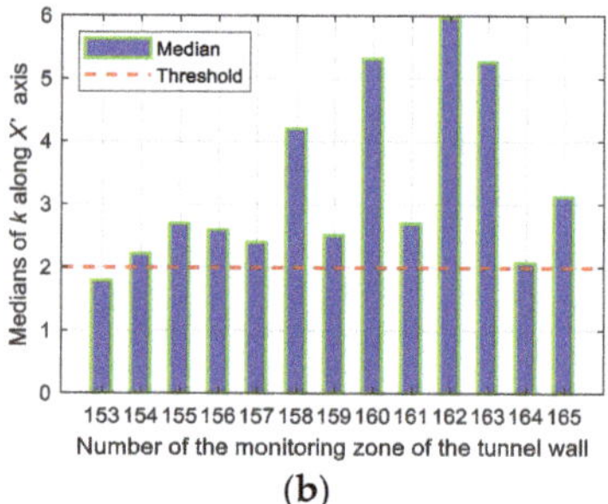

Figure 16. Median distributions along (**a**) the intrusion site and (**b**) the monitoring zone.

As shown in Figure 16a, the medians of k under the perturbation at each site except P5 were greater than 2.5. It can also be observed from the figure that the bar peak occurred at P2 rather than P1. One possibility for this unforeseen situation was the different ground conditions at the simulated intrusion sites. The map in Figure 4 clearly revealed that position P1 was near the edge of the road while positions P2–P5 were all on a path between two plantations, so different ground stiffness may be the cause of the occurrence of the abnormality in Figure 16a. An exponential attenuation process can be found in Figure 16a when fitting the tendency only through four sites located at the same ground condition. Figure 16b shows that the median-based distribution of the 13 monitoring zones at multiple intrusion sites was biased towards the positive direction of the X' axis, which substantially conformed to the response asymmetry observed in Figures 11–15, namely, the offset angle between the actual intrusion path shown in Figure 4 and the defined Y axis in Figure 10 caused the asymmetrical distribution shown in Figure 16b. Compared with the threshold set to two, as depicted in Figure 16b, the median of 12 among 13 monitoring zones was not less than 2.08, although the perturbation responses from the small excavator were tiny during the entire experiment.

5. Conclusions

This study verified that ultra-weak FBG array is a viable method to meet the requirements of distributed dynamic measurement technology in actual engineering. The capability of detecting the ground intrusion in an underground structure based on such measurement technology was reported through field tests. The analysis indicated that the proposed approach has the potential to distinguish the strength and pattern of the typical load in the intrusion process within a certain range. Besides the detectability of the intrusion event based on variations in vibration amplitude, the location of ground intrusion can be inferred by the temporal and spatial distribution characteristics of vibration responses along the tunnel. In view of the approved test time, interferences of subway trains and ground transportation were not considered in the analysis, which seems to be a shortcoming of the study that deserves further attention. However, due to the fact that the load influence generated by the actual intrusion event is often greater than that of the small excavator adopted in this paper, it is believed that the proposed approach can be applied to identify the ground intrusion occurring in the daytime.

Author Contributions: Data curation, S.L.; funding acquisition, S.L.; methodology, W.G. and Z.L.; project administration, Z.L.; supervision, L.S; writing—original draft, W.G. and S.L.; writing—review and editing, L.S.

Funding: This research was funded by the National Natural Science Foundation of China grant numbers 61735013 and 61875155, and the Fundamental Research Funds for the Central Universities (WUT: 2019-III-160CG).

Acknowledgments: The research work reported in this paper was supported by the National Engineering Laboratory for Fiber Optic Sensing Technology, Wuhan University of Technology, and Smart Nanocomposites Laboratory, University of California, Irvine.

Conflicts of Interest: The authors declare no conflicts of interest.

References

1. Nordmark, A. Fire and life safety for underground facilities: Present status of fire and life safety principles related to underground facilities. *Tunn. Undergr. Space Technol.* **1998**, *13*, 217–269. [CrossRef]
2. Liu, Z.; Kim, A.K. Review of recent developments in fire detection technologies. *J. Fire Prot. Eng.* **2003**, *13*, 129–151. [CrossRef]
3. Jiang, D.; Zhou, C.; Yang, M.; Li, S.; Wang, H. Research on optic fiber sensing engineering technology. In Proceedings of the 22nd International Conference on Optical Fiber Sensors, Beijing, China, 15–19 October 2012.
4. Zhang, W.; Zhang, Z.; Qi, D.; Liu, Y. Automatic crack detection and classification method for subway tunnel safety monitoring. *Sensors* **2014**, *14*, 19307–19328. [CrossRef] [PubMed]
5. Ye, X.W.; Ni, Y.Q.; Yin, J.H. Safety monitoring of railway tunnel construction using FBG sensing technology. *Adv. Struct. Eng.* **2013**, *16*, 1401–1410. [CrossRef]
6. Jo, B.W.; Khan, R.M.A.; Lee, Y.S.; Jo, J.H.; Saleem, N. A fiber Bragg grating-based condition monitoring and early damage detection system for the structural safety of underground coal mines using the Internet of things. *J. Sens.* **2018**, *2018*, 9301873. [CrossRef]
7. Hong, C.; Zhang, Y.; Lu, Z.; Yin, Z.A. FBG tilt sensor fabricated using 3D printing technique for monitoring ground movement. *IEEE Sens. J.* **2019**, *19*, 1590. [CrossRef]
8. Chen, X.; Li, X.; Zhu, H. Condition evaluation of urban metro shield tunnels in Shanghai through multiple indicators multiple causes model combined with multiple regression method. *Undergr. Space Technol.* **2019**, *85*, 170–181. [CrossRef]
9. Pamukcu, S.; Cheng, L.; Pervizpour, M. Chapter 1—Introduction and overview of underground sensing for sustainable response. In *Underground Sensing. Monitoring and Hazard Detection for Environment and Infrastructure*, 1st ed.; Pamukcu, S., Cheng, L., Eds.; Academic Press: Cambridge, MA, USA, 2018; pp. 1–42.
10. Dewynter, V.; Rougeault, S.; Magne, S.; Ferdinand, P.; Vallon, F.; Avallone, L.; Vacher, E.; De Broissia, M.; Canepa, C.; Poulain, A. Tunnel structural health monitoring with Brillouin optical fiber distributed sensing. In Proceedings of the 5th European Workshop on Structural Health Monitoring 2010, Naples, Italy, 28 June–4 July 2010.
11. Fajkus, M.; Nedoma, J.; Mec, P.; Novak, M.; Kajnar, T.; Martinek, R.; Jaros, J.; Hruby, D.; Vasinek, V. Monitoring of the structural loads of tunnels using a distributed optical system BOTDR. In Proceedings of the Optical Materials and Biomaterials in Security and Defense Systems Technology XIV 2017, Warsaw, Poland, 11–13 September 2017.
12. Hong, C.; Zhang, Y.; Li, G.; Zhang, M.; Liu, Z. Recent progress of using Brillouin distributed fiber optic sensors for geotechnical health monitoring. *Sens. Actuators A Phys.* **2017**, *258*, 131–145. [CrossRef]
13. Soga, K.; Kechavarzi, C.; Pelecanos, L.; Battista, N.; Williamson, M.; Gue, C.Y.; Murro, V.D.; Elshafie, M.; Monzón-Hernández Sr., D.; Bustos, E.; et al. Chapter 6—Fiber-optic underground sensor networks. In *Underground Sensing. Monitoring and Hazard Detection for Environment and Infrastructure*, 1st ed.; Pamukcu, S., Cheng, L., Eds.; Academic Press: Cambridge, MA, USA, 2018; pp. 287–356.
14. Lei, X.; Xue, Z.; Hashimoto, T. Fiber optic sensing for geomechanical monitoring: (2)-distributed strain measurements at a pumping test and geomechanical modeling of deformation of reservoir rocks. *Appl. Sci.* **2019**, *9*, 417. [CrossRef]
15. Liu, X.; Jin, B.; Bai, Q.; Wang, Y.; Wang, D.; Wang, Y. Distributed fiber-optic sensors for vibration detection. *Sensors* **2016**, *16*, 1164. [CrossRef]

16. Muanenda, Y. Recent advances in distributed acoustic sensing based on phase-sensitive optical time domain reflectometry. *J. Sens.* **2018**, *2018*, 3897873. [CrossRef]

17. He, Z.; Liu, Q.; Fan, X.; Chen, D.; Wang, S.; Yang, G. A review on advances in fiber-optic Distributed Acoustic Sensors (DAS). In Proceedings of the Conference on Lasers and Electro-Optics/Pacific Rim, CLEOPR 2018, Hong Kong, China, 29 July–3 August 2018.

18. He, Z.; Liu, Q.; Fan, X.; Chen, D.; Wang, S.; Yang, G. Fiber-optic distributed acoustic sensors (DAS) and applications in railway perimeter security. In Proceedings of the Advanced Sensor Systems and Applications VIII 2018, Beijing, China, 11–13 October 2018.

19. Cedilnik, G.; Hunt, R.; Lees, G. Advances in train and rail monitoring with DAS. In Proceedings of the Optical Fiber Sensors, OFS 2018, Lausanne, Switzerland, 24–28 September 2018.

20. Miah, K.; Potter, D.K. A review of hybrid fiber-optic distributed simultaneous vibration and temperature sensing technology and its geophysical applications. *Sensors* **2017**, *17*, 2511. [CrossRef]

21. Rao, Y. Recent progress in ultra-long distributed fiber-optic sensing. *WuliXuebao* **2017**, *66*, 074207.

22. He, L.; Tan, Y.; Liu, H.; Zhao, B. UAV-image-based illegal activity detection for urban subway safety. In Proceedings of the 6th International Conference on Remote Sensing and Geoinformation of the Environment, RSCy 2018, Paphos, Cyprus, 26–29 March 2018.

23. Yang, M.; Bai, W.; Guo, H.; Wen, H.; Yu, H.; Jiang, D. Huge capacity fiber-optic sensing network based on ultra-weak draw tower gratings. *Photonic Sens.* **2016**, *6*, 26–41. [CrossRef]

24. Yang, M.; Li, C.; Mei, Z.; Tang, J.; Guo, H.; Jiang, D. Thousand of fiber grating sensor array based on draw tower: A new platform for fiber-optic sensing. In Proceedings of the Optical Fiber Sensors, OFS 2018, Lausanne, Switzerland, 24–28 September 2018.

25. Bartelt, H.; Schuster, K.; Unger, S.; Chojetzki, C.; Rothhardt, M.; Latka, I. Single-pulse fiber Bragg gratings and specific coatings for use at elevated temperatures. *Appl. Opt.* **2007**, *46*, 3417–3424. [CrossRef]

26. Ecke, W.; Schmitt, M.W.; Shieh, Y.; Lindner, E.; Willsch, R. Continuous pressure and temperature monitoring in fast rotating paper machine rolls using optical FBG sensor technology. In Proceedings of the 22nd International Conference on Optical Fiber Sensors, Beijing, China, 15–19 October 2012.

27. Bai, W.; Yang, M.; Hu, C.; Dai, J.; Zhong, X.; Huang, S.; Wang, G. Ultra-weak fiber Bragg grating sensing network coated with sensitive material for multi-parameter measurements. *Sensors* **2017**, *17*, 1509. [CrossRef]

28. Zhou, L.; Li, Z.; Xiang, N.; Bao, X. High-speed demodulation of weak fiber Bragg gratings based on microwave photonics and chromatic dispersion. *Opt. Lett.* **2018**, *43*, 2430–2433. [CrossRef]

29. Guo, H.; Qian, L.; Zhou, C.; Zheng, Z.; Yuan, Y.; Xu, R.; Jiang, D. Crosstalk and ghost gratings in a large-scale weak fiber Bragg grating array. *J. Lightwave Technol.* **2017**, *35*, 2032–2036. [CrossRef]

30. Guo, H.; Liu, F.; Yuan, Y.; Yu, H.; Yang, M. Ultra-weak FBG and its refractive index distribution in the drawing optical fiber. *Opt. Express* **2015**, *23*, 4829–4838. [CrossRef]

31. Gong, H.; Kizil, M.S.; Chen, Z.; Amanzadeh, M.; Yang, B.; Aminossadati, S.M. Advances in fibre optic based geotechnical monitoring systems for underground excavations. *Int. J. Min. Sci. Technol.* **2017**, *29*, 229–238. [CrossRef]

32. Gui, X.; Li, Z.; Wang, F.; Wang, Y.; Wang, C.; Zeng, S.; Yu, H. Distributed sensing technology of high-spatial resolution based on dense ultra-short FBG array with large multiplexing capacity. *Opt. Express* **2017**, *25*, 28112–28122.

33. Li, Z.; Tong, Y.; Fu, X.; Wang, J.; Guo, Q.; Yu, H.; Bao, X. Simultaneous distributed static and dynamic sensing based on ultra-short fiber Bragg gratings. *Opt. Express* **2018**, *26*, 17437–17446. [CrossRef]

34. Tong, Y.; Li, Z.; Wang, J.; Wang, H.; Yu, H. High-speed Mach-Zehnder-OTDR distributed optical fiber vibration sensor using medium-coherence laser. *Photonic Sens.* **2018**, *8*, 203–212. [CrossRef]

35. Chuang, W.H.; Tam, H.Y.; Wai, P.K.A.; Khandelwal, A. Time- and wavelength-division multiplexing of FBG sensors using a semiconductor optical amplifier in ring cavity configuration. *IEEE Photonics Technol. Lett.* **2005**, *17*, 2709–2711. [CrossRef]

36. Luo, Z.; Wen, H.; Guo, H.; Yang, M. A time- and wavelength-division multiplexing sensor network with ultra-weak fiber Bragg gratings. *Opt. Express* **2013**, *21*, 22799–22807. [CrossRef]

Article

Application of a Novel Long-Gauge Fiber Bragg Grating Sensor for Corrosion Detection via a Two-level Strategy

Yuyao Cheng [1], Chenyang Zhao [1], Jian Zhang [1,2,*] and Zhishen Wu [1,3]

[1] School of Civil Engineering, Southeast University, Nanjing 210096, China; chengyy@seu.edu.cn (Y.C.); cyzhao@seu.edu.cn (C.Z.); zswu@seu.edu.cn (Z.W)

[2] Jiangsu Key Laboratory of Engineering Mechanics, Southeast University, Nanjing 210096, China

[3] International Institute for Urban System Engineering, Southeast University, Nanjing 210096, China

* Correspondence: jian@seu.edu.cn

Received: 17 January 2019; Accepted: 19 February 2019; Published: 23 February 2019

Abstract: Corrosion of main steel reinforcement is one of the most significant causes of structural deterioration and durability reduction. This research proposes a two-level detection strategy to locate and quantify corrosion damage via a new kind of long-gauge fiber Bragg grating (FBG) sensor. Compared with the traditional point strain gauges, this new sensor has been developed for both local and global structural monitoring by measuring the averaged strain within a long gauge length. Based on the dynamic macrostrain responses of FBG sensors, the strain flexibility of structures are identified for corrosion locating (Level 1), and then the corrosion is quantified (Level 2) in terms of reduction of sectional stiffness of reinforcement through the sensitivity analysis of strain flexibility. The two-level strategy has the merit of reducing the number of unknown structural parameters through corrosion damage location (Level 1), which guarantees that the corrosion quantification (Level 2) can be performed efficiently in a reduced domain. Both numerical and experimental examples have been studied to reveal the ability of distributed long-gauge FBG sensors for corrosion localization and quantification.

Keywords: long-gauge fiber optic sensor; corrosion detection; strain flexibility; impact test

---

1. Introduction

With the development of society, complex infrastructures such as high-rise buildings and long-span bridges have been widely constructed. During the service life of steel used in these infrastructures, steel corrosion has been considered as one of the main reasons for structural damage and deterioration, especially for those exposed to aggressive environments [1]. Over the past several years, much effort has been devoted to developing a reliable and efficient corrosion monitoring apparatus [1–4]. Zhang et al. designed an innovative Hall-effect magnetic sensor to quantify the corrosion rate for reinforced concrete structures [5]. Sunny et al. utilized a low frequency (LF) RFID sensing system to measure corrosion of steel samples in marine atmosphere and selective transient features are extracted for corrosion characterization [6]. Many conventional methods of corrosion identification, such as sensors based on macrocell measurements, sensors based on in-depth resistivity measurements, and detection based on ultrasonic techniques, have been illustrated in detail in [7]. These techniques are sensitive to structural corrosion, but fail to detect the corrosion unless the apparatus covers the corroded region. Recently, much attention has been paid to the application of fiber optical sensing techniques for corrosion detection [8–15]. As the optical fiber is small and lightweight, it can be easily attached to the surface of concrete or mounted on the steel reinforcement. With high precision and stable sensing capacity, the fiber Bragg grating (FBG)-based strain sensor is the most popular. In some studies, the main approach used to indicate the levels of corrosion is to measure the expansion in the bar diameter due to corrosion deposits [12]; this is accomplished by winding fiber optical strain sensors around the steel reinforcement in the corroded area.

However, the challenging problems described below hinder the development of strain-based corrosion detection approaches. (1) The traditional point strain sensors are unsuitable for large-scale civil engineering because they cannot successfully detect unforeseen corrosion with a short gauge length (around 1–2 cm), unless the sensors are installed in the corrosion domain [16]. Also, as a point sensor, corrosion rust can affect the fixation point [17]. (2) As a kind of structural damage, it has been found that the level of steel corrosion at the early stages of corrosion can be indicated by using a measure of the stiffness [18]. A vibration test, an effective and convenient way to excite the structure, has been widely used to extract the structural modal properties to detect damage caused by stiffness reduction. However, vibration-based corrosion detection methods are seldom studied because corrosion involves a deeper-level damage detection problem: damage quantification. Usually, civil structures with a large scale and complex form include a significant quantity of unknown parameters which will lead to slow convergence and non-uniqueness in inverse and optimal analyses [19]. Most damage detection methods may only have the capacity to locate damage and may fail to quantify the severity. Consequently, the application of strain measurements for effective structural corrosion detection and safety evaluation will be greatly enhanced if these two problems are solved.

Corresponding to the two challenging difficulties discussed above, an effective methodology for structural corrosion detection using a novel strain sensor is proposed in this paper, and two aspects have been included. (1) A kind of novel long-gauge fiber optic sensor has been developed, that can integrate both local and global information by measuring the average strains within a long gauge length (e.g., 1–2 m). Compared with the traditional strain gauge used only for "local" measurements, long-gauge FBG sensors can capture integrated information that covers the entire structure and can thus detect unforeseen damage. (2) The localization and quantification of structural corrosion damage are two different levels of corrosion detection, and they can be considered separately, and the localization and quantification can be conducted step by step. In Level 1, corrosion damage is located according to a damage index with a space resolution of the gauge lengths of the fiber optic sensors. Comprehensive development of damage assessment methodologies has been widely studied based on structural dynamic properties, such as natural frequencies, mode shapes, and their derivatives [20–22]. It well-known that modal flexibility is a more sensitive diagnostic indicator than mode shape and natural frequency [22,23]. In addition, strain flexibility, defined as the strain response of a structure's element to the unit input force, has a direct relationship with structural stiffness and is much more useful for structural safety evaluation [24]. Consequently, strain flexibility is adopted here to construct a damage index. After locating the corrosion damage by a strain flexibility-based index, the quantification of Level 2 corrosion can only be conducted in the detected domain, where the number of structural parameters to be identified is less than the entire structure. Thus, it is helpful for the structural equation, solving the quantification at Level 2.

In this article, after a brief description of long-gauge FBG sensors, a step-by-step corrosion damage detection strategy is illustrated to locate and quantify corrosion; meanwhile, a solid theoretical basis is developed to guarantee accurate detection. Finally, both numerical examples and experimental tests are conducted to verify the robustness of the novel long-gauge FBG sensors and the effectiveness of the proposed method for corrosion detection in in-service structures.

2. Long-Gauge Fiber Optic Sensor

A number of sensors are utilized to measure structural responses during structural performance evaluation. These sensors are of two types: global sensing technology and local sensing technology. The former, including accelerometers and GPS, usually reflects overall structural information and is not a good candidate for detecting structural local damage. The latter, such as strain sensors and corrosion sensors, are sensitive to structural local damage, but they fail to detect local damage unless the sensor covers the damaged region. To overcome the limitation that sensors can only reflect "local" or "global" behavior (but not both), a novel long-gauge FBG sensor, shown in Figure 1, is developed to monitor structure by measuring the average strain within a long gauge length (e.g., 1~2 m), in which both local and global structural information are integrated. The principle of the long-gauge macrostrain is introduced in Figure 1a.

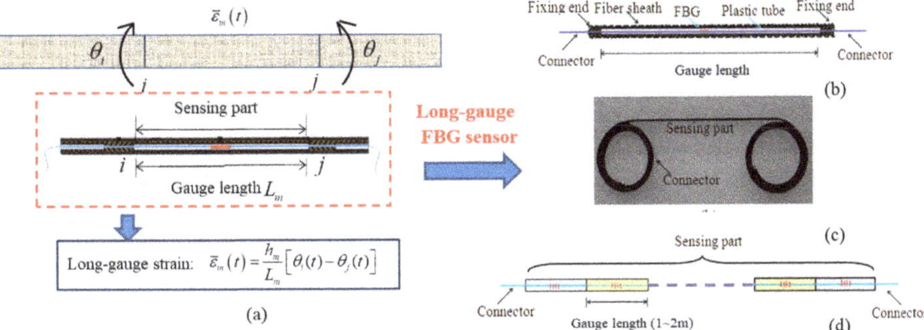

Figure 1. Fiber Bragg grating (FBG) sensor. (**a**) Principle of macrostrain; (**b**) Design of long-gauge FBG sensor; (**c**) Actual sensor; (**d**) Sensor arrays.

The designed FBG sensor with a long gauge length is illustrated in Figure 1b. One significant feature for the designed sensor is the utilization of the outer tube, which makes the in-tube fiber move freely and has the same mechanical behavior as the structure. Thus, when the FBG sensor is mounted on the structure and its two ends fixed, the strain transferred from the shift of Bragg center wavelength represents the average strain over the region the sensor covers. However, since optical fibers are fragile, bare optical fibers cannot be embedded directly into concrete. Therefore, a packaging technique is needed to protect the FBG sensor from some hostile environments, such as high temperatures, corrosion, and humidity. To enhance the measuring sensitivity, composite materials are utilized to package the optical fiber; to ensure the accurate measurements for compressive strain, the fiber is pre-tensioned before packaging to produce an initial pre-tensioned strain. The packaged long-gauge FBG sensor is shown in Figure 1c. Moreover, the long-gauge sensors can be connected in a series to make an FBG sensor array (Figure 1d) for distributed sensing. The above features provide the developed sensor with the advantage of measuring both local and global information about the structure. Therefore, a new set of strain modal identification theory can be developed, and corresponding methods will be studied for damage detection and performance evaluation for structures; this article investigates a novel method for detecting structural corrosion and quantifying the severity by processing monitoring data recorded by FBG sensors in modal space.

3. Two-Level Corrosion Detection Strategy Based on FBG Sensors

Long-gauge fiber optic strain sensors offer an excellent opportunity for developing a macrostrain modal identification theory and accomplishing corrosion detection by vibration test. Based on the macrostrain response measured by long-gauge FBG sensors, the theory of macrostrain flexibility identification is investigated. As the identified strain flexibility directly relates to the structural stiffness, it has clear engineering application potential for the detection of corrosion damage and further structural long-term performance evaluation.

3.1. Framework of the Proposed Method

The proposed two-level corrosion detection method using strain flexibility is illustrated in Figure 2. An impact test is performed on the intact structure. Based on the impacting force and dynamic macrostrain measurements, the strain frequency response function (FRF), strain modal parameter, and scaling factor will be estimated to calculate the macrostrain flexibility in the condition in which the mass is unknown. Other than the undamaged structure, the dynamic macrostrain data for a damaged structure via an impact test is also processed for structural modal identification. Based on these two strain flexibility measures, a flexibility-based damage index can be obtained by extracting the diagonal element of the flexibility difference. Then, the corrosion damage can be located for Level 1 with a space resolution of the gauge lengths of the fiber optic sensors. The strain flexibility estimated from the recorded data can be directly utilized, with no need to construct an analytical model. Here, the "space resolution" represents the region one sensor covers. Once the damage is localized, the number of unknown parameters to be identified can be sharply reduced in the damaged domain, in which the strain flexibility-based sensitivity function will be constructed for Level 2 damage quantification. This step-by-step procedure allows for damage quantification by reducing the significant number of unknown parameters in sensitivity equations.

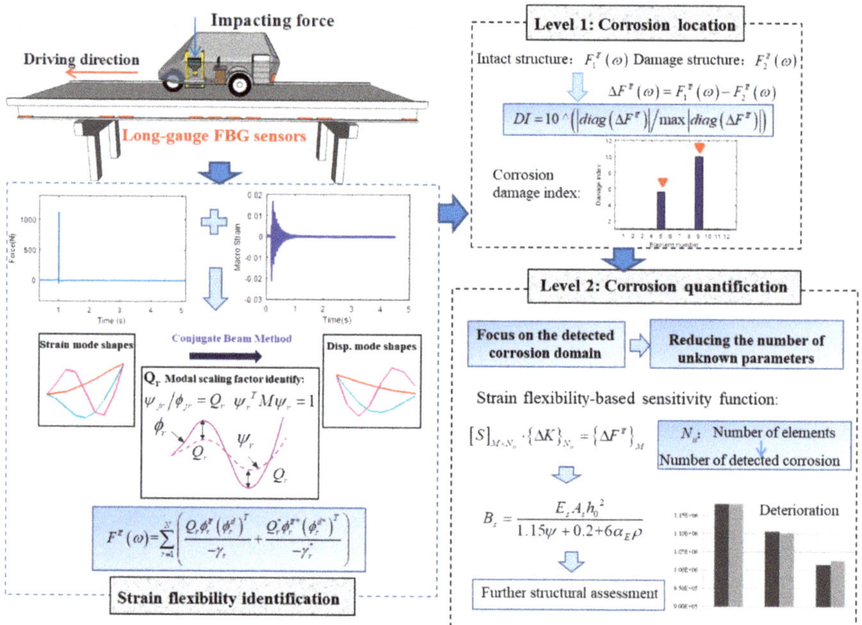

Figure 2. Framework of the proposed method.

3.2. Theoretical Basis of the Proposed Method

3.2.1. Strain Flexibility Identification

The strain flexibility identified by long-gauge FBG sensors has been illustrated in previous research by the authors. In the work of [25], the identification method is briefly introduced as follows. When a force, f_p, is applied at node p, the corresponding long-gauge strain FRF of the mth element is as follows:

$$H_{mp}^{\bar\varepsilon}(\omega) = \frac{\bar\varepsilon_m(\omega)}{f_p(\omega)} = \frac{h_m}{L_m} \cdot \frac{(\theta_i(\omega) - \theta_j(\omega))}{f_p(\omega)}$$
$$= \eta_m \cdot \left(H_{ip}^d(\omega) - H_{jp}^d(\omega) \right) \quad , \tag{1}$$

where $\bar\varepsilon_m(\omega)$ and $f_p(\omega)$ are spectra of the long-gauge strain $\bar\varepsilon_m(t)$ and force $f_p(t)$, i and j are two nodes of the mth element, and θ_i and θ_j represent the rotational displacements of the i and j nodes. $H_{ip}^d(\omega)$

and $H^d_{jp}(\omega)$ are the displacement FRF in the rotation direction. $\eta_m = h_m/L_m$, h_m is the distance from the element bottom surface to the beam neural axis, and L_m is the element length.

A complex mode indicator function (CMIF) method is utilized to identify structural modal parameters. The singular value decomposition (SVD) technique is applied to Equation (1), and it is derived that

$$H^{\bar{\varepsilon}}(\omega_k) = U^{\bar{\varepsilon}}S^{\bar{\varepsilon}}(V^{\bar{\varepsilon}})^T = \eta_m(U_L S_L V_L^T - U_R S_R V_R^T), \tag{2}$$

where $U^{\bar{\varepsilon}} \in \Re^{N_0 \times N_0}$ is the left orthogonal matrix, which relates to the long-gauge strain mode shape; $V \in \Re^{N_i \times N_i}$ is the right orthogonal matrix, V_L and V_R are the same because they consist of information about the displacement modal participation factors corresponding to the same impacting locations; $S \in \Re^{N_0 \times N_i}$ is the singular value matrix, and S_L and S_R are also the same because they consist of the information concerning frequency and damping ratio for the same structure. U_L is the singular matrix including the information of the mode shapes related to the left nodes of all the elements, and U_R is for the right node of all the elements. Thus, they are different. Equation (2) can be rewritten as

$$H^{\bar{\varepsilon}}(\omega_k) = \eta_m(U_L - U_R)SV^T = U^{\bar{\varepsilon}}SV^T. \tag{3}$$

It is also known that the long-gauge strain FRF can be written in the following format:

$$H^{\bar{\varepsilon}}(\omega_k) = \phi^{\bar{\varepsilon}}\left[\frac{1}{j\omega_k - \gamma_r}\right]L^T, \tag{4}$$

where $\phi^{\bar{\varepsilon}}$ is the strain mode shape, $\left[\frac{1}{j\omega_k - \gamma_r}\right] = \begin{bmatrix} \frac{1}{j\omega_k - \gamma_1} & \cdots & 0 \\ \vdots & \ddots & \vdots \\ 0 & \cdots & \frac{1}{j\omega_k - \gamma_N^*} \end{bmatrix}$, $\gamma_r = -\xi_r\omega_r + j\omega_r\sqrt{1 - \xi_r^2}$,

and γ_r^* is the conjugate of γ_r. ω_r is the structural frequency, ξ_r is the damping ratio, and L is the modal participation matrix in which $L_r = Q_r \cdot \phi_{r,drv}$ for the rth mode, $\phi_{r,drv}$ is the mode shape vector of the driving point, and Q_r is the modal scaling factor.

It can be seen that Equation (3) is the real mode form of the strain FRF while Equation (4) is the complex mode form. Basically, they are the same and, thus, the structural modal parameters including natural frequencies, damping ratios, and mode shapes can be identified from Equation (3). The modal scaling factor Q_r can be solved from the least squares estimation formulation as follows:

$$\frac{1}{Q_r} = C_{1r}C_{2r}\left\{\begin{array}{c} eH^{\bar{\varepsilon}}(\omega_1)_r \\ eH^{\bar{\varepsilon}}(\omega_2)_r \\ \vdots \\ eH^{\bar{\varepsilon}}(\omega_k)_r \end{array}\right\}^{+}\left\{\begin{array}{c} 1/(j\omega_1 - \lambda_r) \\ 1/(j\omega_2 - \lambda_r) \\ \vdots \\ 1/(j\omega_k - \lambda_r) \end{array}\right\}, \tag{5}$$

where $C_{1r} = (U_r^{\bar{\varepsilon}})^T\phi_r^{\bar{\varepsilon}}$; $C_{2r} = (\phi_{r,drv}^d)^T V_r$; and $eH^{\bar{\varepsilon}}(\omega_k)_r = (U_r^{\bar{\varepsilon}})^T[H^{\bar{\varepsilon}}(\omega_k)]V_r$, which is the enhanced FRF of the rth model. Then, based on the modal scaling factors and basic modal parameters, the structural long-gauge strain flexibility can be obtained in Equation (6):

$$F^{\bar{\varepsilon}}(\omega) = \sum_{r=1}^{N}\left(\frac{Q_r\phi_r^{\bar{\varepsilon}}(\phi_r^d)^T}{-\gamma_r} + \frac{Q_r^*\phi_r^{\bar{\varepsilon}*}(\phi_r^{d*})^T}{-\gamma_r^*}\right), \tag{6}$$

where $\phi_r^{\bar{\varepsilon}}$ is the long-gauge strain mode shapes identified from $U^{\bar{\varepsilon}}$; ϕ_r^d is the displacement mode shapes calculated from strain mode shapes by using the improved conjugate beam approach, and the symbol * represents the complex conjugate.

Equation (6) indicates that the estimation of the long-gauge strain flexibility does not require one to know the mass of the structure. It just requires the natural frequencies ω_r, damping ratios ξ_r, identified strain modal shapes $\phi_r^{\bar{\varepsilon}}$, displacement mode shapes ϕ_r^d, and modal scaling factors Q_r to be known. For the rth mode, the relationship between the modal scaling factor and the modal mass can be derived as $Q_r = \frac{1}{2j\omega_r M_r}$. See Appendix A for more details. If $M_r = 1$, the mass-normalized mode shape can be denoted as $\psi_r^d = \alpha_r \phi_r^d = \sqrt{2j\omega_r Q_r}\phi_r^d$. Replacing the unscaled mode shape in Equation (6) with the mass normalized mode shape ψ_r^d and the corresponding strain mode shape $\psi_r^{\bar{\varepsilon}}$, the strain flexibility will be derived in another form:

$$F^{\bar{\varepsilon}}(\omega) = \sum_{r=1}^{N} \frac{\psi_r^{\bar{\varepsilon}}(\psi_r^d)^T}{\omega_r^2}. \tag{7}$$

3.2.2. Two-Level Corrosion Detection

Corrosion generally produces reduction of the stiffness of a structure, and these changes will lead to variation in strain flexibility. Thus, the difference between the strain flexibility for intact and corroded structures may be used to detect the corrosion damage in Level 1:

$$\Delta F^{\bar{\varepsilon}}(\omega) = F_1^{\bar{\varepsilon}}(\omega) - F_2^{\bar{\varepsilon}}(\omega), \tag{8}$$

where $F_1^{\bar{\varepsilon}}(\omega), F_2^{\bar{\varepsilon}}(\omega)$ are the strain flexibility of the intact and the damaged structures, respectively. To amplify the difference for easy identification, a damage index is constructed by extracting the diagonal elements of the difference matrix:

$$DI = 10\left(\frac{|\mathrm{diag}(\Delta F^{\bar{\varepsilon}})|}{\max|\mathrm{diag}(\Delta F^{\bar{\varepsilon}})|}\right) \tag{9}$$

In Level 2, the strain flexibility-based sensitivity equation is derived to quantify the damage. From Equation (7), the derivative of the strain flexibility with respect to the ith element stiffness is

$$\frac{\partial F_{mq}^{\bar{\varepsilon}}}{\partial k_i} = \sum_{r=1}^{N}\left\{-\frac{2}{\omega_r^3}\frac{\partial \omega_r}{\partial k_i}\psi_{mr}^{\bar{\varepsilon}}\psi_{qr}^d + \frac{1}{\omega_r^2}\left(\frac{\partial \psi_{mr}^{\bar{\varepsilon}}}{\partial k_i}\psi_{qr}^d + \psi_{mr}^{\bar{\varepsilon}}\frac{\partial \psi_{qr}^d}{\partial k_i}\right)\right\}. \tag{10}$$

The equilibrium equation for the undamaged structural vibration equation is

$$\left(-\omega_r^2[M] + [K]\right)\psi_r^d = \{0\}. \tag{11}$$

Applying the derivative to Equation (11) with respect to element stiffness k_i, it is derived that

$$\left(-2\omega_r\frac{\partial \omega_r}{\partial k_i}[M] - \omega_r^2\frac{\partial[M]}{\partial k_i} + \frac{\partial[K]}{\partial k_i}\right)\psi_r^d + \left(-\omega_r^2[M] + [K]\right)\frac{\partial \psi_r^d}{\partial k_i} = 0. \tag{12}$$

As the mass matrix $[M]$ is independent of k_i, $\partial[M]/\partial k_i = 0$. Multiply $\left(\psi_r^d\right)^T$ on both sides of this equation and the sensitivity coefficient of the rth natural frequency with respect to the ith element flexural stiffness k_i can be derived:

$$\frac{\partial \omega_r}{\partial k_i} = \frac{1}{2\omega_r}\left(\psi_r^d\right)^T\frac{\partial[K]}{\partial k_i}\psi_r^d, (r, i = 1, 2, \ldots, n). \tag{13}$$

Since the mode shapes of a structure are independent from each other and, thus, can form a complete mode shape space, the sensitivity coefficients of the rth mode shape can be expressed in the form of mode shape:

$$\frac{\partial \psi_r^d}{\partial k_i} = \sum_{L=1}^{N} \alpha_L \psi_L^d, \tag{14}$$

where α_L is the weight coefficient. Substitute Equation (14) into Equation (12) and multiply $\left(\psi_s^d\right)^T (s \neq r)$ on both sides. Since the mode shapes are orthogonal, two different modes will yield $\left(\psi_s^d\right)^T [M]\psi_r^d = 0$ and $\left(\psi_s^d\right)^T [K] = \omega_r^2\left(\psi_s^d\right)^T [M]$. Then, Equation (12) can be rewritten as

$$\left(\psi_s^d\right)^T \frac{\partial [K]}{\partial k_i} \psi_r^d + \alpha_s\left(\omega_s^2 - \omega_r^2\right) = 0, \tag{15}$$

solving for α_s:

$$\alpha_s = \frac{1}{\omega_r^2 - \omega_s^2}\left(\psi_s^d\right)^T \frac{\partial [K]}{\partial k_i} \psi_r^d \, (r \neq s). \tag{16}$$

When $s = r$, $\alpha_s = 0$ can be derived.

The long-gauge strain mode shape $\psi_{mr}^{\bar{\varepsilon}}$ can be written as follows:

$$\psi_{mr}^{\bar{\varepsilon}} = \frac{h_m}{L_m}\left(\psi_{or}^d - \psi_{pr}^d\right). \tag{17}$$

The derivative of the rth strain mode shape in terms of k_i can be derived from the displacement mode shape sensitivity:

$$\begin{aligned}\frac{\partial \psi_{mr}^{\bar{\varepsilon}}}{\partial k_i} &= \frac{h_m}{L_m}\left(\frac{\partial \psi_{or}^d}{\partial k_i} - \frac{\partial \psi_{pr}^d}{\partial k_i}\right) = \frac{h_m}{L_m}\left(\sum_{s=1}^{N}\alpha_s\psi_{os}^d - \sum_{s=1}^{N}\alpha_s\psi_{ps}^d\right) = \sum_{s=1}^{N}\alpha_s\left[\frac{h_m}{L_m}\left(\psi_{os}^d - \psi_{ps}^d\right)\right]\\ &= \sum_{s=1}^{N}\alpha_s\psi_{ms}^{\bar{\varepsilon}}\end{aligned} \tag{18}$$

Substituting Equations (13), (14), (16), and (18) into Equation (10) can yield the sensitivity coefficients of the strain modal flexibility:

$$\frac{\partial F_{mq}^{\bar{\varepsilon}}}{\partial k_i} = \sum_{r=1}^{N}\left\{-\frac{1}{\omega_r^4}\left(\psi_r^d\right)^T \frac{\partial [K]}{\partial k_i}\psi_r^d\psi_{mr}^{\bar{\varepsilon}}\psi_{qr}^d + \frac{1}{\omega_r^2}\left(\sum_{s=1}^{N}\alpha_s\psi_{ms}^{\bar{\varepsilon}}\right)\psi_{qr}^d + \frac{1}{\omega_r^2}\left(\sum_{s=1}^{N}\alpha_s\psi_{qs}^d\right)\psi_{mr}^{\bar{\varepsilon}}\right\}. \tag{19}$$

The strain flexibility can be expanded into a first-order Taylor's series, as below:

$$\Delta F_{mq}^{\bar{\varepsilon}} = \sum_{i=1}^{n}\frac{\partial F_{mq}^{\bar{\varepsilon}}}{\partial k_i}\cdot \Delta k_i. \tag{20}$$

The sensitivity function with respect to strain flexibility is obtained:

$$[S]_{M \times Nu}\cdot \{\Delta k\}_{Nu} = \left\{\Delta F^{\bar{\varepsilon}}\right\}_M, S_{mq} = \frac{\partial F_{mq}^{\bar{\varepsilon}}}{\partial k_i}, \tag{21}$$

where $[S]$ is the sensitivity matrix obtained from the initial structural state. Δk is the change of the element stiffness, and $\Delta F^{\bar{\varepsilon}}$ is the variation of strain flexibility. Nu is the number of parameters to be identified, and M is the number of columns for strain flexibility.

Before the corrosion damage is identified in Level 2, an original finite element (FE) model needs to be built to obtain the global stiffness matrix $[K]$, from which the sensitivity matrix $[S]$ can be determined. From Equation (21), it can be found that once the damage caused by corrosion is localized, the unknown parameters have been reduced from the entire structure to detected domain, upon which the quantification process can only focus. Thus, the convergence and reliability of the solution will achieve better results.

4. Numerical Example of a Steel Beam

As shown in Figure 3, a simply supported steel beam model is used to study the validity of the proposed method. The beam has a length of 5.76 m, which is divided into 12 elements. The material is Q235 steel, with the elasticity of modulus 206 GPa, and the unit weight is 7854 kg/m^3. Assume that 12 long-gauge FBG sensors with a gauge length of 0.48 m are mounted at the bottom of the beam to measure the macrostrain response of the structure. The numerical model is simulated in SAP2000 software and a Rayleigh damping matrix is adopted. Other than the baseline structure, three damage patterns are considered: (1) Case 1, single damage scenario for 5% stiffness loss at the fifth element; (2) Case 2, single damage scenario for 10% stiffness loss at the fifth element; (3) Case 3, multiple damage scenarios for 10% and 15% stiffness loss at the fifth and ninth elements. The corrosion damage is simulated by reducing the width of the flange. Impacting forces are applied on the fifth and eighth nodes as excitation. White noises (5%) are added into the response data to act as the observation noise.

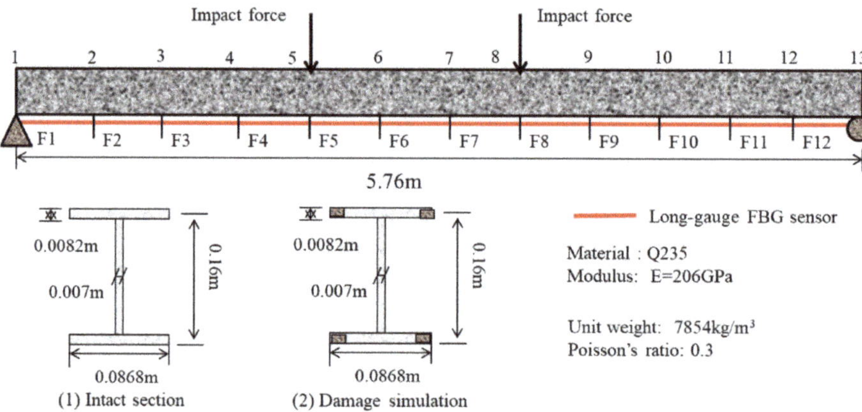

Figure 3. Configuration of the steel beam.

We take the intact structure as an example to illustrate the process of strain flexibility identification. By performing an impact test on the beam, the corresponding macrostrain responses are recorded. The applied impacting force and strain response are plotted in Figure 4a,b. Long-gauge strain FRFs are estimated from the macrostrain responses during the impact. According to Equation (3), the singular value decomposition is applied on the estimated strain FRF to obtain the singular matrix, which is plotted in Figure 4c. As shown in the figure, two curves are plotted due to the fact that two nodes were impacted during the test. Three peaks in the spectral line represent three modes that are identified. The corresponding macrostrain mode shapes and displacement mode shapes are plotted in Figure 4d,e. The natural frequencies in the first three modes are identified as 15.46, 61.2, and 135.37 Hz, and the corresponding damping ratios are identified as 0.5%, 1.1%, and 0.56%, respectively. According to Equation (5), the modal scaling factors can be obtained. Thus, based on the identified basic modal parameters, macrostrain mode shapes, and displacement mode shapes, the macrostrain flexibility is estimated from Equation (6) and plotted in Figure 4f. To verify the accuracy of the estimated strain flexibility, static strains measured from the corresponding static test are plotted for comparison (Figure 4g). The static test is performed by placing four static forces with a magnitude of 100 N each on nodes 3, 5, 8, and 10. It can be seen in each case that the predicted strain has good agreement with the measured strain.

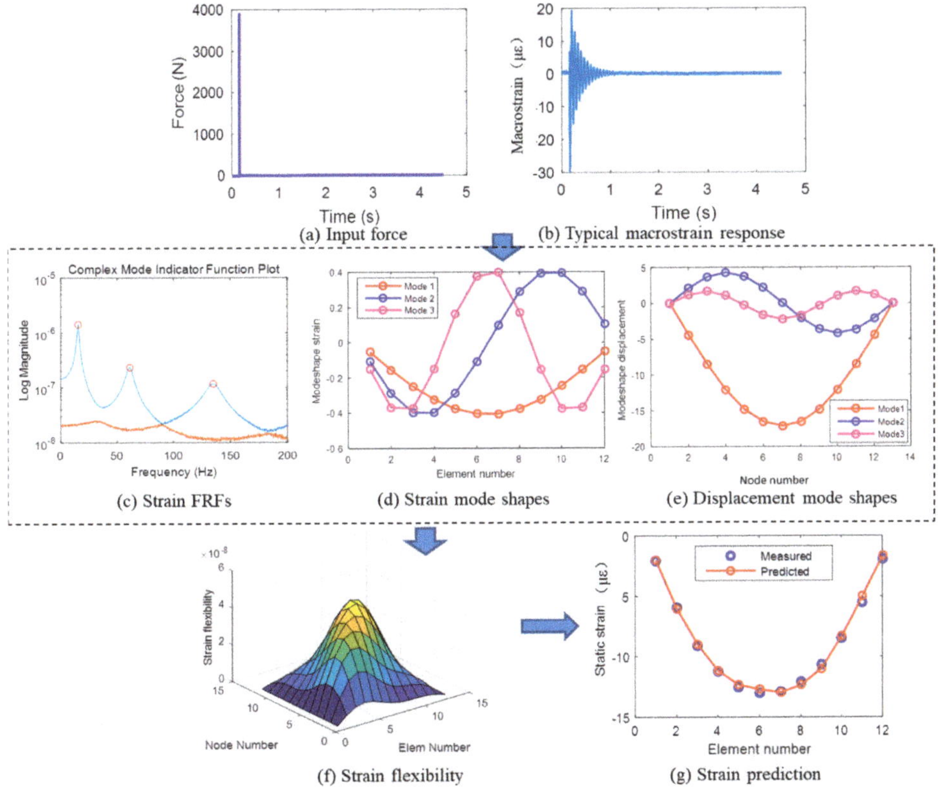

Figure 4. Process of strain flexibility identification for intact structure.

Similar procedures are performed on corroded structures to obtain the corresponding structural strain flexibility, which has been plotted in Figure 5. Structural deformation predicted by the strain flexibility is also plotted to compare with the static test results provided by SAP2000. As shown in the figure, there is a good agreement between the predicted strain from the identified strain flexibility and the measured strain from the static test, demonstrating the accuracy of the strain flexibility identification. Based on the strain flexibility, the difference matrices are calculated, and the diagonal elements are extracted to locate the corrosion damage. The results are plotted in Figure 5a–c for three damage cases. From the figures, it can be found that the location of structural damage can be clearly detected. Therefore, strain flexibility difference is a good indicator for locating corrosion and is suitable for assessing singular and multiple incidents of damage.

Once the corrosion damage is located, the corrosion quantification can be focused on the detected domain and, hence, the number of structural parameters to be identified is greatly reduced. Here, the sectional flexural rigidity of element 5 for cases 1 and 2, and sectional flexural rigidity of element 5 and 9 for case 3, are taken as the objective parameters for damage quantification. According to the scaling factor Q_r identified from Equation (5), the coefficient α_r of the first three modes is calculated as 0.0003, 0.0012, and 0.0022. Thus, the mass normalized shape can be obtained, and the sensitivity coefficients of frequency, displacement mode, and strain mode that correspond to the detected element stiffness can be calculated from Equations (13), (14), and (18). Then, the stiffness reduction will be quantified from Equation (21), and the results are listed in Table 1 for comparison against the theoretical value. It can be found that the damage is quantified effectively using the strain flexibility-based sensitivity function, and the errors are all within an acceptable range. Figure 6 describes the result in a more intuitive way, and the identified value agrees with the theoretical value.

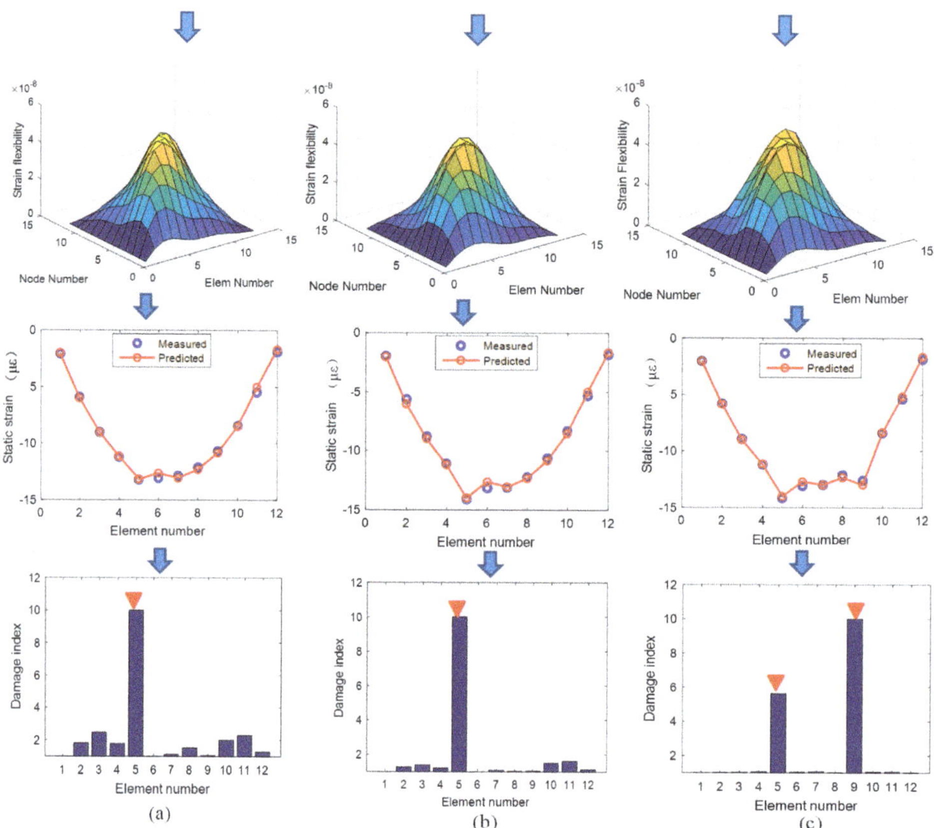

Figure 5. Damage location for Level 1. (**a**) Case 1; (**b**) Case 2; (**c**) Case 3.

Table 1. Damage quantification results.

Damage Location	Element 5			Element 9		
Damage Quantification	Theoretical Value (N·m²)	Experimental Value (N·m²)	Error (%)	Theoretical Value (N·m²)	Experimental Value (N·m²)	Error (%)
Case 1	1.978×10^6	1.955×10^6	1.16	-	-	-
Case 2	1.858×10^6	1.876×10^6	0.97	-	-	-
Case 3	1.858×10^6	1.894×10^6	1.93	1.759×10^6	1.868×10^6	6.2

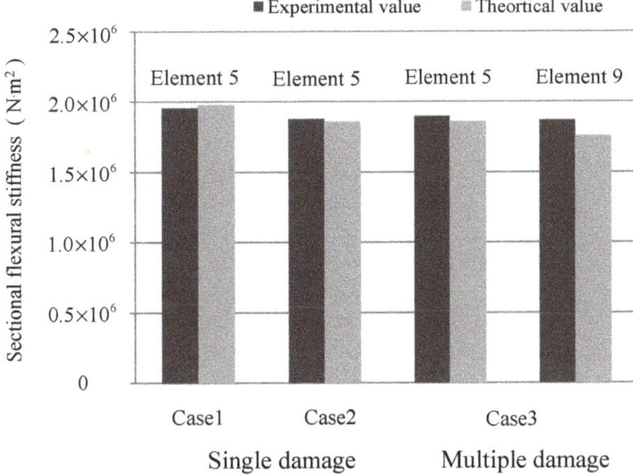

Figure 6. Damage quantification for three cases.

5. Experimental Verification through a Reinforced Concrete (RC) Beam

In Section 4, the applicability of the proposed method was demonstrated by a numerical example. This section is devoted to verifying the robustness of the long-gauge FBG sensors and the effectiveness of the proposed method by an experimental study of a simply supported reinforced concrete (RC) beam.

5.1. Description of the Experimental Setup

The experimental beam has a total size of 2000 mm × 150 mm × 200 mm, and is reinforced with two deformed bars with a diameter of 16 mm and two compression bars with a diameter of 12 mm. The configuration is illustrated in Figure 7. It was been divided into 17 elements, and a visible crack runs through the twelfth element, as shown in Figure 7c.

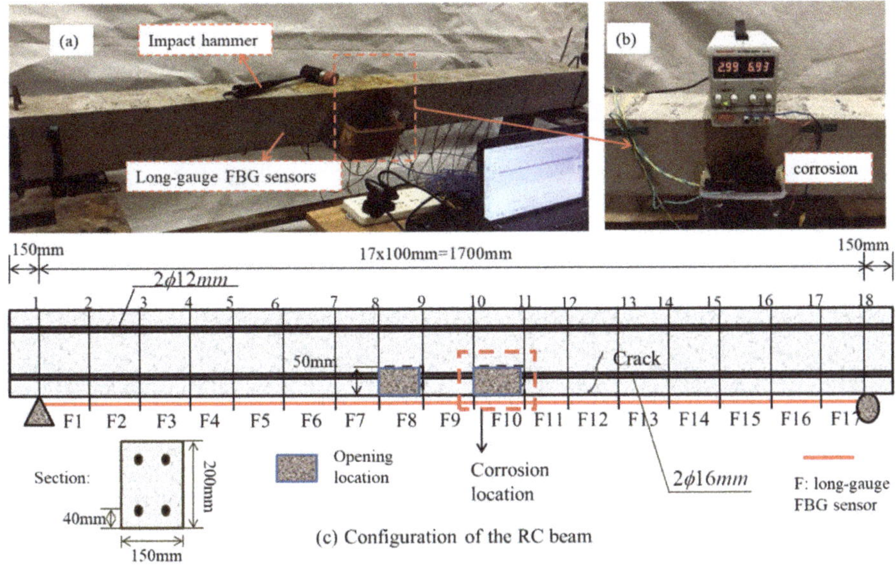

Figure 7. Illustration for the reinforced concrete (RC) beam; (**a**) The tested RC beam; (**b**) Corrosion setup; (**c**) Configuration of the RC beam.

Corrosion is conducted via the accelerated corrosion technique. Seventeen long-gauge FBG sensors with a gauge length of 100 mm were mounted on the bottom of the beam surface, as shown in Figure 7. The long-gauge FBG sensors were fixed using reinforcement glue on the designed locations. All sensors were pre-tensioned during the fabrication to ensure the accurate signals of low-level strains. Before installing the long-gauge FBG sensors, the concrete surface was firstly cleaned by using a sander to remove the cement cover, then alcohol was applied to scrub the concrete surface to make sure that the FBG sensors can be fixed on the concrete surface sustainably by using epoxy primer.

5.2. Accelerated Corrosion Procedure

5.2.1. Calibration Test

To accelerate the corrosion process, the electrochemical corrosion method was applied to a steel bar similar to those used for reinforcing the RC beam. A corrosion calibration test was conducted to determine the theoretical corrosion amount in the specified time. The setup is illustrated in Figure 8a. It can be seen from the figure that the steel bar was immersed in 10% sodium chloride (NaCl) solution to create a path for the current between the anode and cathode terminals. A power supply was set to create a stable 3 A current output. To eliminate the effects of sample length on corrosion, two samples with different lengths (325 and 1000 mm) were tested (Figure 8b–c). However, the length of the corroded part is 78 mm for both samples.

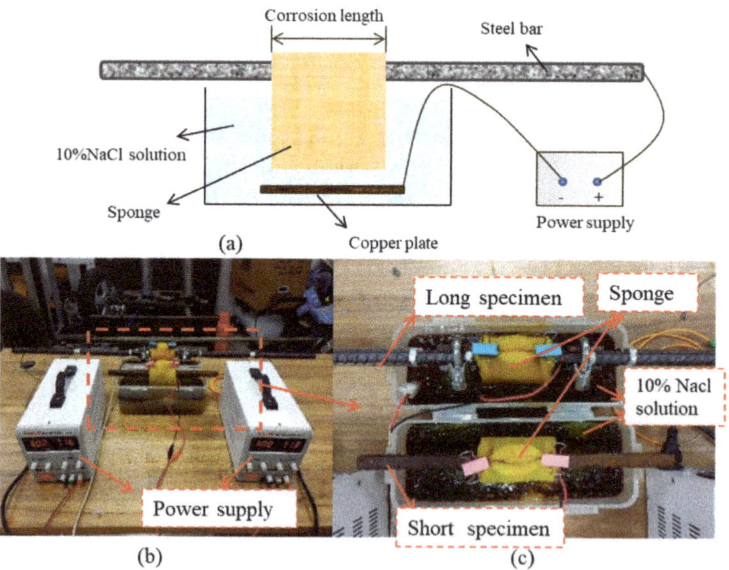

Figure 8. Corrosion calibration test; (**a**) Illustration of the accelerated corrosion; (**b**) Calibration test; (**c**) Details of two specimen.

According to the Faraday's first law of electrolysis, the weight loss of steel within a specified time can be calculated as follows:

$$\Delta W = \frac{\omega \times i \times t}{F},$$ (22)

where ΔW is the weight loss of the steel, $\omega = M/n$ is the constant, n is the number of electrons exchanged during the corrosion process ($n = 2$ for Fe^{2+}), and M is the molar mass of iron (i.e., M = 5584 g/mol). i is the applied current in A, t is the corrosion time, and F is Faraday's constant (96,487 C/mol).

It can be concluded from Equation (22) that the corrosion of the steel bar depends only on the current intensity i. Therefore, the relationship between corrosion weightless rates and time derived from calibration tests is similar to the corrosion that occurs in the RC beam if the electrochemical corrosion can be controlled under the same conditions. To ensure the sustainability of the corrosion progress, the corrosion solution was changed every two hours, the rust adhering to the corroded section was carefully cleaned, then the quality of the remaining bar was measured to obtain the weight loss that occurred during corrosion. The corrosion process lasted for 8 h, and the residual mass of each sample was measured every 2 h; the values are recorded in Table 2. The varying rate of the steel section area can be calculated as follows:

$$r = \frac{\Delta A}{A} = \frac{\Delta W}{W}, \tag{23}$$

where ΔA, ΔW are the loss in area and mass for corroded part; and A, W are the area and mass for the corroded part in the initial state. For long specimens, $W = 78/1000 \times 1517.07 = 118.33$ g; for short specimens, $W = 78/325 \times 454.5 = 109.08$ g.

Table 2. Corrosion calibration result with 3 A current.

Time (h)	Short Specimen (325 mm)			Long Specimen (1000 mm)			Mean of the Ratio (%)
	Residual Mass (g)	Lost Mass (g)	Weight Loss Ratio (%)	Residual Mass (g)	Lost Mass (g)	Weight Loss Ratio (%)	
0	454.5	0	0	1517.07	0	0	0
2	448	6.5	5.96	1510.65	6.42	5.44	5.7
4	441.5	13	11.92	1504.37	12.7	10.76	11.34
6	435	19.5	17.88	1497.89	19.18	16.25	17.1
8	429	25.5	23.38	1491.55	25.52	21.62	22.5

The relationships of the average weight loss ratio with time are described in Table 2. It can be seen that the corrosion level increases linearly with time and the whole length of the steel bar has no effect on the corrosion rate. Therefore, the calibration result can be considered as the theoretical value. In addition, this test confirmed that certain packaging can effectively protect the FBG sensors from electric effects, long-term salt attacks, and temperature variations experienced during the corrosion process. As the Young's modulus of concrete and steel bars is needed to quantify the damage, cube compression and steel tensile tests were conducted to determine the modulus.

5.2.2. Corrosion Setup

To make the beam easier to corrode, two small openings with a size of 100 mm $\times$ 30 mm $\times$ 50 mm are introduced on the bottom surface of the RC beam, and corrosion was applied to one of the main reinforcements through the openings.

The corrosion was introduced on element 10. A plastic bottle with a height of 80 mm, as shown in Figure 7b, was fixed to the RC beam surface with epoxy resin. Then, 10% sodium chloride (NaCl) solution was poured into the plastic bottle to a height of 60 mm to guarantee that the exposed deformed bar was immersed in the solution. The corrosion solution was changed every two hours to ensure the sustainability of the corrosion progress. Two corrosion cases were considered. The first case lasted for 8 h, and the second case lasted for 16 h. They are defined as case 1 and case 2 in the following section.

5.3. Two Level Strategy for Corrosion Damage Quantification

5.3.1. Level 1: Corrosion Damage Localization

Impact testing was performed on the RC beam to obtain the macrostrain flexibility. The impacting force was applied on the nodes from left to right by using the PCB model 086D20 short-sledge impulse hammer and then the corresponding macrostrain responses were measured by the SM130 optical sensing interrogator. The NI PXIe-1082 data acquisition system was used for impacting force measurements and the sampling frequencies are both set as 1000 Hz. The impacting force and macrostrain response were recorded to estimate the strain FRF, and then the basic modal parameters were identified using the CMIF method. For the intact structure, the first natural frequency identified from the impact test data was 56.58 Hz, and the modal scaling factor was 1.45×10^{-5}. The strain mode shapes were identified from Equation (3) and then the improved conjugated beam method was applied to obtain the corresponding displacement mode shapes. As the higher mode frequency of the RC beam was large, its contribution to the flexibility may be neglected. The exact strain flexibility may be obtained by considering only the first mode.

After the basic modal parameters and modal scaling factors are identified in each condition, the structural strain flexibility was estimated. Figure 9 plots the strain flexibility for the intact and corroded structures for case 1 and case 2. A static test was also performed by placing two mass blocks with weights of 61.2 kg and 61.3 kg, respectively, on node 6 and node 13, to demonstrate the accuracy of the strain flexibility calculated from the impact test. For the intact and corroded structures, the identified strain from the flexibility and measured static strain are plotted in Figure 9 for comparison. Owing to the presence of an opening, the strain of elements 8 and 10 increased sharply. Since a crack existed in element 12, the strain was also very large. As shown in the figure, the predicted strain from the flexibility is in accordance with the measured strain from the long-gauge FBG sensors, which demonstrates the accuracy of the estimated strain flexibility. To eliminate the effect caused by temperature, impact and static tests for each condition were all performed at the same time of day.

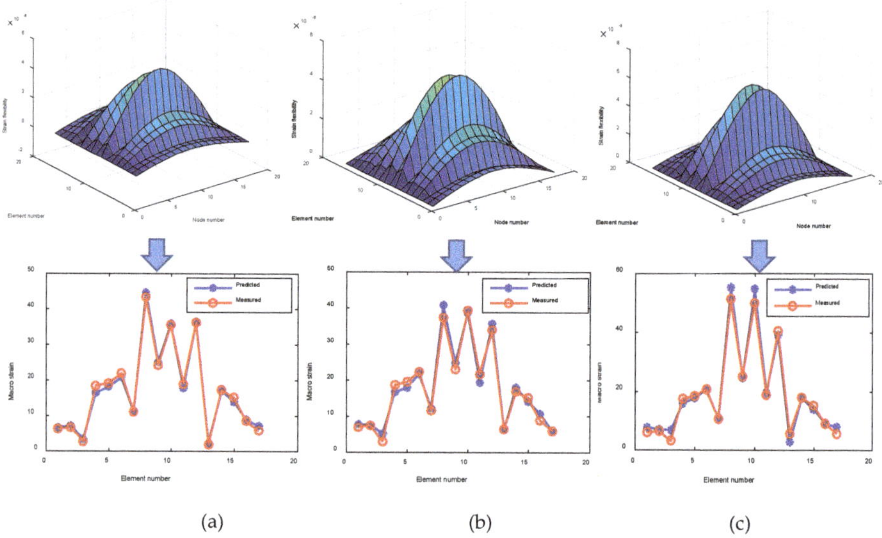

Figure 9. Strain flexibility and static strain prediction for intact structures. (**a**) Intact structure; (**b**) Case 1; (**c**) Case 2.

5.3.2. Level 2: Corrosion Damage Quantification

After the strain flexibility for each condition was obtained, the damage index was calculated by Equation (8) and plotted in Figure 10. In case 1, after 8 h of corrosion, the cross-sectional area of the steel bar was reduced by 22% according to the calibration test. As the contribution of the steel bar to the stiffness of the entire section was very small, a 22% steel area reduction only caused a 7% decrease in cross-sectional stiffness. Although the damage was small, the index located the damage successfully, which demonstrates the robustness of the FBG sensors for perceiving minor damage.

In case 2, according to Faraday's law, an area reduction of about 45% occurred on element 10. As the corrosion degree increased, the damage became easier to locate.

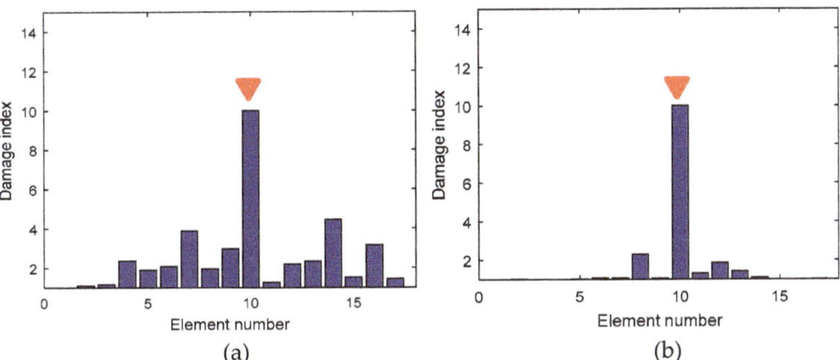

Figure 10. Corrosion detection for Level 1. (**a**) Damage index for Case 1; (**b**) Damage index for Case 2.

Once the damage location was detected in Level 1, the damage severity in terms of stiffness reduction for cross-sections could be quantified according to the Equation (21). The theoretical value for the structural sectional stiffness of the rectangular section under a short-term load can be calculated as follows [26]:

$$B_s = \frac{E_s A_s h_0^2}{1.15\psi + 0.2 + 6\alpha_E \rho},$$ (24)

where $\alpha_E = E_s/E_c$, $\rho = A_s/bh_0$; B_s is the sectional flexural rigidity; and E_s and A_s are the elastic modulus and section areas of the reinforcement, respectively. E_c is the elastic modulus for concrete. h_0 is the effective height of the reinforced concrete section, and $\alpha_E = E_s/E_c$ is the elastic modulus ratio. ψ is the non-uniformity coefficient of strain subjected to longitudinal tension. Comparison between the theoretical value and the experimental results of the stiffness and steel areas are plotted in Figure 11a,b. It can be found that the experimental values agree with the theoretical values.

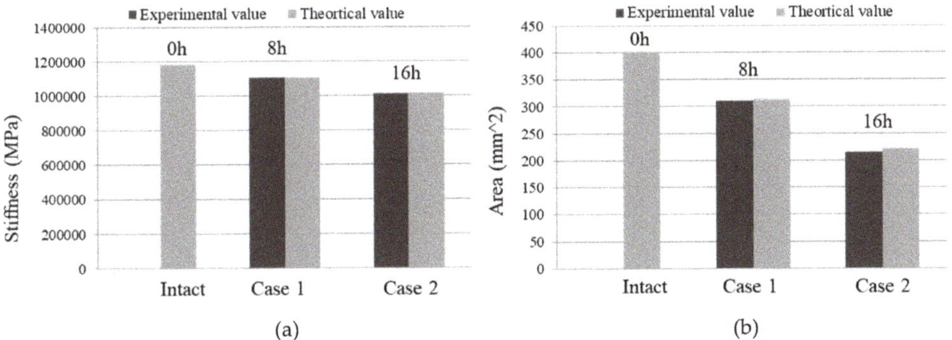

Figure 11. Corrosion quantification for Level 2. (**a**) Identified stiffness; (**b**) Identified steel area.

6. Conclusions

The overall conclusions of this research are that the distributed long-gauge FBG sensor is a potentially promising method for effective and accurate corrosion damage detection and further structural long-term performance evaluation.

(1) Based on the long-gauge FBG strain sensors, a new kind of corrosion detection methodology via impact test is proposed and demonstrated.

(2) The original contribution of this paper is the development of a step-by-step strategy that helps to locate and quantify corrosion damage by using a long-gauge FBG sensor; a solid theoretical basis has been developed to guarantee that this sensor will detect corrosion accurately.

(3) The proposed two-level corrosion detection methodology presents a distinct advantage in that locating Level 1 damage significantly reduces the number of unknown parameters in the sensitivity equations and increases the success of Level 2 corrosion quantification.

Both numerical and experimental examples have been conducted, and they verify the robustness of the novel long-gauge FBG sensors, as well as of the effectiveness of the proposed method for corrosion detection of in-service structures. The proposed two-level corrosion detection strategy that employs distributed long-gauge FBG sensors has a great application potential in civil infrastructural maintenance. In addition, an assumption that the steel of the damaged structure is still in linear elastic state is needed due to the fact that method of the strain flexibility identification is studied based on the theory of elasticity. Taking steel plasticity into consideration is worthy of further study in future works.

7. Patents

The proposed technique for corrosion detection is patent pending.

Author Contributions: The individual contributions of authors are as follows. Y.C. initiated the research and designed the experiments, C.Z. performed the experiments, J.Z. directed the research. The experimental results analysis and performance discussion were performed by Y.C., C.Z. and J.Z., Z.W. designed the long-gauge FBG sensor. The paper was written by Y.C. and revised by J.Z.

Funding: The research presented was financially supported by the National Key R&D Program of China (No.:2018YFC0705601), Jiangsu Distinguished Young Scholars Fund (No.: BK20160002) and National Science Foundation of China (Grant No.: 51578139, 51778134).

Conflicts of Interest: The authors declare no conflict of interest.

Appendix A. Relationship between Modal Mass and Modal Scaling Factor

According to the complex modal theory, the long-gauge strain FRFs can be expanded as

$$H^{\bar{\varepsilon}}(\omega) = \sum_{r=1}^{N} \left(\frac{Q_r \phi_r^{\bar{\varepsilon}} \left(\phi_r^d \right)^T}{j\omega - \gamma_r} + \frac{Q_r^* \phi_r^{\bar{\varepsilon}*} \left(\phi_r^{d*} \right)^T}{j\omega - \gamma_r^*} \right), \tag{A1}$$

where Q_r is the rth scaling factor, $\phi_r^{\bar{\varepsilon}}$ and ϕ_r^d are the rth complex strain mode shape and displacement mode shape, and * denotes the conjugate.

According to real modal theory, the long-gauge strain FRFs is rewritten as below:

$$H^{\bar{\varepsilon}}(\omega) = \sum_{r=1}^{N} \frac{\varphi_r^{\bar{\varepsilon}} \left(\varphi_r^d \right)^T / M_r}{(j\omega)^2 + (j\omega)2\zeta_r\omega_r + \omega_r^2}, \tag{A2}$$

where M_r is the rth modal mass, $\varphi_r^{\bar{\varepsilon}}$ and φ_r^d are the rth real strain mode shape and displacement mode shape.

There is no essential difference between Equation (A1) and Equation (A2) as they are two different forms of long-gauge strain FRFs, thus, the following equation can be obtained:

$$H^{\bar{\varepsilon}}(\omega) = \sum_{r=1}^{N} \left(\frac{Q_r \phi_r^{\bar{\varepsilon}} \left(\phi_r^d \right)^T}{j\omega - \gamma_r} + \frac{Q_r^* \phi_r^{\bar{\varepsilon}*} \left(\phi_r^{d*} \right)^T}{j\omega - \gamma_r^*} \right) = \sum_{r=1}^{N} \frac{\varphi_r^{\bar{\varepsilon}} \left(\varphi_r^d \right)^T / M_r}{(j\omega)^2 + (j\omega)2\zeta_r\omega_r + \omega_r^2}. \tag{A3}$$

For the rth mode, the same contribution is derived from the two representations:

$$\frac{Q_r\phi_r^{\bar{\varepsilon}}\left(\phi_r^d\right)^T}{j\omega-\gamma_r}+\frac{Q_r^*\phi_r^{\bar{\varepsilon}*}\left(\phi_r^{d*}\right)^T}{j\omega-\gamma_r^*}=\frac{\varphi_r^{\bar{\varepsilon}}\left(\varphi_r^d\right)^T/M_r}{(j\omega)^2+(j\omega)2\xi_r\omega_r+\omega_r^2}. \tag{A4}$$

The complex mode shape can be normalized as real mode shape when the structural damping matrix can be orthogonalized, and it is derived that

$$\phi_r^d=\phi_r^{d*}=\varphi_r^d, \tag{A5}$$

$$\phi_r^{\bar{\varepsilon}}=\phi_r^{\bar{\varepsilon}*}=\varphi_r^{\bar{\varepsilon}}. \tag{A6}$$

Substituting Equation (A5) and (A6) into Equation (A4), the following equation is derived:

$$\frac{Q_r\phi_r^{\bar{\varepsilon}}\left(\phi_r^d\right)^T}{j\omega-\gamma_r}+\frac{Q_r^*\phi_r^{\bar{\varepsilon}}\left(\phi_r^d\right)^T}{j\omega-\gamma_r^*}=\frac{\phi_r^{\bar{\varepsilon}}\left(\phi_r^d\right)^T/M_r}{(j\omega)^2+(j\omega)2\xi_r\omega_r+\omega_r^2}, \tag{A7}$$

where $\phi_r^{\bar{\varepsilon}}\left(\phi_r^d\right)^T$ is a matrix and the coefficients must be equal if the matrices on both sides of the above equation are equal, thus, it is derived at any frequency ω:

$$\frac{Q_r}{j\omega-\gamma_r}+\frac{Q_r^*}{j\omega-\gamma_r^*}=\frac{1/M_r}{(j\omega)^2+(j\omega)2\xi_r\omega_r+\omega_r^2}, \tag{A8}$$

when $\omega=0$ and $\omega=\omega_r$, the following equations can be obtained:

$$\begin{cases}\frac{Q_r}{-\lambda_r}+\frac{Q_r^*}{-\lambda_r^*}=\frac{1/M_r}{\omega_r^2}\\[1mm]\frac{Q_r}{j\omega_r-\lambda_r}+\frac{Q_r^*}{j\omega_r-\lambda_r^*}=\frac{1/M_r}{(j\omega_r)^2+(j\omega_r)2\xi_r\omega_r+\omega_r^2}\end{cases}. \tag{A9}$$

The scaling factors Q_r and Q_r^* can be solved as

$$\begin{cases}Q_r=\dfrac{1}{j2M_r\omega_r\sqrt{1-\xi_r^2}}\\[2mm]Q_r^*=\dfrac{1}{-j2M_r\omega_r\sqrt{1-\xi_r^2}}\end{cases}. \tag{A10}$$

Usually, the damping ratio for civil structure is small so that $\sqrt{1-\xi_r^2}\approx1$. Therefore, the relationship between modal mass and scaling factor can be derived as

$$\begin{cases}Q_r=\dfrac{1}{j2M_r\omega_r}\\[2mm]Q_r^*=\dfrac{1}{-j2M_r\omega_r}\end{cases}. \tag{A11}$$

References

1. Li, Z.; Jin, Z.Q.; Zhao, T.J.; Wang, P.G.; Li, Z.J.; Xiong, C.S.; Zhang, K.L. Use of a novel electro-magnetic apparatus to monitor corrosion of reinforced bar in concrete. *Sens. Actuators A* **2019**, *286*, 14–27. [CrossRef]
2. Hong, S.X.; Wiggenhauser, H.; Helmerich, R.; Dong, B.Q. Long-term monitoring of reinforcement corrosion in concrete using ground penetrating radar. *Corros. Sci.* **2017**, *114*, 123–132. [CrossRef]
3. Huang, Y.; Yang, L.J.; Xu, Y.Z.; Cao, Y.Z.; Song, S.D. A novel system for corrosion protection of reinforced steels in the underwater zone. *Corros. Eng. Sci. Technol.* **2016**, *51*, 566–572. [CrossRef]
4. Lu, Y.Y.; Zhang, J.R.; Li, Z.J.; Dong, B.Q. Corrosion monitoring of reinforced concrete beam using embedded cement-based piezoelectric sensor. *Mag. Concr. Res.* **2013**, *65*, 1265–1276. [CrossRef]
5. Zhang, J.R.; Liu, C.; Sun, M.; Li, Z.J. An innovative corrosion evaluation technique for reinforced concrete structures using magnetic sensors. *Constr. Build. Mater.* **2017**, *135*, 68–75. [CrossRef]

6. Sunny, A.I.; Tian, G.Y.; Zhang, J.; Pal, M. Low frequency (LF) RFID sensors and selective transient feature extraction for corrosion characterization. *Sens. Actuators A Phys.* **2016**, *241*, 34–43. [CrossRef]

7. Bertolini, L.; Elsener, B.; Pedeferri, P.; Redaelli, E.; Polder, R.B. *Corrosion of Steel in Concrete—Prevention, Diagnosis and Repair*, 2nd ed.; Wiley: Hoboken, NJ, USA, 2014.

8. Li, W.J.; Xu, C.H.; Ho, S.C.M.; Wang, B.; Song, G.B. Monitoring concrete deterioration due to reinforcement corrosion by integrating acoustic emission and FBG strain measurements. *Sensors* **2017**, *17*, 657. [CrossRef] [PubMed]

9. Mao, J.H.; Chen, J.Y.; Cui, L.; Jin, W.L.; Xu, C.; He, Y. Monitoring the corrosion process of reinforced concrete using BOTBA and FBG sensors. *Sensors* **2015**, *15*, 8866–8883. [CrossRef] [PubMed]

10. Zhao, X.; Gong, P.; Qiao, G.; Lu, J.; Lv, X.; Ou, J. Brillouin corrosion expansion sensors for steel reinforced concrete structures using a fiber optic coil winding method. *Sensors* **2011**, *11*, 10798–10819. [CrossRef]

11. Lee, J.R.; Yun, C.Y.; Yoon, D.J. A structural corrosion-monitoring sensor based on a pair of prestrained fiber Bragg gratings. *Meas. Sci. Technol.* **2009**, *21*, 017002. [CrossRef]

12. Zheng, Z.; Sun, X.; Lei, Y. Monitoring corrosion of reinforcement concrete structures via FBG sensors. *Front. Mech. Eng. China* **2009**, *4*, 316–319.

13. Tan, C.H.; Shee, Y.G.; Yap, B.K.; Adikan, F.M. Fiber Bragg grating based sensing system: Early corrosion detection for structural health monitoring. *Sens. Actuators. A* **2016**, *246*, 123–128. [CrossRef]

14. Gurpreet, K.; Kaler, R.S.; Naveen, K. Experiment on a highly sensitive fiber Bragg grating sensor to monitor strain and corrosion in civil structures. *J. Opt. Technol.* **2018**, *85*, 36–41.

15. Magne, S.; Alvarez, S.A.; Rougeault, S. Distributed corrosion detection using dedicated OFS-based steel rebar within reinforced concrete structures by OFDR. In Proceedings of the 9th European Workshop on Structural Health Monitoring (EWSHM), Manchester, UK, 10–13 July 2018.

16. Zhang, J.; Cheng, Y.Y.; Xia, Q.; Wu, Z.S. Change localization of a steel-stringer bridge through long-gauge strain measurements. *J. Bridge Eng.* **2016**, *21*, 04015057. [CrossRef]

17. Fouad, N.; Saifeldeen, M.A.; Huang, H.; Wu, Z.S. Corrosion monitoring of flexural reinforced concrete members under service loads using distributed long-gauge carbon fiber sensors. *Struct. Health Monit.* **2018**, *17*, 379–394. [CrossRef]

18. Malumbela, G.; Moyo, P.; Alexander, M. Longitudinal strains and stiffness of RC beams under load as measures of corrosion levels. *Eng. Struct.* **2012**, *35*, 215–227. [CrossRef]

19. Wu, Z.S.; Li, S.Z. Two-level damage detection strategy based on modal parameters from distributed dynamic macro-strain measurements. *J. Intell. Mater. Syst. Struct.* **2007**, *18*, 667–676. [CrossRef]

20. Grande, E.; Imbimbo, M. A multi-stage approach for damage detection in structural systems based on flexibility. *Mech. Syst. Signal Process.* **2016**, *76–77*, 455–475. [CrossRef]

21. Cao, M.S.; Radzienski, M.; Xu, W.; Ostachowicz, W. Identification of multiple damage in beams based on robust curvature mode shapes. *Mech. Syst. Signal Process.* **2014**, *46*, 468–480. [CrossRef]

22. Zhao, J.; DeWolf, J.T. Sensitivity study for vibrational parameters used in damage detection. *J. Struct. Eng.* **1999**, *125*, 410–416. [CrossRef]

23. Perera, R.; Ruiz, A.; Manzano, C. An evolutionary multiobjective framework for structural damage localization and quantification. *Eng. Struct.* **2007**, *29*, 2540–2550. [CrossRef]

24. Zhang, J.; Xia, Q.; Cheng, Y.Y.; Wu, Z.S. Strain flexibility identification of bridges from long-gauge strain measurements. *Mech. Syst. Signal Process.* **2015**, *62–63*, 272–283. [CrossRef]

25. Zhang, J.; Zhang, Q.Q.; Guo, S.L.; Xu, D.W.; Wu, Z.S. Structural identification of short/middle span bridges by rapid impact testing: Theory and verification. *Smart Mater. Struct.* **2015**, *24*, 065020. [CrossRef]

26. Cheng, W.R.; Wang, T.C.; Yan, D.H. *Design Theory for Concrete Structure*, 4th ed.; China Architecture & Building Press: Beijing, China, 2008.

Article

Twisted Dual-Cycle Fiber Optic Bending Loss Characteristics for Strain Measurement

Sang-Jin Choi [1], Seong-Yong Jeong [2], Changhyun Lee [1], Kwon Gyu Park [1,*] and Jae-Kyung Pan [2,*]

[1] Petroleum & Marine Division, Korea Institute of Geoscience and Mineral Resources, 124 Gwahak-ro, Yuseong-gu, Daejeon 34132, Korea; sang-jin@kigam.re.kr (S.-J.C.); jetlee@kigam.re.kr (C.L.)

[2] Department of Electrical Engineering and Smart Grid Research Center, Chonbuk National University, 567 Baekje-daero, Deokjin-gu, Jeonju 54896, Korea; jsy50541@jbnu.ac.kr

* Correspondence: kgpark@kigam.re.kr (K.G.P.); pan@jbnu.ac.kr (J.-K.P.); Tel.: +82-42-868-3250 (K.G.P.); +82-63-270-2397 (J.-K.P.)

Received: 17 October 2018; Accepted: 15 November 2018; Published: 16 November 2018

Abstract: The intensity-based fiber optic sensor (FOS) head using twisted dual-cycle bending loss is proposed and experimentally demonstrate. The bending loss characteristics depend on the steel wire radius, number, and distance. To determine the effects of these parameters, two samples in each of seven configuration cases of the proposed FOS head were bonded to fiber reinforced plastics coupons, and tensile and flexural strain tests were repeated five times for each coupon. The bending loss of the manufactured FOS heads was measured and converted to the tensile and flexural strain as a function of configuration cases. The measurement range, sensitivity, and average measurement errors of the tensile load and flexural strain were 4.5 kN and 1760 $\mu\varepsilon$, 0.70 to 3.99 dB/kN and 0.930 to 6.554 dB/mm, and 57.7 N, and 42.6 $\mu\varepsilon$, respectively. The sensing range of FOS head were 82 to 138 mm according to configuration cases. These results indicate that it is possible to measure load, tensile strain, and flexural strain using the proposed FOS head, and demonstrate that the sensitivities, the operating ranges, and the sensing range can be adjusted depending on the deformation characteristics of the measurement target.

Keywords: fiber optic sensor; intensity-based fiber optic sensor; fiber reinforced plastics coupon strain; strain sensor; bending loss

---

1. Introduction

Over the last decade, aging structures have necessitated the development of techniques to ensure their safety, constituting a prominent aspect of structure monitoring. Assessing a structure by using a health-monitoring system can reduce the cost of maintenance and ensure safety [1]. Measurement factors for stability diagnosis in structural health-monitoring systems include cracks, surface degradation, acceleration, temperature, humidity, displacement, and strain. Excessive strain on a structure causes cracking and degradation of structural integrity, ultimately leading to collapse. Thus, measurement of the remaining fatigue lifetime of a structure is necessary to ensure its safety and can be achieved by conducting continuous strain monitoring [1].

Fiber optic sensor (FOS) technology has developed along with the growth of the optoelectronic and optical fiber telecommunication industry. An optical network can be used to obtain real-time measurement information. In addition, it can measure various types of measurement factors with arbitrary spatial distribution [2]. Generally, FOSs can be used to measure physical quantities, such as deformation, temperature, humidity, corrosion, and vibration. As structures have become larger, FOS-based schemes have received increasing attention in monitoring construction conditions.

Among fiber optic-based sensors, the intensity-based FOSs were the first to be developed due to their simplicity and potentially low cost. They continue to offer an attractive option in many sensing applications because they can measure a wide variety of parameters using inexpensive light sources and non-sophisticated detection schemes, while still benefiting from the intrinsic advantages of photonic sensors, i.e., their low weight, small size, and electromagnetic immunity [3,4]. An intensity-based FOS needs self-referencing characteristics to minimize the influences of long-term aging of source characteristics and to overcome short-term fluctuations. In addition, the sensor head needs to convert measurements such as strain, pressure, or force into the corresponding optical intensity change [4–11].

It is well known that radiation loss occurs when an optical fiber is bent [12–16]. Many polymer optical fibers (POFs) have utilized the attenuation of optical power caused by bending-induced mode conversion [16–21]. Lu et al. presented a POF displacement sensor that enhanced sensitivity using a bent and elongated grooved POF [16]. Kuang et al. presented a POF coupon subjected to dual cyclic bending that improved the sensitivity of the POF displacement sensor [17]. Wang et al. presented a novel means of transducing plantar pressure and shear stress using a sensor based on fiber optic bending loss with an FOS array [18]. Abe et al. reported a strain sensor that used a twisted optical fiber that measured the optical loss due to fiber curvature, in which the distributed strain along the sensor axis was converted into a distributed optical loss [19]. Zendehnam et al. investigated the characteristics of bending loss by studying the effect of bending radius, and the influence of the number of wrapping turns [20].

In this paper, an intensity-based FOS head, consisting of high carbon steel wires and a standard single-mode optical fiber, is proposed and applied to the measurement of fiber reinforced plastics (FRP) coupon tensile strain and three-point flexural strain. In Section 2, the FOS head design is presented, along with the twisted dual-cycle bending loss characteristics and a theoretical analysis. Section 3 presents the process for manufacturing the FRP coupon and the FOS head embedded on the coupon. Section 4 describes the experimental validation of the proposed FOS head, in which the bending loss characteristics were obtained via experimental measurements according to the distance between steel wires, wire radius, and the number of steel wires. The results of a tensile test and three-point bending test of the FRP coupon using the proposed FOS head are then provided, and can be used to manufacture FRP intensity-based sensing elements. Finally, we conclude the paper in Section 5.

2. Twisted Dual-Cycle Bending Loss Characteristics

The twisted dual-cycle bending structure for an intensity-based FOS head was designed using a standard single-mode optical fiber. The fiber optic bending loss characteristics of the designed twisted dual-cycle bending structure were then analyzed and experimentally demonstrated.

2.1. Twisted Dual-Cycle Bending Structure

The proposed twisted dual-cycle bending structure for an FOS head that measures fiber optic bending loss is shown in Figure 1. The single-mode optical fiber is twisted around itself and is supported by high carbon steel wires mounted on an FRP coupon. The entire structure consists of the optical fiber, steel wires, a round structure, and a spring. The left side of the twisted optical fiber is fixed, and the right side consists of a turnaround structure where an external force (F_s) is applied to the optical fiber using the spring deflection length (z) shown in Figure 1. In the figure, the first cycle of the optical fiber bending route is from the input (P_{in}) to the round structure. The second cycle of the optical fiber bending route is from the round structure to the output (P_{out}), forming the dual-cycle bending loss structure.

The bending loss of an optical fiber includes both macro- and micro-bending losses. Macro-bending loss dominates when the curvature radius of the optical fiber is much larger than the diameter of the optical fiber [21]. The macro-bending loss increases exponentially when the curvature radius (R_b) is smaller than the critical radius (R_c) of the optical fiber [21]. Therefore, the critical radius plays a significant role in the design and application of optical fibers in sensing.

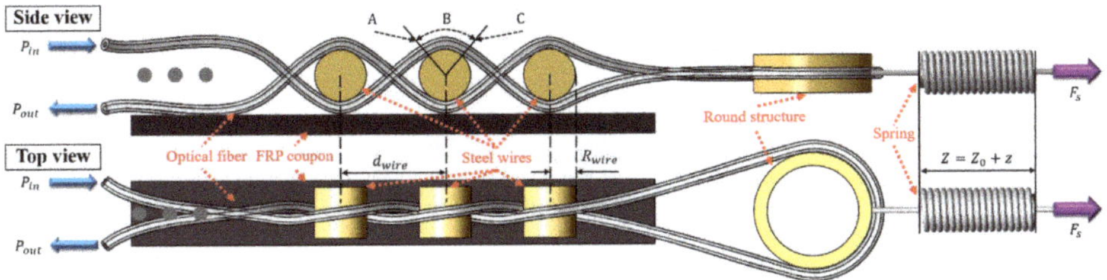

Figure 1. Schematic of the experimental setup for measuring the twisted dual-cycle bending loss characteristics (d_{wire}: distance between steel wires; R_{wire}: steel wire radius; Z: length of spring; Z_0: free length of spring; z: spring deflection length; F_s: spring tension load).

The critical radius (R_c) of a single mode optical fiber is given by [21]:

$$R_c \approx 20 \frac{\lambda}{(\Delta)^{3/2}} \left(2.748 - 0.996 \frac{\lambda}{\lambda_{cutoff}} \right)^{-3} \tag{1}$$

where λ is the wavelength of the light, λ_{cutoff} is the cutoff wavelength of an optical fiber, and Δ is the relative difference between the refractive index of the optical fiber core and cladding. A single mode optical fiber with a λ_c of 1260 nm and a Δ of 0.00529 was used in our experiments. The calculated R_c using Equation (1) is 22.85 mm. In the experimental setup shown Figure 1, the macro-bending loss dominates because the radii of the steel wires are smaller than the critical radius of the optical fiber.

The total loss of a bent fiber includes the pure bending loss in the bent section and the transition loss caused by the mismatch in the propagation mode between the bent and straight sections. Figure 1 illustrates the macro-bending loss of the twisted optical fiber, which includes the pure bending loss in region B and the transition loss at positions A and C. When the inherent attenuation of the optical fiber is neglected, the optical power ratio of the output optical power, $P_{wire-out}$, to the input optical power, $P_{wire-in}$, for the passage of one steel wire in Figure 1 can be expressed as follows [21]:

$$\frac{P_{wire-out}}{P_{wire-in}} = \left(\left. \frac{P_{wire-out}}{P_{wire-in}} \right|_A \left. \frac{P_{wire-out}}{P_{wire-in}} \right|_B \left. \frac{P_{wire-out}}{P_{wire-in}} \right|_C \right) \tag{2}$$

where losses A and C are the transition losses and loss B is the pure bending loss. Because a transition loss is much smaller than a pure bending loss for a certain wavelength, Equation (2) can be simplified as follows [21]:

$$\frac{P_{wire-out}}{P_{wire-in}} = \left. \frac{P_{wire-out}}{P_{wire-in}} \right|_B = \exp(-2\alpha l_{b-wire}) \tag{3}$$

where α is the coefficient of the pure bending loss and l_{b-wire} is the bending length of the optical fiber around one steel wire, as shown in Figure 2. The bending loss in the dB scale at the l_{b-wire} section, L_{single}, using Equation (3) can be defined as follows [21]:

$$L_{single} = 10 \log_{10} \left(\frac{P_{wire-in}}{P_{wire-out}} \right) = 10 \log_{10} \left[\frac{1}{\exp(-2\alpha l_{b-wire})} \right] = 4.342(2\alpha l_{b-wire}) \tag{4}$$

The bending loss coefficient, α, of a single mode optical fiber under weak guiding conditions is given by [22]:

$$2\alpha = \frac{\sqrt{\pi}\kappa^2}{2\gamma^{3/2} V^2 \sqrt{R_b} K_{+1}^2(\gamma a)} \exp\left(-\frac{2\gamma^3 R_b}{3\beta^2} \right) \tag{5}$$

where κ is the normalized radial phase constant, γ is the normalized radial attenuation constant, V is the normalized frequency, $K_{+1}^2(\gamma a)$ is the modified Hankel function, a is the radius of the fiber core, and β is the axial propagation constant. These parameters are fixed values once the single-mode optical fiber and the wavelength of incident light are chosen. According to Equations (4) and (5), the bending loss of an optical fiber passing over one steel wire, L_{single}, is determined as follows [21]:

$$L_{single} = 4.342\left(\frac{A_{strc}}{\sqrt{R_b}}\exp(-B_{strc}R_b)\right)l_{b-wire} \tag{6}$$

where $A_{strc} = \sqrt{\pi}\kappa^2/2\gamma^{3/2}V^2 K_{+1}^2(\gamma a)$ and $B_{strc} = 2\gamma^3/3\beta^2$.

The bending length, l_b, of the twisted dual-cycle bending structure when the number of steel wires, N_{wire}, is can be expressed as follows:

$$l_b = 2l_{b-wire}N_{wire} = 2R_b\theta N_{wire} = 4R_b\sin^{-1}\left(\frac{2R_b}{d_{wire}}\right)N_{wire} \tag{7}$$

where R_b is the sum of the steel wire radius, R_{wire}, and the optical fiber radius, R_{fiber}; d_{wire} is the distance between two steel wires; and $\theta = 2\sin^{-1}\left(\frac{2R_b}{d_{wire}}\right)$. Further, the bending loss, L_{strc}, given by the twisted dual-cycle bending structure using Equations (6) and (7) can be expressed as follows:

$$L_{strc} = 17.368N_{wire}A_{strc}\sqrt{R_b} \times \exp(-B_{strc}R_b)\sin^{-1}\left(\frac{2R_b}{d_{wire}}\right) \tag{8}$$

Equation (8) shows that the bending loss, L_{strc}, can be adjusted by changing N_{wire}, R_b, and/or d_{wire}.

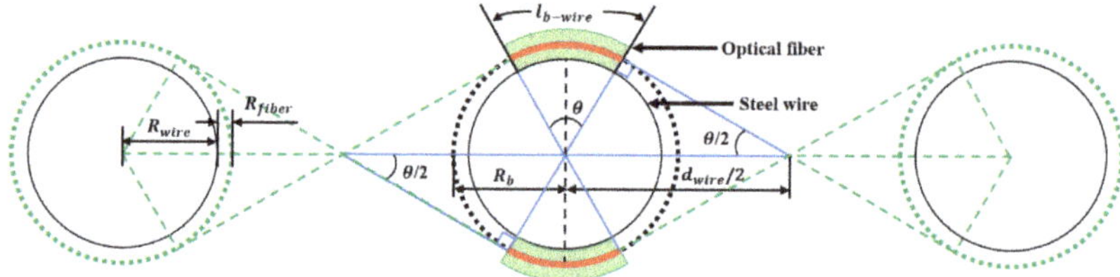

Figure 2. Schematic of the twisted dual-cycle bending length l_{b-wire}.

2.2. Bending Loss Characteristics

The twisted dual-cycle bending loss characteristics were evaluated by measuring the optical power loss (dB) corresponding to the spring tension load (F_s) or the spring deflection length (z) using the experimental setup shown in Figure 1. In the experiments, we used a tunable laser (MG9638A, Anritsu, Atsugi-shi, Japan) with an input optical power of 0 dBm, wavelength of 1550 nm, and full width at half maximum (FWHM) of 0.2 nm; an optical power and energy meter (PM320E, Thorlabs, Newton, NJ, USA) with a fiber photodiode power sensor (S154C, Thorlabs); a single-mode optical fiber (SMF-28, Heesung Cable, Seoul, Korea) with an optical fiber radius, R_{fiber}, including polymer coating, of 0.1225 mm, an attenuation of 0.190 dB/km, and a polarization mode dispersion of 0.049 ps/$\sqrt{\text{km}}$ at a wavelength of 1550 nm; and a tension spring (AWY8-60, Misumi, Tokyo, Japan) with an initial tension of 2.35 N and a reference spring constant of 0.2 N/mm.

Figure 3 shows the measured results of the spring tension load according to the spring deflection length in stepwise increments from 0.1 mm up to 30 mm via a universal testing machine (UTM) (INSTRON, 5982). The spring tension load increased linearly as the spring deflection length increased from 1.5 to 30 mm. The measured initial tension was 2.088 N, and the spring constant was 0.2781 N/mm. Therefore, the optical power loss (dB) versus spring deflection length (z) was determined using a spring tension load (F_s) within a linearly increasing range. An increase in the spring deflection length from 3 to 15 mm in 3 mm increments increased the spring tension load to 2.92, 3.76, 4.59, 5.43, and 6.26 N, respectively.

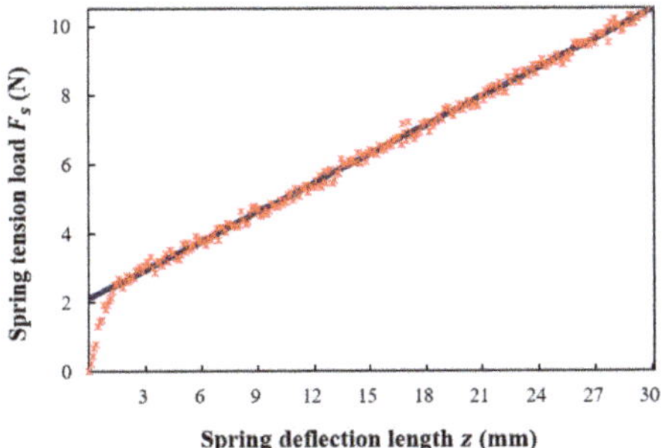

Figure 3. Measured spring tension load F_s (N) versus spring deflection length z (mm).

Figure 4 shows the results measured for the optical power loss versus the spring tension load and spring deflection length using five values of the bending radius (R_b of 0.6225, 0.8725, 1.1225, 1.6225, and 2.1225 mm) and six distances between the steel wires (d_{wire} of 10, 12, 14, 16, 18, and 20 mm) when N_{wire} is 3. The relationship between the total optical power loss (L_{total}) and the spring tension load is linear. According to the Abe et al. the loss variation is linear when strain satisfies $\Delta\alpha\varepsilon \ll 1$ and a larger lateral rigidity of the primary coating (k), a smaller core radius (a) and a smaller refractive index deference (Δ) are necessary to achieve linearity for larger strains [19]. Because we used standard single mode optical fiber (SMF-28) that satisfies $\Delta\alpha\varepsilon \ll 1$, the optical power loss linearly increases with the increase of the tensile load.

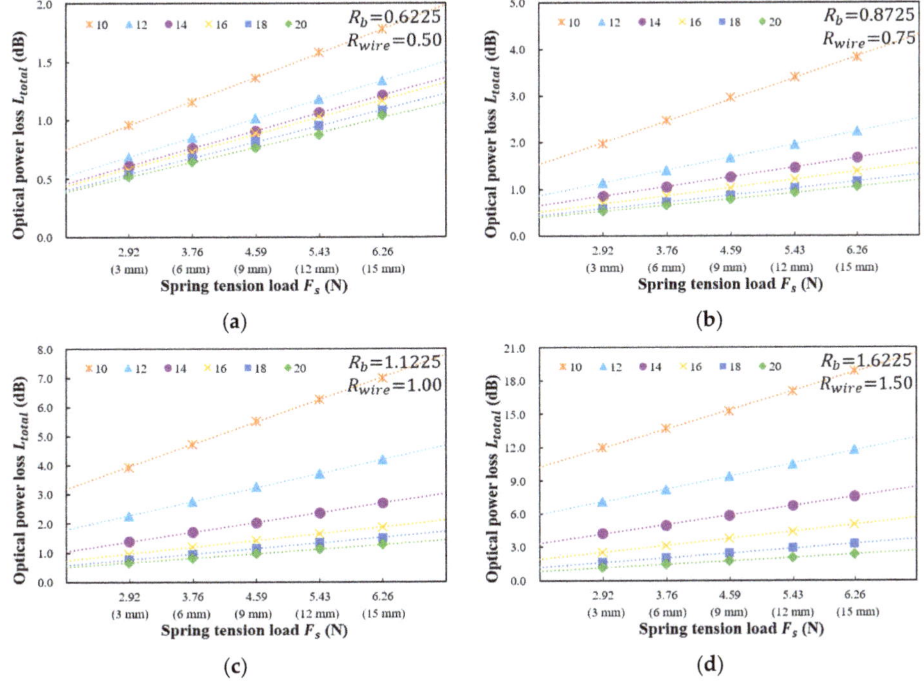

Figure 4. *Cont.*

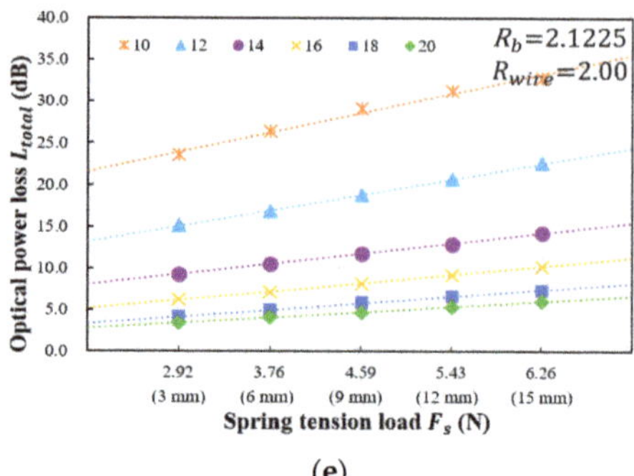

(e)

Figure 4. Measured optical power loss (dB) versus spring tension load F_s (N) and spring deflection length z (mm) for various distances between the steel wires. The bending radii R_b are (**a**) 0.6225 mm; (**b**) 0.8725 mm; (**c**) 1.1225 mm; (**d**) 1.6225 mm; (**e**) 2.1225 mm.

The total optical power loss, L_{total}, for the proposed twisted dual-cycle bending structure can be expressed as:

$$L_{total} = L_{strc} + F_s \times L_{force} \tag{9}$$

where L_{force} is the optical power loss per unit Newton of the spring tension load. In Figure 4, the L_{force} according to the sensor head structure can be obtained by using L_{total} according to F_s. The value of L_{strc} due to the twisted dual-cycle bending structure can be obtained by subtracting the loss due to F_s from the measured value of L_{total}.

For the proposed twisted dual-cycle bending structure, the calculated optical power loss, L_{strc}, versus the bending radius, R_b, for six different distances between the steel wires, d_{wire}, is shown in Figure 5.

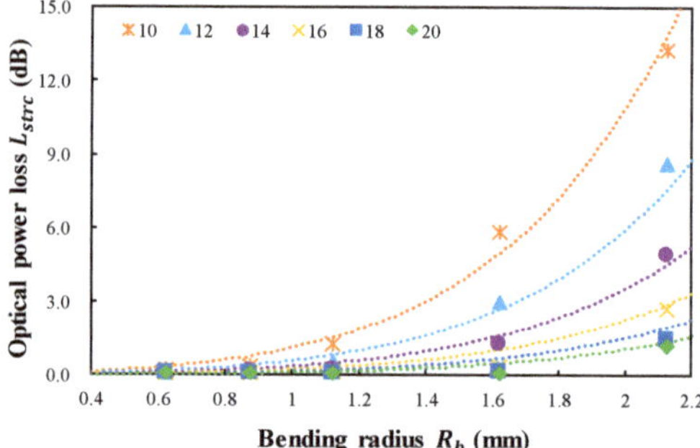

Figure 5. Calculated optical power loss (L_{strc}) versus bending radius (R_b) for six distances between steel wires (d_{wire}) of the twisted dual-cycle bending structure.

The optical power loss is increased by increasing R_b and decreasing d_{wire}. When d_{wire} is decreased, the bending length (l_b) in Equation (7) is increased. Ordinarily, increasing R_b decreases the total optical loss because the optical power loss per unit length is decreased. However, our experiment demonstrated an increase in L_{strc} when increasing R_b due to the increase in l_b in Equation (8). In this equation, A_{strc} and B_{strc} are the calculation results of the twisted dual-cycle bending loss for five different bending radii (R_b) and six distances between the steel wires (d_{wire}). Using the MATLAB exponential fitting model, A_{strc} was determined to be 0.030 and B_{strc} was determined to be -2.380. The exponential model has the largest multiple correlation coefficient with a value of 0.9816, and the root mean square deviation is 0.4064. Both A_{strc} and B_{strc} are constant values depending on the parameters of the twisted dual-cycle bending loss structure. Substituting these constant values for A_{strc} and B_{strc} in Equation (8) enables L_{strc} to be predicted according to d_{wire}, R_{wire}, and N_{wire}.

Figure 6 plots the calculated results of the optical power loss per unit Newton (L_{force}) determined by the spring tension load (F_s) for five bending radii (R_b) and six distances between the steel wires (d_{wire}). The value of L_{force} is also increased by increasing R_b and decreasing d_{wire}. Furthermore, the optical power loss per unit newton, L_{force}, can be expressed as in Equation (8) as follows:

$$L_{force} = 17.368 N_{wire} A_{force} \sqrt{R_b} * \exp\left(-B_{force} R_b\right) \sin^{-1}\left(\frac{2R_b}{d_{wire}}\right) \tag{10}$$

Using the MATLAB exponential fitting model, A_{force} was calculated to be 0.492 and B_{force} was calculated to be -0.332. The exponential model has the largest multiple correlation coefficient with a value of 0.9932, and the root mean square deviation is 0.0642. Applying the constant value of A_{force} and B_{force} to Equation (10), L_{force} can be predicted according to the d_{wire}, R_{wire}, and N_{wire}.

Figures 5 and 6 indicate that the operating range of the twisted dual-cycle bend structure can be improved by decreasing R_{wire} and/or increasing d_{wire} while its sensitivity can be improved by decreasing d_{wire} and/or increasing R_{wire}.

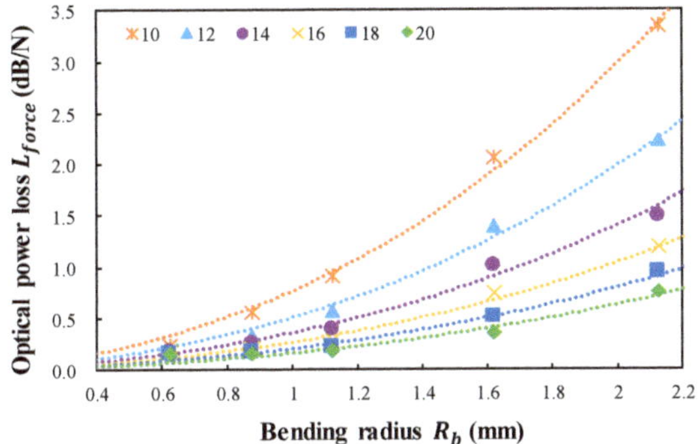

Figure 6. Calculated optical power loss per unit Newton (L_{force}) determined by the spring tension load (F_s) versus bending radius (R_b) for six distances between steel wires (d_{wire}) of the twisted dual-cycle bending structure.

3. FOS Head Bonded to the FRP Composite Coupon

Different FRP composites have different physical characteristics depending on the constitution of the fiberglass mat lay-up and manufacturing method [23]. Typically, the laminated structure of an FRP consists of chopped strand mat fabric and roving cloth fabric combined in an orderly manner. The FRP coupon used in this study was produced by the vacuum infusion bagging method using the infusion set up and glass fiber lay-up shown in Figure 7. In this lay-up, two roving cloth fabric sheets, constituting an anisotropic material, are located between three chopped strand mat fabric sheets that constitute an isotropic material. Thus, the composite material is composed of five sheets: two chopped strand mat fabric sheets with a thickness of 0.65 mm and a weight per unit area of 300 g/m^2, and the three roving cloth fabric sheets with a thickness of 0.5 mm and a weight per unit area of 570 g/m^2. The epoxy resin matrix material (RESOLTECH 1050, resoltech, Rousset, France) was mixed in a volume ratio of 7:3 with RESOLTECH 1056 hardener to manufacture the FRP composite coupon. The epoxy resin was injected using the vacuum infusion bagging method and cured at room temperature for 16 h.

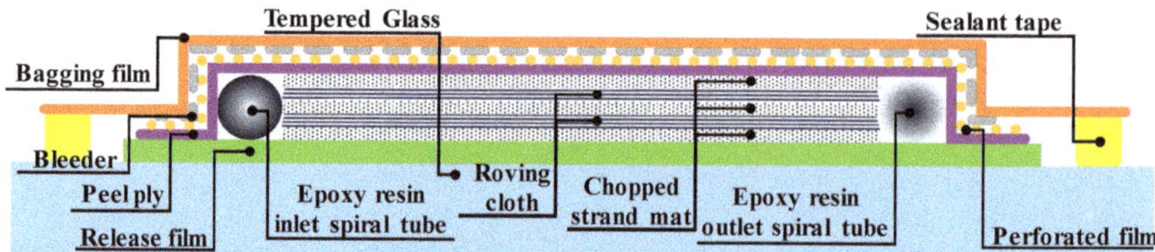

Figure 7. Schematic of the vacuum infusion bagging method.

In Figure 7, epoxy resin is injected through the inlet spiral tube located on the left side of the fiberglass mat, and a negative pressure of 95 kPa is applied by a vacuum pump connected to the outlet spiral tube located on the right side of the fiberglass mat. The vacuum pump removes air and unnecessary epoxy resin inside the compartment, which is otherwise sealed with bagging film and sealant tape. As a result, the prepared FRP composite material has a uniform thickness and surface quality, and the interlayer adherence, tensile strength, and impact strength are improved. Once produced, the manufactured FRP plate was cut into FRP coupons with a length of 200 mm and a width of 40 mm using an automatic cut-off machine (Brillant 220, ATM, Mammelzen, Germany) for testing using the UTM. The average thickness of the final FRP coupons was 2.2 mm.

The strain in the twisted dual-cycle bending structure was measured by attaching the proposed FOS head to the FRP coupon surface. The twisted dual-cycle bending structure shown in Figure 1 contains a standard single-mode optical fiber, the strands of which are twisted around each other as they travel between steel wires. This FOS head was installed on the surface of the FRP coupon using the procedure shown in Figure 8a–f and described as follows:

(a)   Attach three sheets of double-sided tape (93015LE, 3M, Maplewood, MN, USA) in two strips to the surface of the FRP coupon. Each strip of double-sided tape fixes one end of the steel wire, leaving the middle part of the coupon without tape for the passage of the optical fiber.

(b)   Place the steel wire on top of the double-sided tape under the conditions listed in Table 1. The sensor length depends on the number of steel wires and the distance between these wires.

(c)   Fix both ends of the steel wire to the surface of the FRP coupon.

(d)   Weave the single-mode optical fiber over and under the steel wires as they pass across the specimen.

(e)   Fix the optical fiber to the FRP coupon surface with epoxy resin 20 mm from the left-most steel wire. After the epoxy resin is fully cured, connect the spring to the round structure for returning the optical fiber. Pull the spring to apply force to the optical fiber for setting the operating point of the sensor head that the optical power loss can be linearly operated according to the strain applied to the FRP coupons.

(f)   Fix the optical fiber to the FRP coupon surface with epoxy resin 20 mm from the right-most steel wire. Remove the round structure and the spring after the epoxy resin has fully cured.

Table 1. Seven kinds of intensity-based FOS heads implemented with different conditions.

FOS Heads	R_{wire} (mm)	N_{wire}	d_{wire} (mm)	Sensor Length (mm)	Insertion Loss (dB)	FOS Heads	R_{wire} (mm)	N_{wire}	d_{wire} (mm)	Sensor Length (mm)	Insertion Loss (dB)
Case 1-1	1	6	14	110	2.43	Case 1-2	1	6	14	110	2.89
Case 2-1	0.75	6	14	110	1.17	Case 2-2	0.75	6	14	110	1.65
Case 3-1	1.5	6	14	110	9.93	Case 3-2	1.5	6	14	110	9.98
Case 4-1	1	4	14	82	2.43	Case 4-2	1	4	14	82	2.56
Case 5-1	1	8	14	138	2.80	Case 5-2	1	8	14	138	3.35
Case 6-1	1	6	12	100	5.24	Case 6-2	1	6	12	100	5.27
Case 7-1	1	6	16	120	1.42	Case 7-2	1	6	16	120	1.61

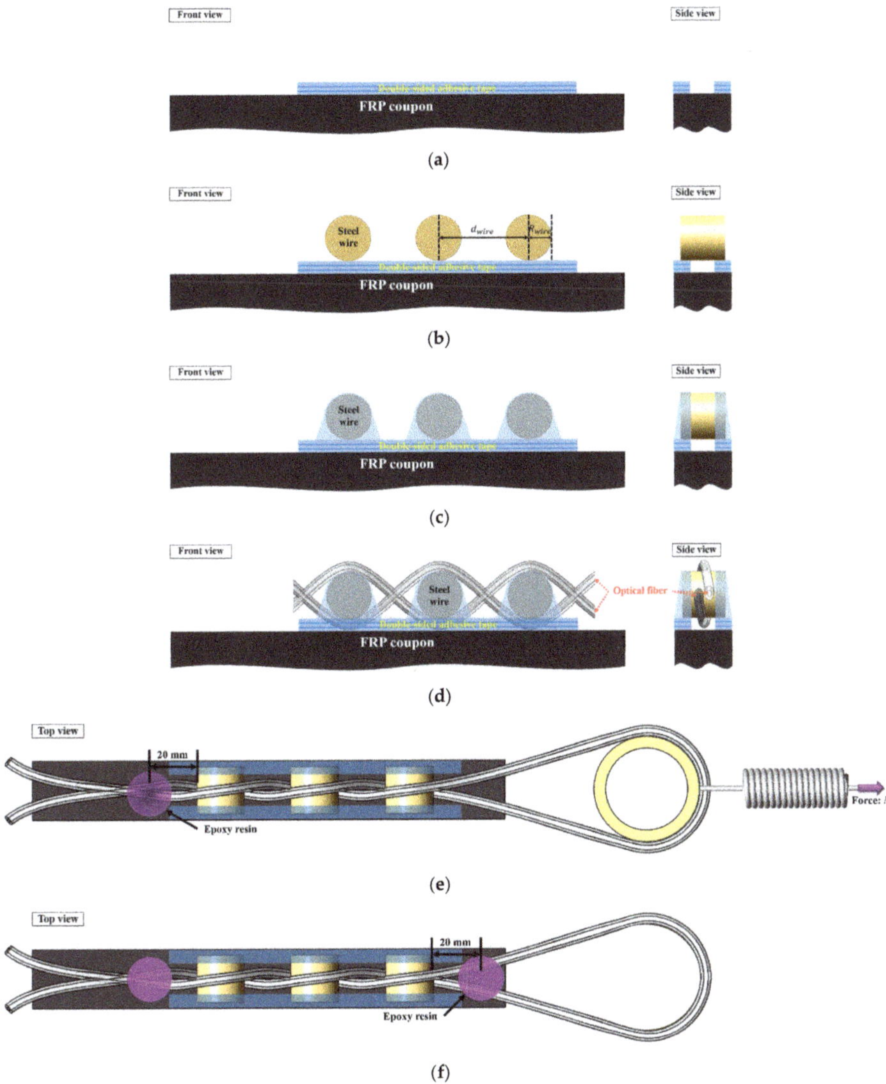

Figure 8. Intensity-based FOS head fabrication process.

The force applied at the manufacturing process shown in Figure 8e is 3 N, which was derived from the analysis/investigation on the relation between power loss and tensile force shown in Figure 4. Actually, the magnitude of force applied during fabrication is important. If no force is applied at all or the force is too small, it is difficult to measure the optical power loss. On the other hand, if too much force is applied, insertion loss become larger.

Table 1 contains seven types of intensity-based FOS heads that were implemented using the above procedure and specifies the steel wire radius, number of steel wires, and distance between steel wires used in each head variant. As determined theoretically in Section 2, as the distance between steel wires and the number of steel wires increases, the sensing area increases, while as the steel wire radius and the number of steel wires increase, and the distance between steel wires decreases, the insertion loss increases.

4. Measurement of FRP Coupon Strain

The strain measurement experiment was designed and conducted using the configuration shown in Figure 1 with the proposed FOS heads in Table 1 based on the bending loss characteristics of the twisted dual-cycle configuration. The proposed FOS heads were used to collect measurements by bonding the twisted dual-cycle bending structure to the FRP coupon using epoxy resin, as described in Section 3, to conduct tensile strain and three-point bending tests. In these experiments, we used a tunable laser (MG9638A, Anritsu) with an input optical power of 0 dBm at 1550 nm and a full width at half maximum (FWHM) of 0.2 nm, an optical power and energy meter (PM320E, Thorlabs) with a fiber photodiode power sensor (S154C, Thorlabs), and a single mode optical fiber (SMF-28, Heesung Cable).

4.1. Tensile Strain Test

To determine the relationship between the FRP coupon tensile strain under the load F_t versus the optical power loss of the FOS head, seven different tensile strength test cases were investigated using a UTM in load control mode. The optical power and FRP coupon tensile strain data from the UTM were recorded using a separate data acquisition system at a sampling rate of 2 Hz. The measurement started at an initial load of 50 N. The loading rate was 20 N/s and the maximum load was 4.5 kN with a measurement interval of 10 N. The measured load-strain relationships of the subject FRP coupons were linear: a tensile load of 4.5 kN is equivalent to an approximate tensile strain of 2000 $\mu\varepsilon$.

The bending loss characteristics depend on the steel wire radius, numbers, and distance. To determine the effects of these parameters, two FRP coupons were tested for each of the seven cases and each coupon test was repeated five times. Table 1 provides the details of the twisted dual-cycle bending structure test cases for each of the coupons. Figure 9 shows the measured optical power loss (left) and error (right) of the FOS head versus the applied tensile load (F_t) for each of the seven cases. The black solid line is the calculated reference curve plotted using the MATLAB curve-fitting tool for one-term power series models ($s \times F_t^l$). Table 2 provides the results of the tensile tests, in which the parameters s and l represent the loss characteristics of the FOS head corresponding to the load or strain. These are correction variables for converting the optical power loss measured by the optical power meter into the measured load or strain applied to the FRP coupon upon which the sensor head is installed. When the s and l parameters are applied to the optical power loss occurring in the sensor, the applied tensile load, F_t, can be expressed as follows:

$$F_t = (L_{tensile}/s)^{1/l} \text{ [kN]} \tag{11}$$

where $L_{tensile}$ is the optical power loss of the FOS head according to the tensile load applied to the FRP coupon.

The tensile load calculated using Equation (11) was compared with that applied by the UTM and is indicated by the error axis in Figure 9. The average error of the load measured by the FOS heads is 57.7 N and the average error of the tensile strain is less than 26.1 $\mu\varepsilon$. Clearly, it is possible to measure the load and tensile strain by using the proposed intensity-based FOS head.

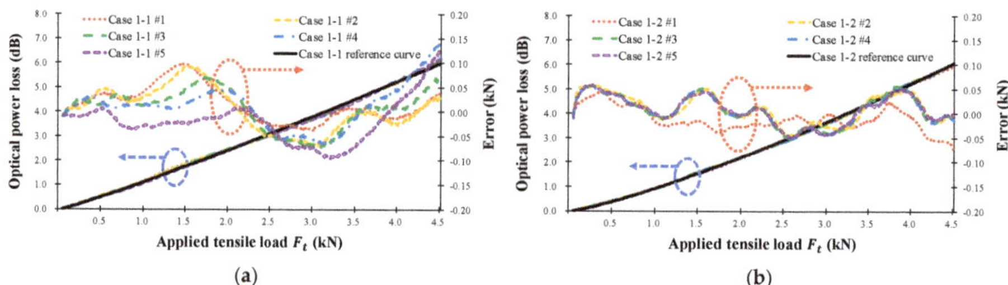

(a)   (b)

Figure 9. *Cont.*

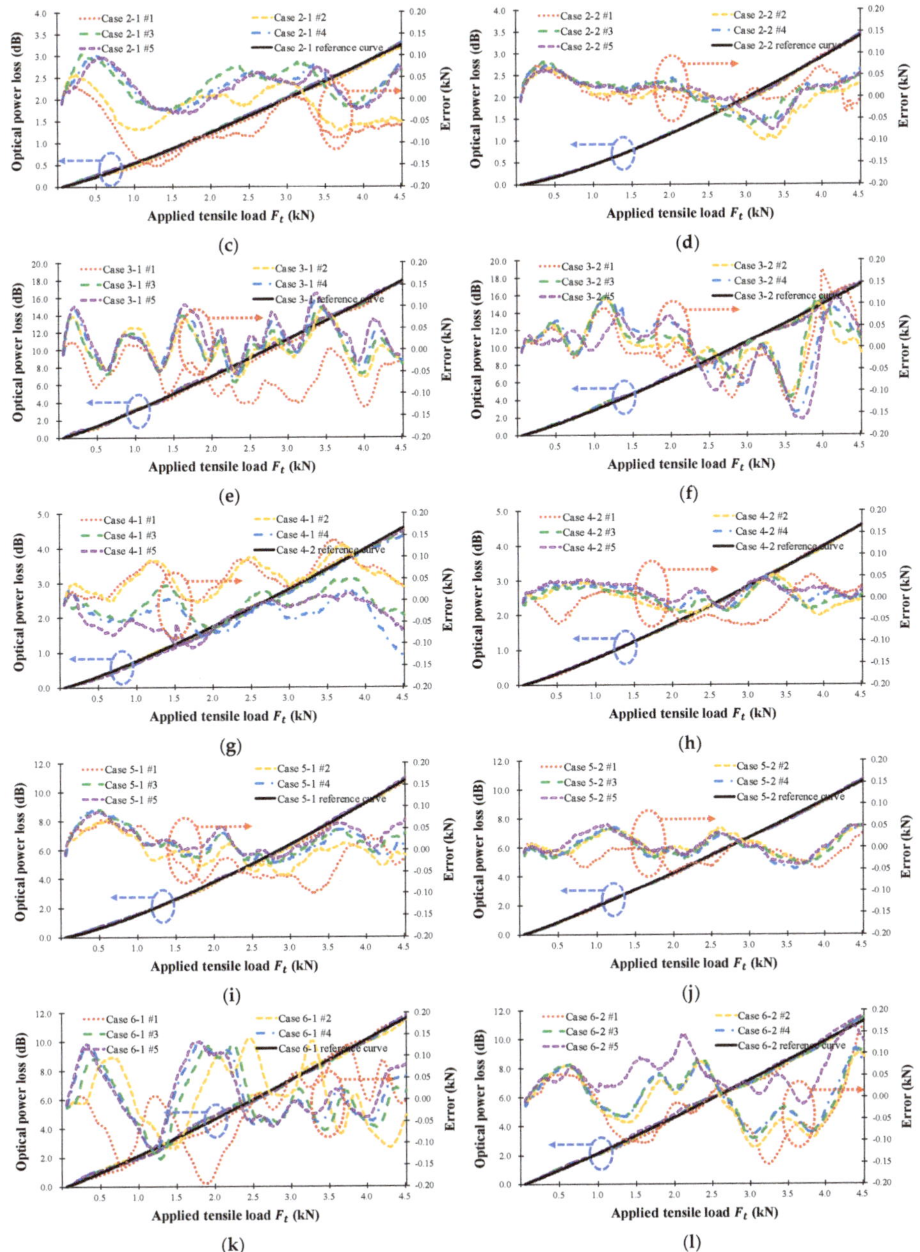

Figure 9. *Cont.*

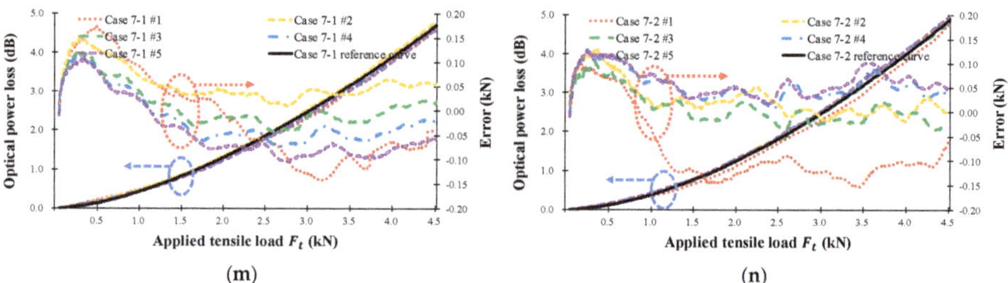

Figure 9. Measured FOS head optical power loss (dB) according to the applied tensile load F_t (kN) for the seven types of FRP sensor head in Table 1. (**a**) Case 1-1; (**b**) Case 1-2; (**c**) Case 2-1; (**d**) Case 2-2; (**e**) Case 3-1; (**f**) Case 3-2; (**g**) Case 4-1; (**h**) Case 4-2; (**i**) Case 5-1; (**j**) Case 5-2; (**k**) Case 6-1; (**l**) Case 6-2; (**m**) Case 7-1; (**n**) Case 7-2.

Table 2. Measurement results of tensile strain test.

FOS Heads	Average Sensitivity		Curve Fitting Results ($s \times F_t^l$)				Error
	(dB/kN)	(dB/$\mu\varepsilon$)	s	l	R-Square	RMSE	Average (dB, N, $\mu\varepsilon$)
Case 1-1	1.33	0.00294	1.1530	1.101	0.9988	0.06000	0.046, 34.2, 15.5
Case 1-2	1.35	0.00299	0.9624	1.231	0.9995	0.04020	0.034, 24.9, 11.3
Case 2-1	0.70	0.00155	0.5698	1.161	0.9983	0.03962	0.032, 45.1, 20.4
Case 2-2	0.76	0.00168	0.4716	1.319	0.9992	0.02768	0.023, 30.1, 13.6
Case 3-1	3.99	0.00883	3.2550	1.140	0.9984	0.21370	0.164, 41.1, 18.6
Case 3-2	3.83	0.00847	3.0000	1.168	0.9976	0.24810	0.180, 46.9, 21.2
Case 4-1	1.00	0.00221	0.7599	1.191	0.9975	0.06675	0.042, 42.0, 19.0
Case 4-2	1.02	0.00226	0.7939	1.171	0.9996	0.02755	0.022, 21.9, 9.9
Case 5-1	2.40	0.00531	1.6120	1.269	0.9992	0.09092	0.077, 32.1, 14.5
Case 5-2	2.35	0.00520	2.0260	1.103	0.9996	0.06102	0.048, 20.4, 9.2
Case 6-1	2.58	0.00571	2.2290	1.102	0.9971	0.18360	0.149, 57.7, 26.1
Case 6-2	2.50	0.00553	2.2310	1.081	0.9976	0.16090	0.129, 51.8, 23.4
Case 7-1	1.05	0.00232	0.4587	1.556	0.9976	0.06938	0.060, 57.5, 26.0
Case 7-2	1.09	0.00241	0.4109	1.655	0.9973	0.07569	0.060, 55.1, 24.9

Figure 10 shows the calculated reference curve for the seven test cases, while Table 2 presents the calculated results of the one-term power series models (s and l), regression analysis (R-square), and Root Mean Squared Error (RMSE). In Table 2, the R-square values were measured five times, producing a minimum R-square value of 0.997; therefore, the calculated reference curve can be considered a good representation of the characteristics of the seven sensor head cases. Table 2 also indicates that increasing the sensor head sensitivity increases the R-square, RMSE, and average error values. Figure 10a shows the optical power loss as a function of the wire radius, R_{wire}, in Equation (8): as the wire radius increases, the sensitivity of the sensor head increases, and the average error is not affected. Figure 10b displays the loss depending on the number of wires, N_{wire}, in Equation (8). As the number of wires increases, the sensitivity of the sensor head increases, and the average error is nearly constant. Additionally, as the number of wires increases, the sensor length increases, and thus the measurement range also increases. Figure 10c shows that the sensitivity of the sensor head decreases as the distance between wires, d_{wire}, in Equation (8) increases because of the bending length (l_b) in Equation (7), and because the measurement range is increased by increasing the sensor length. In summary, the sensitivity of the twisted dual-cycle bending structure can be adjusted by changing the radius of the steel wires, the number of steel wires, and the distance between these steel wires.

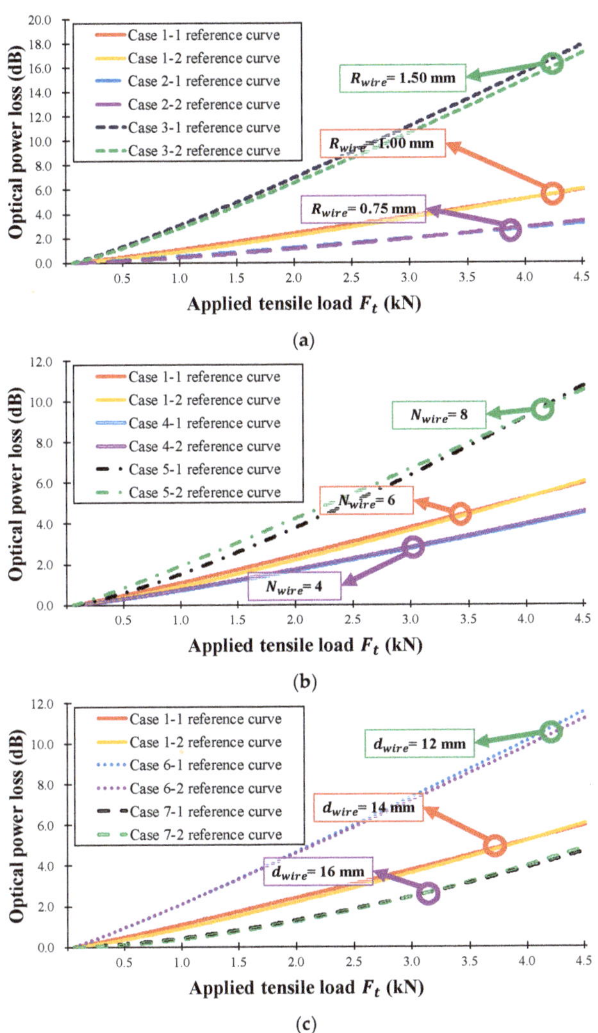

Figure 10. Measured optical power loss (dB) versus applied tensile load F_t (kN) for the manufactured FOS heads in Table 1. (**a**) R_{wire} = 0.75, 1.00, and 1.50 mm, N_{wire} = 6, d_{wire} = 14 mm; (**b**) R_{wire} = 1.00 mm, N_{wire} = 4, 6, and 8, d_{wire} = 14 mm; (**c**) R_{wire} = 1.00 mm, N_{wire} = 6, d_{wire} = 12, 14, and 16 mm.

4.2. Three-Point Bending Test

In the three-point bending test, strain was applied to an FRP coupon embedded with the proposed twisted dual-cycle bending structure to investigate the relationship between the optical power loss of the FOS head and the applied flexural strain. The bending loss characteristics depend on the steel wire radius, numbers, and distance. To determine the effects of these parameters, two FRP coupons were tested for each of the seven cases detailed in Table 1, in which the steel wire radius, number of steel wires, distance between steel wires, total embedment length, and insertion loss (dB) is varied. Figure 11 shows image of the FOS head installed in the UTM under unloaded and loaded condition and schematic of the three-point bending test setup where the FOS head adhered to the center of the FRP coupon as described in Figure 8. The FRP coupon was subjected to increasing cyclic loads following a triangularly shaped deflection pattern applied by the UTM. The diameter of the vertically moving loading roller and horizontally moving supporting rollers was 10 mm. The experiment measured the FRP coupon deflection, which we plotted it against the optical power loss measured by the proposed FOS head. The FRP coupon was loaded at a vertical displacement rate of 0.04 mm/s in measurement intervals of 0.02 mm until the deflection reached 3.5 mm from the initial 0.5 mm deflection.

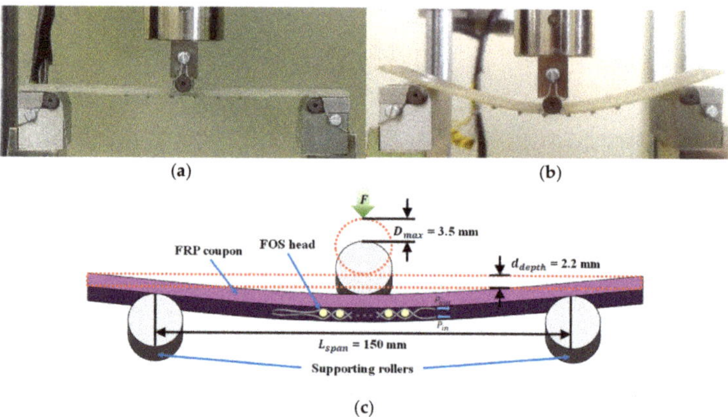

Figure 11. Three-point bending test setup with (**a**) unloaded image of the FOS head installed in the UTM; (**b**) loaded image of the FOS head installed in the UTM; (**c**) Schematic of the FOS head placement.

The theoretical flexural strain in the outer surface of the FRP coupon, ϵ_f, can be expressed as [24]:

$$\epsilon_f = \frac{6Dd_{thickness}}{L_{span}^2} \tag{12}$$

where D is the deflection of the center of the FRP coupon under increasing cyclic loading following a triangularly shaped deflection from 0.5 mm to 3.5 mm, $d_{thickness}$ is the depth or thickness of the tested FRP coupon ($\cong$2.2 mm), and L_{span} is the support span ($\cong$150 mm). From Equation (12), we can see that the flexural strain and deflection have a linear relationship. The change in ϵ_f between the maximum and initial deflection was 1760 $\mu\varepsilon$ and the deflection sensitivity was 586.7 $\mu\varepsilon$/mm. The calculated flexural strain in the outer surface of the FRP coupon was then used as a reference value to verify the performance of the FOS head.

The correction parameters h and k are used to represent the loss characteristics of the FOS head corresponding to the flexural strain, allowing the optical power loss measured by the optical power meter to be converted into flexural strain. The experimental flexural strain can be obtained by applying the parameters h and k to the measured optical loss occurring in the sensor head as follows:

$$\epsilon_f = \left(L_{flex}/h\right)^{1/k} \tag{13}$$

where L_{flex} is the optical power loss of the FOS head due to the flexural strain applied to the FRP coupon.

Figure 12 shows the measured optical power loss of the FOS head (left vertical axis, red dotted line) and theoretical ϵ_f in the outer surface (right vertical axis, black solid line) versus the deflection of the center of the FRP coupon for the seven cases in Table 1. The experimentally determined flexural strain values (right vertical axis, green dotted line) using Equation (13) and the theoretical flexural strain calculated using Equation (12) was compared and is indicated by the flexural strain error (right vertical axis, purple double solid line) in Figure 12, where it can be seen that the experimental flexural strain is nearly equivalent to the theoretical flexural strain, regardless of the characteristics of the sensor head. Therefore, it can be concluded that the optical loss occurring in each sensor head can be accurately converted into flexural strain by using Equation (13). The error in the flexural strain increases abruptly when the deformation of the FRP coupon changes direction, i.e., from increasing to decreasing deflection or from decreasing to increasing deflection, as indicated by the experimental error. The average error in the flexural strain measured by the FOS heads was less than 42.6 $\mu\varepsilon$. Thus, it is possible to accurately measure the flexural strain in FRP by using the proposed intensity-based FOS head.

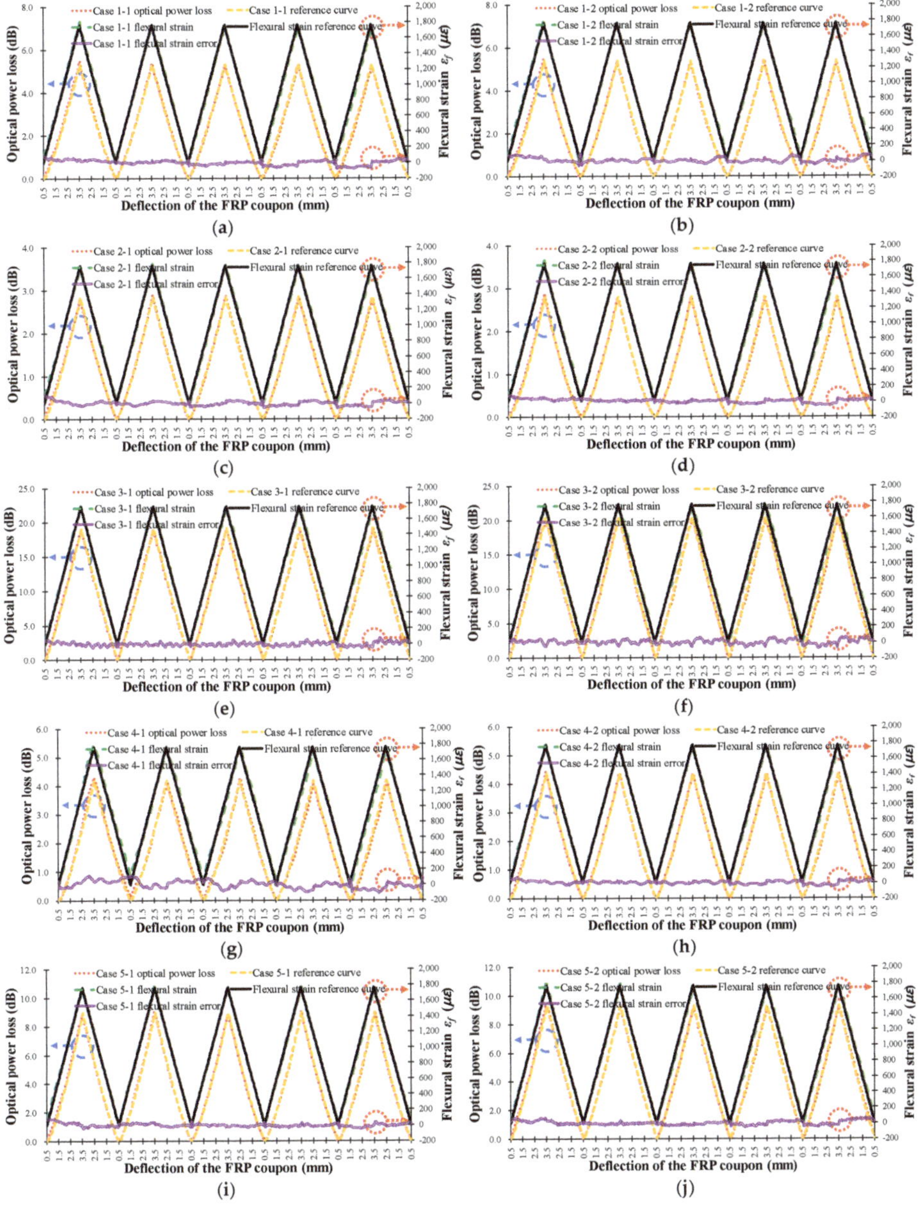

Figure 12. *Cont.*

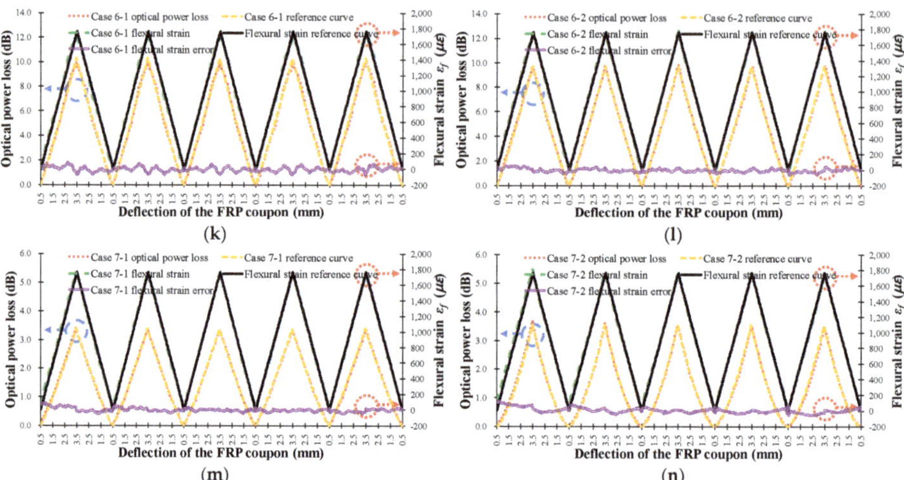

(k)

(l)

(m)

(n)

Figure 12. Measured FOS head optical power loss (**left**), measured flexural strain, and calculated flexural strain error (**right**) according to the deflection of the FRP coupon for the seven types of FRP sensor head in Table 1. (**a**) Case 1-1; (**b**) Case 1-2; (**c**) Case 2-1; (**d**) Case 2-2; (**e**) Case 3-1; (**f**) Case 3-2; (**g**) Case 4-1; (**h**) Case 4-2; (**i**) Case 5-1; (**j**) Case 5-2; (**k**) Case 6-1; (**l**) Case 6-2; (**m**) Case 7-1; (**n**) Case 7-2.

Figure 13 shows the optical power loss of the sensor head as a function of the steel wire radius, the number of steel wires, and the distance between steel wires. Figure 13a shows the optical power loss according to the wire radius, R_{wire}, in Equation (8). As the wire radius increases, the sensitivity of the sensor head increases because the angle of contact between the steel wire and the optical fiber has increased. The sensitivity of the FOS head can therefore be adjusted by varying the radius of the steel wires without changing the length of the sensor head. Both the tensile strain test and the flexural strain test displayed an increase in the average sensitivity when the steel wire radius was increased. Figure 13b shows the dependence of optical power loss on the number of wires, N_{wire}, in Equation (8). As the number of wires increases, the sensing area and the sensitivity increase because the total bending length (l_b) of the optical fiber increases. Figure 13c shows the decrease in sensitivity of the sensor head as the wire interval, d_{wire}, in Equation (8) is increased and the angle of contact between the steel wire and the optical fiber accordingly decreases.

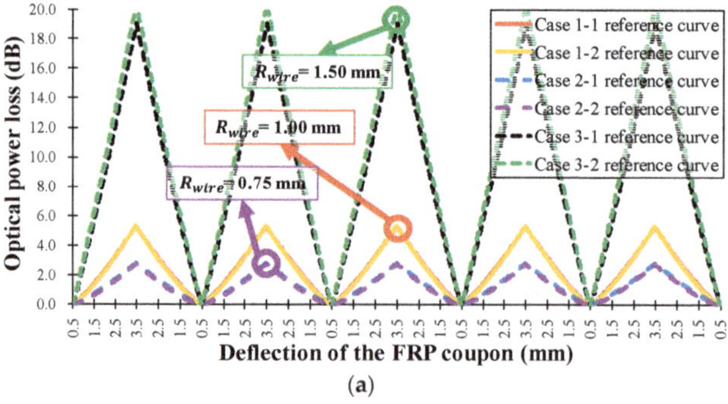

(a)

Figure 13. *Cont.*

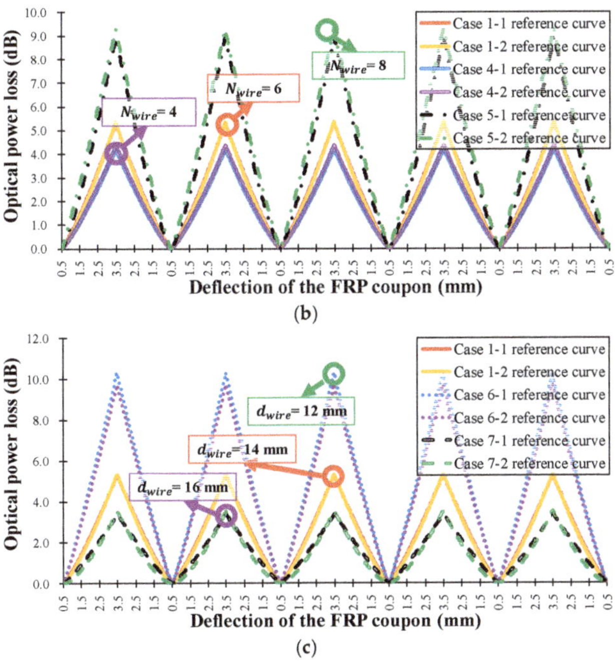

Figure 13. Measured FOS head optical power loss according to the three-point bending deflection of the FRP coupon of the seven types of FRP sensor head in Table 1. (**a**) R_{wire} = 0.75, 1.00, and 1.50 mm, N_{wire} = 6, d_{wire} = 14 mm; (**b**) R_{wire} = 1.00 mm, N_{wire} = 4, 6, and 8, d_{wire} = 14 mm; (**c**) R_{wire} = 1.00 mm, N_{wire} = 6, d_{wire} = 12, 14, and 16 mm.

The results in Figure 13 and Table 3 indicate that the average sensitivities and operation ranges of the proposed twisted dual-cycle fiber optic bending loss sensor head can be increased by increasing the number of steel wires and the radius of the steel wires, and by decreasing the distance between steel wires. Therefore, the sensitivities, sensing range, and the operating range of the proposed FOS head can be readily adjusted depending on the deformation characteristics of the measurement target.

Table 3. Measurement results of three-point flexural test.

FOS Heads	Average Sensitivity		Curve Fitting Results ($h \times \varepsilon_f^{\,k}$)				Error
	(dB/mm)	(dB/$\mu\varepsilon$)	h	k	R-Square	RMSE	Average (dB, $\mu\varepsilon$)
Case 1-1	1.758	0.002996	43.95	1.216	0.9958	0.10200	0.080, 26.59
Case 1-2	1.777	0.003029	49.14	1.276	0.9972	0.08437	0.070, 23.11
Case 2-1	0.938	0.001599	23.55	1.217	0.9973	0.04419	0.036, 22.62
Case 2-2	0.930	0.001585	29.78	1.362	0.9975	0.04135	0.032, 20.44
Case 3-1	6.267	0.010682	129.8	1.105	0.9981	0.24650	0.194, 18.17
Case 3-2	6.554	0.011172	131.1	1.070	0.9969	0.33200	0.256, 22.88
Case 4-1	1.377	0.002347	36.81	1.251	0.9902	0.12330	0.100, 42.58
Case 4-2	1.448	0.002469	36.80	1.227	0.9984	0.05245	0.042, 16.90
Case 5-1	2.976	0.005073	70.79	1.181	0.9983	0.10980	0.083, 16.30
Case 5-2	3.085	0.005258	66.04	1.129	0.9974	0.13950	0.112, 21.30
Case 6-1	3.294	0.005615	70.91	1.109	0.9964	0.18170	0.134, 23.92
Case 6-2	3.243	0.005529	69.4	1.131	0.9983	0.11960	0.097, 17.58
Case 7-1	1.118	0.001905	45.84	1.505	0.9981	0.04317	0.033, 17.09
Case 7-2	1.187	0.002024	50.45	1.529	0.9963	0.06420	0.050, 24.83

5. Conclusions

A twisted dual-cycle bending structure for ingenerating optical fiber bending loss was proposed for strain measurement. The bending loss characteristics according to the distance between steel wires, wire radius, and number of steel wires were experimentally evaluated to verify the proposed method. The results showed that the twisted dual-cycle bending structure exhibits a linear relationship between detected optical power loss and the spring deflection length (z) corresponding to the applied force. These experimental results can be used for manufacturing an FRP intensity-based sensing element in an intensity-based FOS measurement system.

To apply the proposed FOS head to an experimental measurement, the twisted dual-cycle bending structure was bonded to the FRP, and tensile and three-point bending tests were conducted. Two FRP coupons were tested in each of seven cases and each coupon was tested five times. The measurement range, sensitivity, and average measurement errors of the tensile load and flexural strain were 4.5 kN and 1,760 $\mu\varepsilon$, 0.70 to 3.99 dB/kN and 0.930 to 6.554 dB/mm, and 57.7 N, and 42.6 $\mu\varepsilon$, respectively. The sensing range of FOS head were 82 to 138 mm according to configuration cases.

The advantage of the proposed FOS head structure is that the sensitivity and measurement area can be readily adjusted by changing the radius of the steel wires, the number of steel wires, and the distance between steel wires. The sensitivity can also be improved by reducing the distance between steel wires and/or increasing the radius of the steel wires. In addition, the measurement area of the proposed FOS head can be adjusted by changing the number of steel wires and the distance between steel wires. The bending loss characteristics depend on the FOS head steel wire radius, numbers, and distance. To determine the effects of these parameters, two samples in each of seven configuration cases of the proposed FOS head were bonded to FRP coupons and tensile and flexural strain tests were repeated five times. The manufactured sensor head with same configuration has similar characteristics of the sensitivity and the operating ranges. An additional advantage of FOS implemented with standard single-mode optical fibers are definite: An optical network can be used to obtain real-time measurement information. Many devices developed for optical communications like light source, optic circulator, optical coupler, fiber Bragg grating, wavelength division multiplexing can be used with FOS head for multi-point sensing. In addition, it can measure various types of measurands with arbitrary spatial distribution. We already proposed and demonstrated the self-referencing, intensity-based FOS [3]. The manufactured FOS head can be applied together with the self-referencing intensity-based fiber optic sensor interrogator presented in 2014 [3] that features functionality such as self-referencing, remote sensing, and multiple sensor heads with cascade and/or parallel forms. The proposed FOS head can be manufactured into a patch-type sensor that can be easily installed or removed, which enables more convenient and accurate measurement of structural behavior in the field.

Author Contributions: Conceptualization, S.-J.C. and J.-K.P.; methodology, S.-J.C.; validation, K.G.P. and J.-K.P.; formal analysis, S.-Y.J. and C.L.; investigation, S.-J.C., S.-Y.J. and J.-K.P.; resources, J.-K.P.; data curation, S.-J.C. and S.-Y.J.; writing—original draft preparation, S.-J.C. and S.-Y.J.; writing—review and editing, S.-J.C., C.L. and K.G.P.; supervision, K.G.P. and J.K.P.; project administration, J.-K.P.; funding acquisition, K.G.P. and J.-K.P.

Funding: This work was supported by the Basic Science Research Program through the National Research Foundation of Korea (NRF) funded by the Ministry of Education (2016R1D1A1A09917117) and the Basic Research Programs (GP2017-033) of the Korea Institute of Geoscience and Mineral Resources (KIGAM).

Conflicts of Interest: The authors declare no conflict of interest.

References

1.   López-Higuera, J.M.; Cobo, L.R.; Incera, A.Q.; Cobo, A. Fiber optic sensors in structural health monitoring. *IEEE J. Lightwave Tech.* **2011**, *29*, 587–608. [CrossRef]

2.   Yin, S.S.; Ruffin, P.B.; Yu, F.T.S. Overview of fiber optic sensors. In *Fiber Optic Sensors*, 2nd ed.; CRC Press: Boca Raton, FL, USA, 2008; pp. 1–2. ISBN 9781420053654.

3.   Choi, S.J.; Kim, Y.C.; Song, M.; Pan, J.K. A self-referencing intensity-based fiber optic sensor with multipoint sensing characteristics. *Sensors* **2014**, *14*, 12803–12815. [CrossRef] [PubMed]

4. Silva, R.M.; Baptista, J.M.; Santos, J.L.; Lobo Ribeiro, A.B.; Araújo, F.M.; Ferreira, L.A.; Frazão, O. A simple, self-referenced, intensity-based optical fibre sensor for temperature measurements. *Opt. Commun.* **2013**, *291*, 215–218. [CrossRef]

5. Spillman, W.B.; Lord, J.R. Self-referencing multiplexing technique for fiber-optic intensity sensors. *IEEE J. Lightwave Technol.* **1987**, *5*, 865–869. [CrossRef]

6. García, C.V.; Montalvo, J.; Lallana, P.C. Radio-frequency ring resonators for self-referencing fibre-optic intensity sensors. *Opt. Eng. Lett.* **2005**, *44*, 040502. [CrossRef]

7. Vázquez, C.; Montalvo, J.; Montero, D.S.; Pena, J.M.S. Self-referencing fiber-optic intensity sensors using ring resonators and fiber Bragg gratings. *IEEE Photon. Technol. Lett.* **2006**, *18*, 2374–2376. [CrossRef]

8. Montalvo, J.; Frazão, O.; Santos, J.L.; Vázquez, C.; Baptista, J.M. Radio-frequency self-referencing technique with enhanced sensitivity for coarse WDM fiber optic intensity sensors. *IEEE J. Lightwave Technol.* **2009**, *27*, 475–482. [CrossRef]

9. Montero, D.S.; Vázquez, C. Remote interrogation of WDM fiber-optic intensity sensors deploying delay lines in the virtual domain. *Sensors* **2014**, *13*, 5870–5880. [CrossRef] [PubMed]

10. Perez-Herrera, R.A.; Pereira, D.A.; Frazão, O.; Castro Ferreira, J.M.; Santos, J.L.; Araújo, F.M.; Ferreira, L.A.; Baptista, J.M.; Lopez-Amo, M. Optimization of the frequency-modulated continuous wave technique for referencing and multiplexing intensity-based fiber optic sensors. *Measurement* **2011**, *44*, 230–237. [CrossRef]

11. Montalvo, J.; Araújo, F.M.; Ferreira, L.A.; Vázquez, C.; Baptista, J.M. Electrical FIR filter with optical coefficients for self-referencing WDM intensity sensors. *IEEE Photon. Technol. Lett.* **2008**, *20*, 45–47. [CrossRef]

12. Wang, Q.; Farrell, G.; Freir, T. Theoretical and experimental investigations of macro-bend losses for standard single mode fibers. *Opt. Express* **2005**, *13*, 4476–4484. [CrossRef] [PubMed]

13. Harris, A.J.; Castle, P.F. Bend loss measurements on high numerical aperture single-mode fibers as a function of wavelength and bend radius. *IEEE J. Lightwave Technol.* **1986**, *4*, 34–40. [CrossRef]

14. Faustini, L.; Martini, G. Bend loss in single-mode fibers. *IEEE J. Lightwave Technol.* **1997**, *15*, 671–679. [CrossRef]

15. Renner, H. Bending losses of coated single-mode fibers: A simple approach. *IEEE J. Lightwave Technol.* **1992**, *10*, 544–551. [CrossRef]

16. Lu, W.H.; Chen, L.W.; Xie, W.F.; Chen, Y.C. A sensing element based on a bent and elongated grooved polymer optical fiber. *Sensors* **2012**, *12*, 7485–7495. [CrossRef] [PubMed]

17. Kuang, J.H.; Chen, P.C.; Chen, Y.C. Plastic optical fiber displacement sensor based on dual cycling bending. *Sensors* **2010**, *10*, 10198–10210. [CrossRef] [PubMed]

18. Wang, W.C.; Ledoux, W.R.; Sangeorzan, B.J.; Reinhall, P.G. A shear and plantar pressure sensor based on fiber-optic bend loss. *J. Rehabil. Res. Dev.* **2005**, *42*, 315–325. [CrossRef] [PubMed]

19. Abe, T.; Mitsunaga, Y.; Koga, H. A strain sensor using twisted optical fibers. *IEEE J. Lightwave Technol.* **1989**, *7*, 525–529. [CrossRef]

20. Zendehnam, A.; Mirzaei, M.; Farashiani, A.; Farahani, L.H. Investigation of bending loss in a single-mode optical fibre. *Pramana-J. Phys.* **2010**, *74*, 591–603. [CrossRef]

21. Qiu, J.; Zheng, D.; Zhu, K.; Fang, B.; Cheng, L. Optical fiber sensor experimental research based on the theory of bending loss applied to monitoring differential settlement at the earth-rock junction. *J. Sens.* **2015**, *2015*, 346807. [CrossRef]

22. Marcuse, D. Curvature loss formula for optical fibers. *J. Opt. Soc. Am.* **1976**, *66*, 216–220. [CrossRef]

23. Song, H.C.; Yum, J.S. A study of the mechanical properties of fiberglass reinforcements with constitution of lay-up, manufacturing method, and resins. *J. Ocean. Eng. Technol.* **2010**, *24*, 75–80.

24. Čapek, J.; Vojtěch, D.; Oborná, A. Microstructural and mechanical properties of biodegradable iron foam prepared by powder metallurgy. *Mater. Des.* **2015**, *83*, 468–482. [CrossRef]

Article

Comparative Analysis of Deformation Determination by Applying Fiber-optic 2D Deflection Sensors and Geodetic Measurements

Marko Z. Marković [1,*], Jovan S. Bajić [2], Mehmed Batilović [1], Zoran Sušić [1] and Goran M. Stojanović [2]

[1] Department of Civil Engineering and Geodesy, Faculty of Technical Sciences, University of Novi Sad, Novi Sad 21101, Serbia; mehmed@uns.ac.rs (M.B.); zsusic@uns.ac.rs (Z.S.)

[2] Department of Power, Electronic and Telecommunication Engineering, Faculty of Technical Sciences, University of Novi Sad, Novi Sad 21101, Serbia; bajic@uns.ac.rs (J.S.B.); anajoza@uns.ac.rs (A.J.); sgoran@uns.ac.rs (G.M.S.)

* Correspondence: marko_m@uns.ac.rs; Tel.: +381-65-6056-053

Received: 13 December 2018; Accepted: 31 January 2019; Published: 18 February 2019

Abstract: In the paper the description of an experiment for a comparative analysis of two different methods for deformation determination, geodetic and 2D deflection sensors based on fiber-optic curvature sensors (FOCSs) is given. The experiment is performed by a using specially designed assembly which makes it possible to apply both methods. For performing geodetic measurements, a geodetic micro-network is established. Measurements by applying a 2D deflection sensor and three total stations are carried out for comparison. The data processing comprises graphical and numerical analysis of the results. Based on the presented results the potential of 2D deflection sensor application in structural health monitoring (SHM) procedures is indicated. The analysis of the measurement results also indicates the importance of integrating various types of sensors for obtaining more accurate and more reliable deformation measurements results.

Keywords: fiber-optic curvature sensor; total station; deformation analysis; measurements; displacement

1. Introduction

Civil engineering objects (bridges, tunnels, dams, etc.) are exposed to deformations under the influence of various factors, such as changes of ground water level, tectonic phenomena, landslides, etc. The development and integration of interdisciplinary methods of measuring and data processing have undergone a revolutionary transformation from the mere noting and describing a deformation towards the analysis of what causes the phenomenon of deformations. There are several measurement methods aimed at deformation detection. They can be divided into geodetic and non-geodetic methods.

The conventional geodetic measurements offer a high precision in the relative positioning of the discrete (control) points and yield a global picture of deformations affecting the object under observation. However, such measurements are slow and their adaptation to continuous and automatic monitoring is complicated and relatively expensive [1].

Geotechnical sensors provide exceptionally precise information about deformations and they are easily adapted to continuous, completely automatic and telemetric data acquisition. Compared to conventional and satellite geodetic methods, geotechnical sensors are mostly independent of some outdoor conditions, such as snow cover and weak visibility. However, the information provided by them is only local, concerning discrete points [1].

The geodetic methods are based on establishing a geodetic micro-network of points which are related through different types of measurements (measurements of angles, lengths, altitude differences,

GNSS vectors). In order to calculate statistical quality estimation and errors identification, measurements are performed with optimal redundancy. By applying such methods, global information about the behavior of an engineering object exposed to deformations can be obtained. Short periodic measurements of objects (dams, bridges, etc.) deformations by geodetic methods are performed based on points of the geodetic micro-network. This concept is also shown in this paper.

Depending on the type of optical fibers used and the change of the optical signal properties due to environmental influences, there are various fiber-optic sensor (FOS) configurations. In practice, the most widely used FOS are based on one of the four principles: change of light intensity (intensiometric FOS), spectrum, polarization or phase (interferometric FOS) [2,3]. In all four cases the change of physical quantity interacts with the light in the optical fiber or with a non-fiber sensor which is connected to the optical fiber so that it can register changes of intensity, spectrum, polarization or phase [4]. The main advantages of most sensors based on FOS technology are the use of low-power energy sources, immunity to strong electromagnetic fields, corrosion resistance, small size, high sensitivity and large bandwidth.

Optical fibers with fiber Bragg gratings (FBGs) are increasingly being used as photonic sensors for various applications in SHM [4]. These highly precise systems require however expensive light sources and spectrum analyzers. Also, they cannot distinguish between concave and convex bending. In addition, their accuracy is affected by external factors, such as temperature, and because of that their application is limited [5]. The application of FBGs, similar to the 2D deflection sensor application in deformation measurement which will be explained in more detail in the following text, is described in [6]. Deformation measurements based on integrated geodetic and FBG sensors system are presented in [7].

Fiber-optic interferometric sensors and curvature analysis using parallel sensor topology are explained in details in [8–10]. The main advantage of interferometric FOSs is their great potential in practical applications, such as monitoring deformations of airplanes, ships and constructions in real time [10]. Nevertheless, these high-accuracy systems require relatively complicated measurement systems and because of that they are often considered expensive [5].

The FOSs may be designed so that they can discriminate in the spatial mode, and in this way, the measurand can be determined along the length of the fiber itself, in a process normally termed distributed sensing. This principle has been employed widely in the temperature measurement using non-linear effects in fibers, such as Brillouin or Raman scattering or in some types of strain sensing [11–13].

By using FOS based on a change of light intensity it is possible to measure bending deformations of the structure. Depending on the mechanical configuration of an intensiometric FOS many physical quantities, such as strain, torsion, position, can be calculated on the basis of bending measurements [14]. The sensitivity of an optical fiber to bending can be increased by applying various types of structural imperfections on the surface of an optical fiber. By applying such imperfections on the optical fiber, besides increasing the bending sensitivity, it is also possible to determine the bending direction (positive or negative) [15–18]. Also, in the case of FOCS a simple system for signal demodulation is sufficient, in contrast to expensive techniques of processing and analyzing signals from interferometric/polarimetric sensors with problems, such as signal fading, interrupt and sign ambiguity, nonlinearity and multi-valued response [19]. The FOCS operation principle is given in more detail in the study by Fu and Di [14].

The main characteristics of the 2D deflection sensor are the ability to monitor both static and dynamic deformations with high observation frequency, high resolution, accuracy and reliability of measurements, long-term preservation and stability, low cost and simple implementation. Compared to other similar sensor solutions, the 2D deflection sensor is characterized by a robust and low-cost design making it an economical and suitable solution for application in the SHM process. This is supported by the fact that to produce 2D deflection sensor robust and low-cost plastic (PMMA) optical fibers were used. In this paper, a new approach, developed for monitoring of engineering structures, is considered, which, in addition to geodetic measurements, applies 2D deflection sensor in detecting local deformations [20].

2. Geodetic Deformation Analysis Based on Robust Iterative Weighted Similarity Transformation (IWST) Method

The models of geodetic deformation analysis are divided into descriptive and cause-response models [21,22]. Congruence and kinematic models may be classified as descriptive ones, whereas static and dynamic models belong to the cause-response models. The most important task in deformation analysis is to correctly identify unstable points and to isolate them from the set of stable ones. In the literature, various approaches can be found based on congruence models [23], on common adjustment of two measuring epochs based on assumed stable points determined by geoengineering research [24], robust estimates [25–27], finite element strain analysis method [22], and others. Most conventional models are based on the method of least squares at all measuring epochs (congruence model, common-adjustment model, finite element strain analysis method). However, a model based on the least squares method is not always realistic. If the data follow a normal distribution, the least squares will yield the most probable values for the estimated parameters. If the assumptions concerning the model are incorrect, due to non-modelled systematic influences (even of a small magnitude), or to correlated observations, the chosen distribution must be modified. Thus, robust variants of standard estimates have been created through which an estimation of parameters is attempted without the influence of the deviation model. This is namely the reason why in the present paper the method of robust deformation estimating is chosen for the needs of analyzing the displacements detected by geodetic measurements.

After the publication of Huber's paper [28], robust methods have been applied more frequently in deformation analysis. Their basic characteristic is that the parameter estimate is done without a priori assumption about a normal distribution, i.e. the nature of the deviations, which is the main characteristic of the congruence model. The estimates should be close to the true values, even when the data contain gross errors, so that bearing in mind the correct model and the data free of errors they yield almost optimal results [29].

Frequently used robust methods are Iterative Weighted Similarity Transformation (IWST), which was developed at the University of New Brunswick in Canada [26], and Least Absolute Sum [30]. Both methods are based on the application of the S-transformation for detecting the trend of points displacement.

The deformation analysis procedure based on the performed geodetic measurements using the robust estimates is a classic approach based on the IWST method. This method is used for the purpose of estimating the trend of the point displacement and it finds a significant application in numerous studies, among which the best known are Tevatron atomic particle accelerator complex at the Fermilab Laboratory in the USA and automated ALERT monitoring system developed by the Canadian Centre for Geodetic Engineering [31].

The IWST method satisfies the condition of minimum sum of the component moduli of the displacement vector [25–27,31,32]. The method is based on the S transformation (Helmert's similarity transformation):

$$\left.\begin{aligned}
\overset{\wedge(k)}{\mathbf{d}} &= \mathbf{S}^{(k)}\mathbf{d} \\
\mathbf{Q}_{\overset{\wedge}{\mathbf{d}}}^{(k)} &= \mathbf{S}^{(k)}\mathbf{Q_d}\left(\mathbf{S}^{(k)}\right)^T \\
\mathbf{W}^{(k+1)} &= diag(\ \cdots\ , w_{S_i}^{(k+1)},\ \cdots\ \ \cdots\ , 0,\ \cdots\ )
\end{aligned}\right\}_{k=1,\dots} \qquad (1)$$

where $\mathbf{d} = \overset{\wedge}{\mathbf{x}}_2 - \overset{\wedge}{\mathbf{x}}_1$ is the displacement vector, $\mathbf{Q_d} = (\mathbf{Q}_{\overset{\wedge}{\mathbf{x}}_1} + \mathbf{Q}_{\overset{\wedge}{\mathbf{x}}_2})$ cofactor displacement matrix, $\mathbf{S}^{(k)} = \mathbf{I} - \mathbf{H}(\mathbf{H}^T\mathbf{W}^{(k)}\mathbf{H})^{-1}\mathbf{H}^T\mathbf{W}^{(k)}$ S transformation matrix, $\mathbf{I}$ unit matrix, $\mathbf{H}$ matrix of datum conditions and $\mathbf{W}$ weight matrix.

In the case of two-dimensional geodetic networks, the matrix of datum conditions **H** has the following form:

$$\mathbf{H} = \begin{bmatrix} \vdots & \vdots & & \vdots & \vdots \\ 1 & 0 & & -\bar{y}_i & \bar{x}_i \\ 0 & 1 & & \bar{x}_i & \bar{y}_i \\ \vdots & \vdots & & \vdots & \vdots \end{bmatrix}_{2m \times h}$$

where m is the number of points in the network, h the number of all datum network parameters, and $\bar{y}_i$ and $\bar{x}_i$ are adjusted coordinates of the point from the zero-epoch reduced to the network centroid. The first two columns of the matrix represent the translation along the coordinate axes y and x, the third column represents the rotation about the z axis, whereas the fourth one defines the scale [25–27,31,33].

Only the points of the basic network can participate in the optimization process (1). Thus, the weights of the points on the object must be zero, because then the points will not be corrected and in this way, they will not participate in the optimization process. In the first iteration of the transformation ($k = 1$) the weight matrix is the unit matrix ($\mathbf{W} = \mathbf{I}$). In the subsequent iterations the weights of the basic network points are determined in the following way:

$$w_{S_i}^{(k+1)} = 1/|\hat{d}_i^{(k)}| \tag{2}$$

where $\hat{d}_i$ is the corresponding component of the displacement vector for a point ($\hat{d}_{y_i}$ or $\hat{d}_{x_i}$). During the iterative optimization process (1), some values of $\hat{d}_i$ can be very close to zero, causing numerical instability when forming a weight matrix $\mathbf{W}$. Because of this (2) is modified in the following way:

$$w_{S_i}^{(k+1)} = 1/(|\hat{d}_i^{(k)}| + c)$$

where c is the assumed value of the tolerance (for instance, $c = 0.1$ mm). Iterative process (1) is performed until the differences between successively transformed displacement vectors $|\hat{d}^{(k+1)} - \hat{d}^{(k)}|$ are less than the assumed tolerance c.

For the purposes of testing the stability of the network points, the displacement vector and the corresponding cofactor matrix from the last iteration are used. The stability examination of the network points is performed by applying the single-point test. For this purpose, hypotheses are set:

$$H_0 : E(\hat{\mathbf{d}}_i) = 0 \text{ against } H_a : E(\hat{\mathbf{d}}_i) \neq 0$$

where $\hat{\mathbf{d}}_i$ is the displacement vector of the i-th point. The test statistics is formed according to the following equation:

$$T_i = \frac{\hat{\mathbf{d}}_i^T \mathbf{Q}_{\hat{\mathbf{d}}_i}^{-1} \hat{\mathbf{d}}_i}{h_i \hat{\sigma}_0^2} \sim F_{1-\alpha,\, h_i,\, f} \tag{3}$$

where $\hat{\mathbf{d}}_i$ is the displacement vector, $\mathbf{Q}_{\hat{\mathbf{d}}_i}$ cofactor matrix of the displacement vector, $h_i = rank(\mathbf{Q}_{\hat{\mathbf{d}}_i})$, $f = f_1 + f_2$ total number of degrees of freedom from two measuring epochs and $\hat{\sigma}_0^2 = (f_1 \hat{\sigma}_{0_1}^2 + f_2 \hat{\sigma}_{0_2}^2)/f$ total a posteriori dispersion coefficient from two measuring epochs.

If $T_i \leq F_{1-\alpha,\, h_i,\, f}$, the null hypothesis is not rejected, and the point can be regarded as stable. When $T_i > F_{1-\alpha,\, h_i,\, f}$, the null hypothesis is rejected, and it can be concluded that the point is significantly displaced. Detailed explanations concerning the IWST method can be found in the aforementioned publications [25–27,31].

3. 2D Deflection Sensor Design

The basis of the 2D deflection sensor is a polyamide beam (1020 mm long with 30 mm diameter). On the polyamide beam surface, by using precise tools, five trenches are engraved where four of them have triangular cross sections and are positioned with 90° spacing between them (Figure 1). The fifth trench having a rectangular cross section is positioned in the middle, between two neighboring, arbitrarily selected trenches, and within it the thermistor for temperature monitoring and compensation is placed.

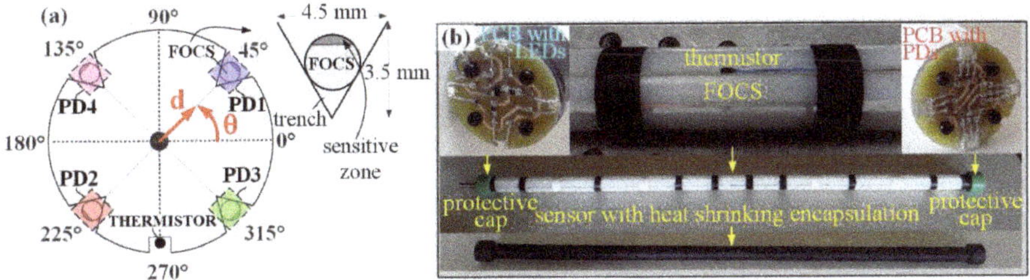

Figure 1. (**a**) Arrangement of the FOCSs and PDs at beam cross section in respect to the reference coordinate system; (**b**) Photo of the fabricated sensor.

The four trenches contain four plastic optical fibers, each of 1.5 mm diameter, which constitute 2D deflection sensor (Figure 2). On each optical fiber structural imperfections (teeth) are applied to increase its sensitivity to bending (Figure 2).

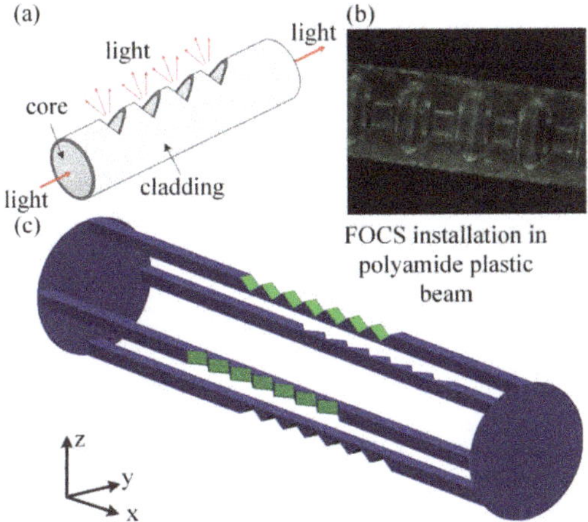

Figure 2. 2D deflection sensor based on FOCS, (**a**) Light propagation; (**b**) Photo of teeth; (**c**) FOCSs installation within a polyamide beam.

Plastic optical fibers are chosen because of their robustness and low price, as well as simplicity of applying teeth on their surface. The cutting is done by using a precise tool, a Protomat S100 device, produced by LPKF Laser & Electronics AG (Grabsen, Germany). The total number of teeth is 50. Spacing between individual teeth is 1.1 mm, and each of them is 0.35 mm deep (Figure 3). Optical fibers 1 and 3, as well as 2 and 4, are installed mutually parallel within the beam. In this way it is possible that during the deformation of a 2D deflection sensor, by observing the parallel, diametrically opposite optical fibers, one of them detects positive and the other one negative bending.

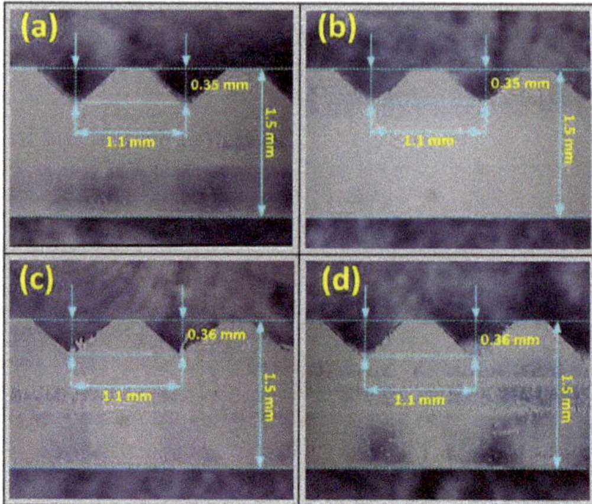

Figure 3. Teeth of the FOCS.

Considering the size of the engraved trenches any uncontrolled displacement of FOCSs is impossible. To prevent any FOCSs moving outside of the trenches (falling out) all optical fibers are additionally fixed by applying silicon. Also, the whole 2D deflection sensor is protected with the heat-shrinkable encapsulation. FOCSs are installed in such a way that in all trenches their teeth are oriented towards the trench top. FOCSs are, on one side, connected to a light source (LED) and, on the other side, to photodetector (PD). LEDs and PDs are mounted on the circular shaped (30 mm in diameter) printed circuits boards (PCBs) that are screwed to the beam ends (Figure 1b). Both, the LEDs and PDs, are connected to PC via NI USB-6351 card within which the electronics provided for operation of the entire system including FOCSs is integrated and programmed. In this way, the readings of the light intensity are provided in real time. FOCSs are connected to independent LEDs and PDs [20]. The described method of 2D deflection sensor installation makes it possible to perform differential measurements, as well as a higher sensitivity in the measurements of deformations.

2D deflection sensor development, design, calibration and characterization were performed by some of the authors of this manuscript. The values obtained for positional and angular accuracies are ±0.15 mm and ±2.5°, respectively, for the resolution (1σ) of 0.01 mm and 0.33°, respectively. The calibration and characterization were done in a laboratory of the Faculty of Technical Sciences in Novi Sad. Results are given in detail in the study by Bajic et al. [20].

4. Test description

The experiment is aimed at comparing two methods of deformation determination, i.e. at comparison of 2D deflection sensor with the geodetic deformation analysis model. To carry out the geodetic measurements a geodetic micro-network consisting of four points is formed (Figure 4). Along a straight line, at distances of about 44 m from point 2 and 3 m from point 4 a setup containing a 2D deflection sensor was placed (Figure 5).

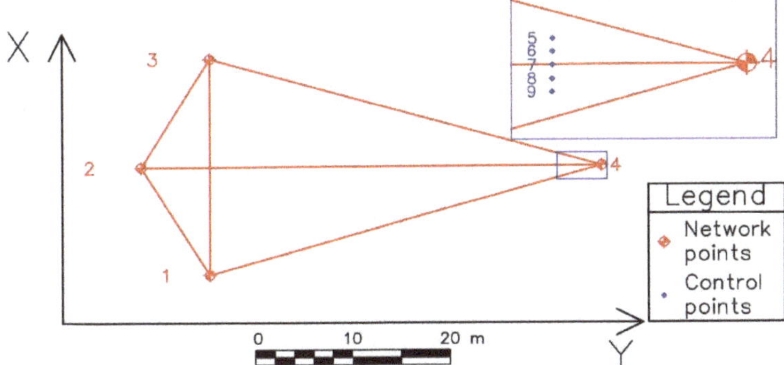

Figure 4. Geodetic micro-network.

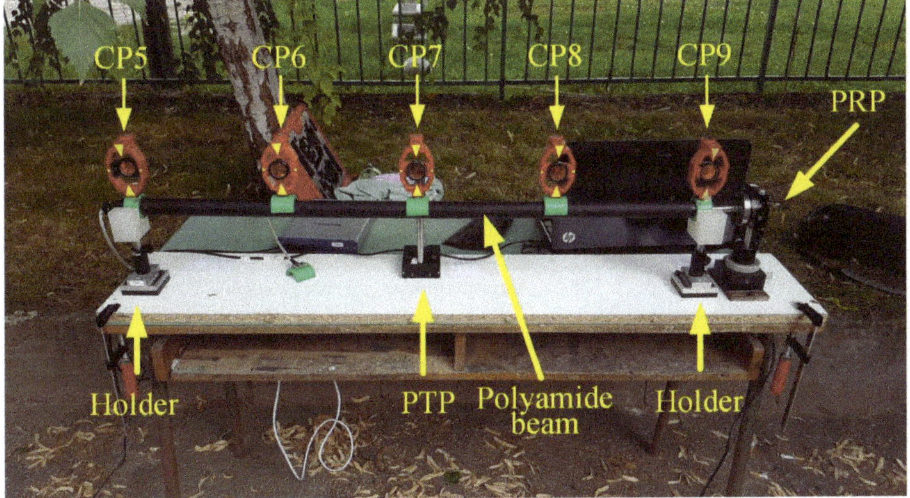

Figure 5. Experimental setup.

On the 2D deflection sensor there are five mini prisms (control points (CP) 5, 6, 7, 8 and 9) placed at approximately equal mutual distances (Figures 4 and 5). The setup is arranged in the way that the direction of the 2D deflection sensor on which the prisms are is parallel to that defined by points 1 and 3 of the geodetic micro-network. The mini prisms can be observed from points 1, 2 and 3 of the geodetic micro-network, whereas their position is perpendicular to point 2. On a wooden beam, at a 900 mm distance, two holders are tightly fixed. Their role is to prevent any movement of the 2D deflection sensor in the negative direction of the Y axis (Figure 4) and to enable its setting to the optimal height above the wooden beam which is placed on the table. Undesirable 2D deflection sensor movement in the positive direction of the Y axis is not possible because on the other side is placed a precise translation positioner (PTP), PT1/M by Thorlabs (Newton, NJ, USA), that performs the loading directed to the point in the middle of the 2D deflection sensor. The 2D deflection sensor is also at one end tightly fixed to a precise rotation positioner (PRP), PR01/M by Thorlabs, which enables the 2D deflection sensor to rotate and reach a desirable position.

The experiment was performed for comparative determination of the simulated (by PTP) displacement values by applying geodetic measurements and 2D deflection sensor. During the performing geodetic measurements, the following measured quantities were registered: horizontal and vertical angles and oblique lengths. The measurements within the geodetic micro-network are performed in three gyruses with forced centering of the instrument and signal on the tripods, whereas the measurements of the CPs on the sensor are carried out in six independent series, in two gyruses. The CPs are measured simultaneously with three total stations, from points 1, 2 and 3 of the geodetic micro-network. In the zero series the PTP is only leaned on the 2D deflection sensor, without applying additional force, to prevent any movement of the sensor in the positive direction of the Y axis. In every subsequent series the PTP is moved in the negative direction of the Y axis by 1 mm whereby a force action is applied to the middle of the 2D deflection sensor. The 2D deflection sensor approximate position is regulated before the measurements begin by using the PRP so that fibers 3 and 4 are in the horizontal plane, and fibers 1 and 2 are in the vertical plane (Figure 1a). The 2D deflection sensor readings are registered twice, at the beginning and at the end of the geodetic measurements, i.e. after and before the PTP action. Considering the coordinate system of the geodetic micro network, the 2D deflection sensor provides the registration of displacements along the Y and Z axes. The equipment used in the geodetic measurements is a Leica Geosystems set (Heerbrugg, Switzerland). From points 1 and 3 the measurements are performed with a TCRP1201+ total station with declared accuracies of 1″ for direction measurement and 1 mm + 1.5 ppm for length measurement. From point 2 a TS06 total station is used with declared accuracies of 2″ for direction measurement and 1.5 mm + 2 ppm for length measurement.

5. Measurements results and discussion

The geodetic deformation analysis is performed by applying the IWST method. As already said, iteration process (1) is continued until the differences of the successively transformed displacement vectors are under a value specified for the tolerance c. The value assumed for the tolerance is equal to 0.001 mm.

The results of simulated (by PTP) displacement measurements detected by 2D deflection sensor are presented in Table 1, whereas in Table 2 the measurement results of rotation angle values in the Y-Z plane for each individual series are given. The results of the geodetic measurements are presented in Table 3. It should be noted that the applied geodetic measurements measure both, deformed shape and rigid body movements, while 2D deflection sensor measures only deformed shape. As explained earlier, the experiment is performed in five variants with simulated displacements from 1 to 5 mm in steps of 1 mm. Accordingly, in Table 3 the results of the estimated components of the displacement vector in the Y and X axes are presented in five columns, as well as the values of the test statistics (3). In Table 3 in each column the points identified as unstable are indicated (test statistics for (3) exceeds the tabular value of the Fisher distribution depending on the chosen probability and number of degrees of freedom). In all calculations the standard value for the probability is used $(1 - \alpha = 0.95)$. The measurements in the Z axis are not analyzed, both for 2D deflection sensor and for the geodetic measurements, because the deviations in the Z axis are negligible compared to the registered displacements. Standard deviations of displacement vector components are presented in Table 4. In the first measurement series, using IWST method, significant displacements were not identified because simulated displacements reach the limit of measurement accuracy.

Table 1. Results of measuring displacements by applying 2D deflection sensor.

Point Number	0-1	0-2	0-3	0-4	0-5
	$\hat{d}(\text{mm})$	$\hat{d}(\text{mm})$	$\hat{d}(\text{mm})$	$\hat{d}(\text{mm})$	$\hat{d}(\text{mm})$
7	0.77	1.69	2.62	3.50	4.43

Table 2. Results of reading the values of angles by applying 2D deflection sensor.

Point Number	0	1	2	3	4	5
	Angle (°)	Angle (°)	Angle (°)	Angle (°)	Angle (°)	Angle (°)
7	309.2964	306.2564	309.5777	310.8704	311.2057	311.9849

Table 3. Geodetic measurement results.

Point Number	0-1		0-2		0-3		0-4		0-5	
	$\hat{d}_y$ / $\hat{d}_x$ (mm)	T_i	$\hat{d}_y$ / $\hat{d}_x$ (mm)	T_i	$\hat{d}_y$ / $\hat{d}_x$ (mm)	T_i	$\hat{d}_y$ / $\hat{d}_x$ (mm)	T_i	$\hat{d}_y$ / $\hat{d}_x$ (mm)	T_i
1	0.00 / −0.12	0.14	0.00 / 0.00	0.00	0.01 / −0.07	0.03	0.00 / 0.05	0.02	0.00 / 0.00	0.00
2	−0.05 / 0.00	0.05	0.00 / 0.01	0.00	−0.02 / 0.00	0.00	0.03 / 0.00	0.02	0.01 / 0.00	0.00
3	0.00 / 0.12	0.12	0.00 / −0.01	0.00	0.02 / 0.07	0.03	0.02 / −0.04	0.02	0.00 / −0.01	0.00
4	0.07 / 0.00	0.01	−0.03 / 0.00	0.00	−0.01 / 0.00	0.00	−0.12 / 0.00	0.04	−0.04 / 0.00	0.00
5	−1.04 / 0.29	1.60	−1.13 / −0.09	1.43	−1.33 / 0.55	2.80	−1.28 / 0.10	1.71	−1.70 / 0.13	3.23
6	−1.25 / 0.39	2.43	−1.78 / −0.04	3.50[a]	−2.22 / 0.49	5.96[a]	−3.12 / 0.19	10.15[a]	−3.90 / 0.14	16.89[a]
7	−1.23 / 0.13	1.90	−1.95 / −0.45	4.83[a]	−3.09 / 0.40	10.71[a]	−3.85 / 0.13	15.48[a]	−4.75 / −0.05	24.90[a]
8	−0.74 / 0.25	0.87	−1.03 / −0.29	1.42	−1.95 / 0.38	4.49[a]	−2.50 / −0.08	6.58[a]	−3.35 / 0.20	12.52[a]
9	−0.28 / 0.45	0.78	−0.32 / −0.02	0.11	−0.30 / 0.36	0.47	−0.46 / 0.20	0.31	−0.95 / 0.32	1.31

[a] The test statistics exceeds the critical value $F_{0.95,\,2,70} = 3.13$.

Figure 6 presents the values of the relative displacements expressed in mm registered by 2D deflection sensor and geodetic measurements. It can be noticed that for the simulated displacement of 1 mm (by PTP), the value determined by applying 2D deflection sensor in the first series is equal to 0.77 mm. The mean displacement value in all other series is 0.915 mm with a standard deviation of 0.017 mm which is in accordance with the positional accuracy of 2D deflection sensor equal to ±0.15 mm obtained in the sensor calibration. The average values of the standard deviations of the direction and length measurements with total stations to the CPs are 1.7″ and 0.1 mm, respectively. The average values of the standard deviations of the direction and length measurements with total stations in geodetic network are 1.5″ and 0.05 mm, respectively. These values are consistent with the declared accuracies of used total stations. Maximum values of displacement vector standard deviations for CPs 5 to 9 are 0.69 mm and 0.42 mm, for Y and X axes respectively. Therefore, measurements using 2D deflection sensor yield more precise results, but it can also be concluded that both technologies have a sub-millimeter accuracy. It can be also noticed, by inspecting Table 2, that only during the first measurement series the value of the measured angle by applying 2D deflection sensor was decreased compared to the zero series. In all other series the value of the measured angle gradually increased to reach a value of 311.9849° in the fifth series.

Table 4. Standard deviations of displacement vector components.

Point Number	0-1	0-2	0-3	0-4	0-5
	$\sigma_{\hat{d}_y}$	$\sigma_{\hat{d}_y}$	$\sigma_{\hat{d}_y}$	$\sigma_{\hat{d}_y}$	$\sigma_{\hat{d}_y}$
	$\sigma_{\hat{d}_x}$	$\sigma_{\hat{d}_x}$	$\sigma_{\hat{d}_x}$	$\sigma_{\hat{d}_x}$	$\sigma_{\hat{d}_x}$
	(mm)	(mm)	(mm)	(mm)	(mm)
1	0.09	0.12	0.20	0.03	0.09
	0.25	0.11	0.27	0.27	0.17
2	0.16	0.11	0.22	0.20	0.16
	0.05	0.23	0.01	0.02	0.14
3	0.14	0.14	0.23	0.26	0.14
	0.25	0.28	0.27	0.27	0.26
4	0.46	0.49	0.33	0.50	0.48
	0.04	0.04	0.01	0.01	0.05
5	0.64	0.67	0.68	0.69	0.67
	0.38	0.40	0.41	0.42	0.40
6	0.64	0.67	0.68	0.69	0.67
	0.38	0.40	0.41	0.42	0.40
7	0.64	0.67	0.68	0.69	0.67
	0.38	0.40	0.41	0.42	0.40
8	0.64	0.67	0.68	0.69	0.67
	0.38	0.40	0.41	0.42	0.40
9	0.64	0.67	0.68	0.69	0.67
	0.38	0.40	0.41	0.42	0.40

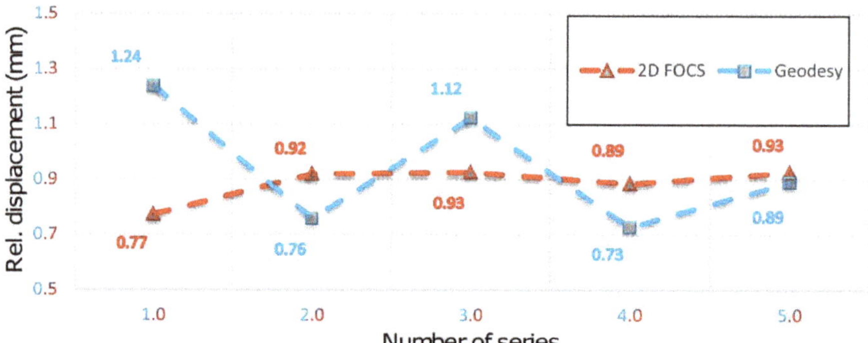

Figure 6. Values of relative displacements registered by 2D deflection sensor and geodetic measurements.

Based on the above facts, it can be concluded that during the simulation of displacement (by PTP) in the first measurement series there was movement of the beam, which was not registered by 2D deflection sensor. This event is result of the fact that during the action of the PTP the beam was not leaned on the holder which resulted in an "idle motion" of about 0.15 mm in the X-Y plane and in a negative rotation of about $3°$ around the beam rotation axis. In addition, the insight into the movements of the CPs 5 and 9 determined by applying geodetic measurements indicates that the beam also rotated in the X-Y plane. This happened because the beam was rigidly bound to the PRP in the immediate neighborhood of CP 9 so that the base rotation point was also a contact between the PTP and the beam. In the first measurement series an error in the experimental setup was evidently manifested. The error occurred first due to the beam rotation, but there is a possibility that the beam was also influenced to a small extent by the stability of the holders, PRP and PTP, basis of the mini prisms on which the PTP was leaned and by the thickness and irregular shape of the heat-shrinkable encapsulation. Since the stability of the entire measurement system was achieved after the first measurement series, the analysis of the collected data is done with and without the zero series. Figure 7 concerns all measurement series, whereas in Figure 8 the zero series is excluded. G (1-5) represent geodetic measurements and F (1-5) represent 2D deflection sensor measurements. An inspection of Figures 7 and 8 shows that by eliminating the zero series the error which appeared during the measurements in the first series is also eliminated to a large extent. Measured results and differences for point 7 by applying 2D deflection sensor and geodetic measuring are presented in Figure 9 (all series included) and Figure 10 (zero series excluded).

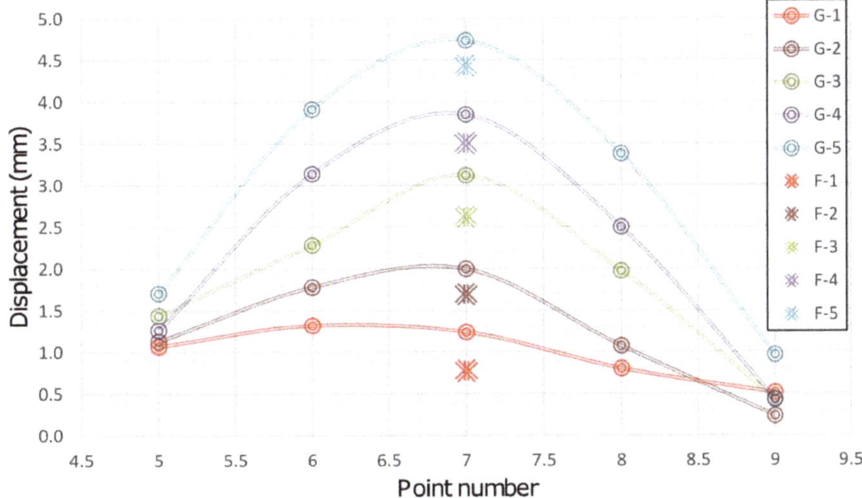

Figure 7. Results of all geodetic and 2D deflection sensor measurements.

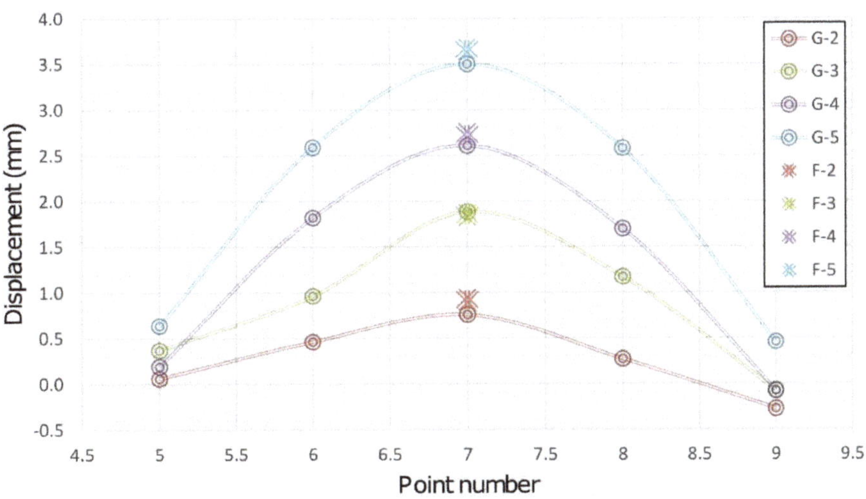

Figure 8. Results of geodetic and 2D deflection sensor measurements without zero series.

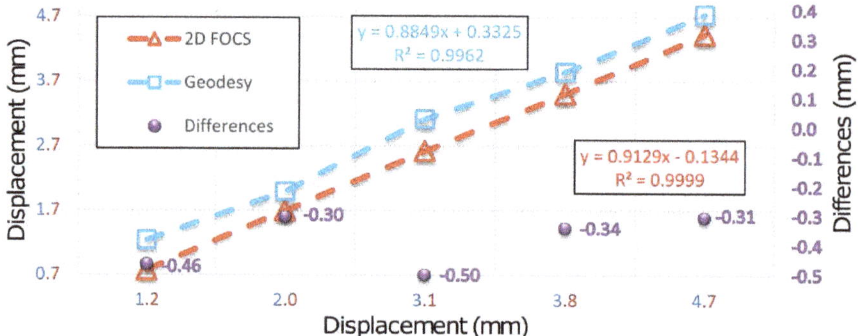

Figure 9. Results and differences for all series of geodetic and 2D deflection sensor measurements for point 7.

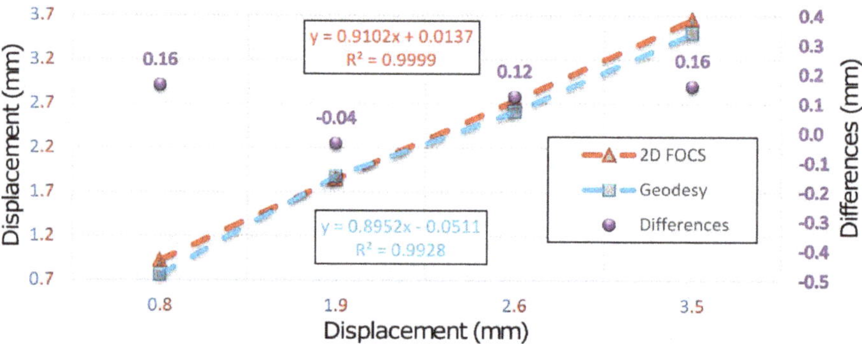

Figure 10. Results and differences of geodetic and 2D deflection sensor measurements for point 7 without zero series.

6. Conclusions

In the present paper a direct comparison of a simple, low-cost 2D deflection sensor for deformation determination and geodetic measurements is done. If only the measurements performed with 2D deflection sensor were considered, it would not be possible to establish that during the measurement process the beam rotation occurred which is confirmed by the geodetic measurements. However, an inspection of the 2D deflection sensor measurements clearly indicates a displacement of the sensor by about 0.15 mm during the motion in the first measurement series, which was not registered by 2D deflection sensor. This is an additional information whereby the effect of the 2D deflection sensor rotation, detected by the geodetic measurements, is also confirmed. The method of measuring deformations by applying an integrated system, consisting of 2D deflection sensor and total stations, would be applicable in the continuous monitoring of arch dams or bridges. It would be necessary to install an optimal number of 2D deflection sensors evenly distributed over the longitudinal section of the object and based on their measurements it would be possible to approximate the geometry of the entire span of the bridge or arch dam. Rigidly stabilized the geodetic reflectors (prisms) would be necessary to place at the support points of the bridge span (pillars, dilatations, etc.), or in the case of an arch dam, on the dam, nearby the touch point of the dam with the surrounding terrain, to monitor absolute deformations of the mentioned building structures. Based on the study described in the present paper it can be concluded that the proposed system with a 2D deflection sensor can be successfully used in the monitoring of deformations. In addition, based on all facts presented in the paper it is possible to reach the conclusion that by integrating several distinct sensor types more complete and more reliable results of deformation measurements can be obtained and, consequently, make the correct decisions in the potential SHM process.

Author Contributions: Conceptualization, M.Z.M., J.S.B. and G.M.S.; Methodology, M.Z.M., J.S.B., M.B. and G.M.S.; Formal Analysis, M.Z.M., J.S.B., M.B. and Z.S.; Investigation, M.Z.M., J.S.B., M.B. and A.J.; Data Curation, M.Z.M., J.S.B., M.B. and A.J.; Writing-Original Draft Preparation, M.Z.M., J.S.B., M.B. and Z.S.; Writing-Review & Editing, M.Z.M., J.S.B., M.B., Z.S., A.J. and G.M.S.; Visualization, M.Z.M., J.S.B. and A.J.; Supervision, Z.S. and G.M.S.; Funding Acquisition, G.M.S.

Funding: The authors gratefully acknowledge the funding provided by the Ministry of Education, Science and Technological Development of Republic of Serbia under project "Development of methods, sensors and systems for monitoring quality of water, air and soil", number III43008. This work is also partly supported by AQUASENSE project, no. 813680.

Conflicts of Interest: The authors declare no conflict of interest.

References

1. Chrzanowski, A.; Chen, Y.; Romero, P.; Secord, J.M. Integration of geodetic and geotechnical deformation surveys in the geosciences. *Tectonophysics* **1986**, *130*, 369–383. [CrossRef]
2. Yin, S.; Ruffin, P.B.; Yu, F.T.S. *Fiber Optic Sensors*, 2nd ed.; Taylor & Francis Group: New York, NY, USA, 2008.
3. Agarwal, T. Introduction to Fiber Optic Sensors and their Types with Applications. Available online: https://www.elprocus.com/diffrent-types-of-fiber-optic-sensors/ (accessed on 9 August 2016).
4. Stupar, D. Elektronski sistem za merenje deformacija pri savijanju pomoću polimernog optičkog vlakna sa osetljivom zonom. Ph.D. Thesis, University of Novi Sad, Novi Sad, Serbia, April 2016.
5. Marković, M.Z.; Bajić, J.S.; Vrtunski, M.; Ninkov, T.; Vasić, D.D.; Živanov, M.B. Application of fiber-optic curvature sensor in deformation measurement process. *Measurement* **2016**, *92*, 50–57. [CrossRef]
6. Metje, N.; Chapman, D.N.; Rogers, C.D.F.; Henderson, P.; Beth, M. An Optical Fiber Sensor System for Remote Displacement Monitoring of Structures—Prototype Tests in the Laboratory. *Struct. Health Monit.* **2008**, *7*, 51–63. [CrossRef]
7. Lackner, S.; Lienhart, W.; Supp, G.; Marte, R. Geodetic and fibre optic measurements of a full-scale bi-axial compressional test. *Surv. Rev.* **2016**, *48*, 86–93. [CrossRef]
8. Glišić, B.; Inaudi, D. *Fibre Optic Methods for Structural Health Monitoring*; John Wiley & Sons, Ltd.: Chichester, UK, 2007.
9. Inaudi, D.; Vurpillot, S.; Casanova, N.; Kronenberg, P. Structural monitoring by curvature analysis using interferometric fiber optic sensors. *Smart Mater. Struct.* **1998**, *7*, 199–208. [CrossRef]
10. Lee, B.H.; Kim, Y.H.; Park, K.S.; Eom, J.B.; Kim, M.J.; Rho, B.S.; Choi, H.Y. Interferometric fiber optic sensors. *Sensors* **2012**, *12*, 2467–2486. [CrossRef]
11. Liehr, S.; Wendt, M.; Krebber, K. Distributed perfluorinated POF strain sensor using OTDR and OFDR techniques. *Proc. SPIE* **2009**, *7503*. [CrossRef]
12. Inaudi, D.; Glišić, B. Integration of distributed strain and temperature sensors in composite coiled tubing. *Proc. SPIE* **2006**. [CrossRef]
13. Grattan, K.T.V.; Sun, T. Fiber optic sensor technology: An overview. *Sens. Actuators A Phys.* **2000**, *82*, 40–61. [CrossRef]
14. Fu, Y.; Di, H. Fiber-optic curvature sensor with optimized sensitive zone. *Opt. Laser Technol.* **2011**, *43*, 586–591. [CrossRef]
15. Babchenko, A.; Weinberger, Z.; Itzkovich, N.; Maryles, J. Plastic optical fibre with structural imperfections as a displacement sensor. *Meas. Sci. Tehnol.* **2006**, *17*, 1157–1161. [CrossRef]
16. Babchenko, A.; Maryles, J. A sensing element based on 3D imperfected polymer optical fibre. *J. Opt. A Pure Appl. Opt.* **2007**, *9*, 1–5. [CrossRef]
17. Lomer, M.; Arrue, J.; Jauregui, C.; Aiestaran, P.; Zubia, J. Lateral polishing of bends in plastic optical fibres applied to a multipoint liquid-level measurement sensor. *Sens. Actuators A Phys.* **2007**, *137*, 68–73. [CrossRef]
18. Lomer, M.; Quintela, A.; López-Amo, M.; Zubia, J.; López-Higuera, J.M. A quasi-distributed level sensor based on a bent side-polished plastic optical fibre cable. *Meas. Sci. Technol.* **2007**, *18*, 2261–2267. [CrossRef]
19. Ma, J.; Asundi, A. Structural health monitoring using a fiber optic polarimetric sensor and a fiber optic curvature sensor—Static and dynamic test. *Smart Mater. Struct.* **2001**, *10*, 181–188. [CrossRef]

20. Bajić, J.S.; Marković, M.Z.; Joža, A.; Vasić, D.D.; Ninkov, T. Design calibration and characterization of a robust low-cost fiber-optic 2D deflection sensor. *Sens. Actuators A Phys.* **2017**, *267*, 278–286. [CrossRef]

21. Heunecke, O. *Zur Identifikation und Verifikation von Deformationsprozessen Mittels Adaptiver KALMAN-Filterung (Hannoversches Filter)*; Fachrichtung Vermessungswesen der University: Hannover, Germany, 1995.

22. Welsch, W.M. Finite element analysis of strain patterns from geodetic observations across a plate margin. *Tectonophysics* **1983**, *97*, 57–71. [CrossRef]

23. Pelzer, H. *Zur Analyse Geodätischer Deformationsmessungen*, 164th ed.; Verlag der Bayer Akad D Wiss: München, Germany, 1971.

24. Heck, B. Das Analyseverfahren des Geodätischen Instituts der Universität Karlsruhe—Stand 1983. In *Geometrische Analyse und Interpretation von Deformationen Geodätischer Netze*; Wissenschaftlicher Studiengang Vermessungswesen, Hochschule der Bundeswehr München: München, Germany, 1983; pp. 153–182.

25. Chen, Y.Q.; Chrzanowski, A.; Secord, J.M. A strategy for the analysis of the stability of reference points in deformation surveys. *CISM J.* **1990**, *44*, 141–149.

26. Chen, Y.Q. *Analysis of Deformation Surveys—A Generalized Method*; University of New Brunswick: Fredericton, NB, Canada, 1983.

27. Setan, H.; Singh, R. Deformation analysis of a geodetic monitoring network. *Geomatica* **2001**, *55*, 333–346.

28. Huber, P.J. Robust Estimation of a Location Parameter. *Ann. Math. Stat.* **1964**, *35*, 1, 73–101. [CrossRef]

29. Niemeier, W. *Ausgleichungsrechnung: Eine Einführung für Studierende und Praktiker des Vermessungs- und Geoinformationswesens*; de Gruyter: Berlin, Germany, 2001.

30. Caspary, W.F.; Borutta, H. Robust estimation in deformation models. *Surv. Rev.* **1987**, *223*, 29–45. [CrossRef]

31. Nowel, K.; Kamiński, W. Robust estimation of deformation from observation differences for free control networks. *J. Geod.* **2014**, *88*, 749–764. [CrossRef]

32. Gökalp, E.; Taşçı, L. Deformation Monitoring by GPS at Embankment Dams and Deformation Analysis. *Surv. Rev.* **2009**, *41*, 86–102. [CrossRef]

33. Caspary, W.F. *Concepts of Network and Deformation Analysis*, 3rd ed.; School of Geomatic Engineering, The University of New South Wales: Kensington, Australia, 2000.

 *sensors*

Article

Research on Optical Fiber Sensor Based on Underwater Deformation Measurement

Jiawang Chen, Chen Cao *, Yue Huang, Yonglei Zhang and Yongqiang Ge

Ocean College, Zhejiang University, Zhoushan 316021, China; arwang@zju.edu.cn (J.C.); 21834170@zju.edu.cn (Y.H.); yonglei@zju.edu.cn (Y.Z.); Ge_yongqiang@zju.edu.cn (Y.G.)
* Correspondence: cc666@zju.edu.cn; Tel.: +86-1336-287-9082

Received: 18 January 2019; Accepted: 2 March 2019; Published: 5 March 2019

Abstract: With the increase in the scale and complexity of underwater engineering, safety problems caused by underwater deformation have become increasingly prominent. Although the intensity fiber curvature sensor can be used for curvature monitoring on the ground, its sensing mechanism is still under investigation. This paper establishes the mathematical model of optical power relative loss and bending radius during deformation of the fiber sensitive region and uses the optical power meter to measure light intensity loss in the sensitive region, which verifies the correctness of the model and reveals the sensing mechanism of the intensity fiber curvature sensor, then optimizes the sensor signal conditioning circuit, applies the sensor to the single-point deformation curvature measurement, and analyzes its measurement error and accuracy. It is proved that the linear measurement range of the sensor is improved when compared with existing similar products.

Keywords: underwater deformation; sensitive region; optical power loss; fiber curvature sensor; measurement range

---

1. Introduction

With the vigorous development of the marine industry in China, the quantity and scale of underwater engineering have been increasing, and underwater deformation detection is imperative [1,2]. For example, the stability of the subsea gas hydrate structure is so susceptible to temperature and pressure changes that its exploitation may result in drastic changes in the seabed topography, destroying subsea engineering facilities such as surrounding oil pipelines. The underwater topographic survey determines the plane coordinates and depth of the underwater topographic point [3]. With the development of underwater acoustic measurement, GPS positioning and computer technology [4], underwater topographic surveys have gone into a new period from traditional optical positioning, single-beam sounding, manual data processing to make use of GPS positioning and multiple sounding depths with automated data processing and diversification of results [5]. Compared with the commonly used single beam, multi-beam, side-scan sonar, GPS [6,7], MEMS and other measurement methods, the intensity type fiber curvature sensor not only has better characteristics than traditional sensor technology, such as large broadband information capacity, long-distance transmission, strong anti-electromagnetic interference capability, strong security and confidentiality [8], but also possesses the advantages of simple structure, low cost, wide measuring range and easy implementation. At present, the monitoring technology of domestic underwater engineering is still in its infancy [9,10], so it is of great significance to apply fiber-optic sensors to the measurement of underwater deformation.

Although the fiber curvature sensor has been widely used in the field of aquatic intelligent health monitoring, the sensing mechanism is still being explored, including the assumptions of Lee Danisch and optical loss, KSC Kuang and the region and R Philip-Chandy and the sensitive region shape [11].

However, most of these explanations lack the support of theoretical and experimental results. They are in the guessing stage which cannot accurately and effectively explain the working mechanism of the intensity-type fiber curvature sensor. Therefore, it is necessary to establish a mathematical model to quantitatively analyze the sensor to reveal the sensing mechanism of this type of sensor.

In recent years, domestic scholars have begun to study this type of sensor. Fu Yili, Liu Renqiang, Di Haiting and others have done some research on the principle characteristics and application of intensity-modulated fiber-optic sensors. Liu Renqiang developed a buried fiber curvature sensor based on the sensitive region of the long-length fiber. It can be used for shape detection of intelligent structures such as bridges [12]. Di Haiting developed a new quasi-distributed intelligent sandwich sensing system based on the sensitive region for the sawtooth fiber curvature sensor, which can be used to monitor the bending deformation of composite structures [13]. However, the above sensor measurement systems are complex and the measurement range is narrow. Therefore, this paper improves and optimizes the sensor measurement system so that it can be better used for underwater deformation measurement.

2. Optical Power Loss in Bending of the Fiber Sensitive Region

Traditional fiber-optic sensors which use fiber macro-bend loss to measure curvature have no surface treatment [14], so sensitivity is very low and the bending direction cannot be distinguished, which is difficult to apply in practical engineering. As shown in Figure 1, a strip-shaped light leakage region of length l and depth h is processed on the surface of the optical fiber to increase the amount of light leakage when the optical fiber is bent, thereby increasing the sensitivity of the optical fiber to bending. Generally, when the positive direction of the fiber sensitive region (the sensitive region is located on the convex side of the curved fiber) is bent and the negative (the sensitive region is located on the concave side of the curved fiber sensor) is bent, the amount of light leakage is different; when the positive direction is bent, the light leakage is larger; while the light is bent in the negative direction, the light leakage is less [15]. This paper will use the theory of light to establish a mathematical model between the relative loss of optical power and the bending radius of the fiber sensitive region to reveal the working mechanism of the intensity modulated fiber sensor [16,17]. Since the measurement range of the fiber curvature sensor is relatively wide, the plastic fiber SH-4001 is used.

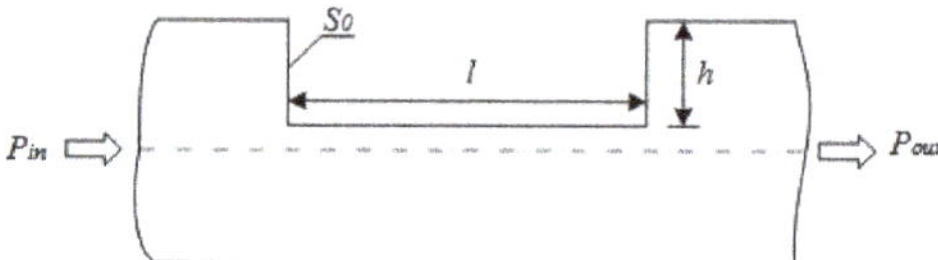

Figure 1. Shape of the fiber sensitive region.

2.1. Transmission Power of the Optical Fiber Sensitive Region during Bending

2.1.1. Positive Bending

For the light propagating in the core, part of the light reaches the sensitive region and leaks out of the core, and another part of the light continues to propagate along the fiber. The optical power of the micro-cell ds_0 on the s_0 plane leaking to the outside of the core at the $d\Omega$ angle is dp_0, and in the positive bending state (the sensitive region is on the convex side), the optical power p_0 leaking from the surface s_0 can be expressed as [11]:

$$\begin{aligned}
P_0 &= \iint_{S_0} dS \int_0^{2\pi} d\theta_\varphi \int_0^{\theta_b} I_0 \sin\theta_0 \cos\theta_0 d\theta_0 \\
&= 2\pi I_0 \int_{a-h}^{a} 2\sqrt{a^2 - h_1^2} dh_1 \int_0^{\theta_b} \sin\theta_0 \cos\theta_0 d\theta_0 \\
&= 2\pi I_0 \int_{a-h}^{a} J(h_1) dh_1
\end{aligned} \tag{1}$$

where h is the depth of the sensitive region and when the light passes through the surface s_0, the optical power ratio α_0 of the leakage is:

$$\alpha_0 = P_0/P_b = \int_{a-h}^{a} J(h_1)dh_1 / \int_{-a}^{a} J(h_1)dh_1, \tag{2}$$

Assuming that the depth h of the processed fiber sensitive region is uniform and the light in the core leaks out of the core at the sensitive region, it is completely absorbed by the medium outside the core and is no longer returned to the core. According to the assumption, the fiber sensitive region of length l is regarded as N s_0 planes, and the leakage ratio of optical power of each surface is α_0, so the relationship between the output optical power and the bending radius R of the fiber and the depth h of the sensitive region, that is, the mathematical model of the positive direction bending resulting in optical power loss is:

$$P_{out} = P_b(1 - \alpha_0)^N, \tag{3}$$

The relative loss of optical power in the sensitive region can be expressed as:

$$\delta P = (P_b - P_{out})/P_b, \tag{4}$$

The relative loss of optical power in the sensitive region is obtained when the optical fiber sensor is in a positive bending state.

$$\delta P = 1 - (1 - \alpha_0)^N, \tag{5}$$

2.1.2. Negative Bending

Similar to the case of positive bending, in the negative bending state (the sensitive region is on the concave side), the optical power leakage from the surface s_0 can be expressed as [11]:

$$
\begin{aligned}
P_0' &= \iint_{S_0} dS \int_0^{2\pi} d\theta_\varphi \int_0^{\theta_b} I_0 \sin\theta_0 \cos\theta_0 d\theta_0 \\
&= 2\pi I_0 \int_{-a}^{-a+h} 2\sqrt{a^2 - h_1^2} dh_1 \int_0^{\theta_b} \sin\theta_0 \cos\theta_0 d\theta_0 \\
&= 2\pi I_0 \int_{-a}^{-a+h} J(h_1)dh_1
\end{aligned}
\tag{6}
$$

Then the optical power factor of the leakage at the surface is:

$$\alpha_0' = P_0'/P_b = \int_{-a}^{-a+h} J(h_1)dh_1 / \int_{-a}^{a} J(h_1)dh_1, \tag{7}$$

Similarly, in the negative bending state of the fiber sensor, the relative loss of optical power in the sensitive region can be expressed as:

$$\delta P' = 1 - (1 - \alpha_0')^N, \tag{8}$$

According to the formula (5) and the formula (8), the relationship between the relative loss of the optical power in the sensitive region and the bending radius is plotted. As shown in Figure 2, for the long sensitive region, the relative loss of optical power decreases with the increase of the bending radius in the positive bending; while the relative loss of the optical power increases when the bending radius in the negative bending increases.

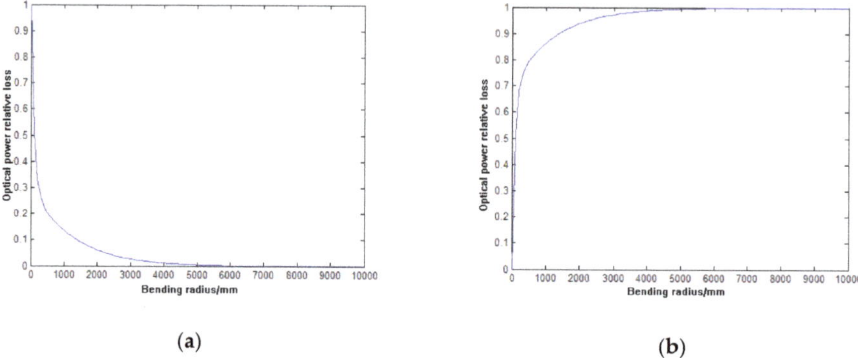

Figure 2. (**a**) relationship between positive bending relative optical power loss and bending radius; (**b**) relationship between negative optical bending relative optical power loss and bending radius.

2.2. Transmission Power of the Optical Fiber Sensitive Region during Bending

According to the directional nature of the bending, a certain degree of deformation is performed on the sensitive regions of the optical fiber, respectively. The optical power meter THORLABS PM200 is used to measure the output intensity of the fiber in different bending radii of the fiber sensitive region. Finally, the output light intensity and bending of the fiber are drawn. The curve relationship of the radius is shown in Figure 3.

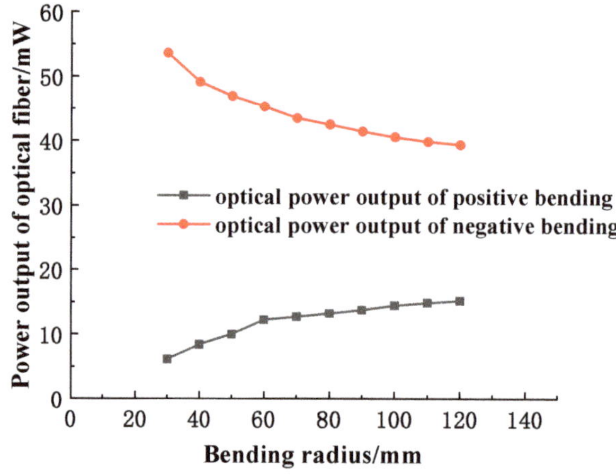

Figure 3. The relation curve of output light intensity and bending radius of fiber.

According to the curve, the relationship between the output power of the fiber and the bending radius can be known as follows: When the sensitive region of the fiber is bent positively, the output optical power of the fiber gradually increases with the increase of the bending radius; when bending is negative, the opposite is true.

Being aware of the output intensity of the fiber when the fiber sensitive region is not bent, according to the measured output optical power of the fiber under different bending radii, the relationship between the relative loss of the optical power of the fiber curvature sensor and the bending radius of the fiber would be obtained. As shown in Figure 4, when the bending radius is large ($\geq$120 mm), since the output voltage of the fiber curvature sensor has no obvious relationship with the bending radius, the second half of the curve cannot be proved in Figure 2. However, within the linear measurement range of the sensor from 20–120 mm, the measured curve is basically consistent with the simulation results in the mathematical model, indicating that the established mathematical model is effective.

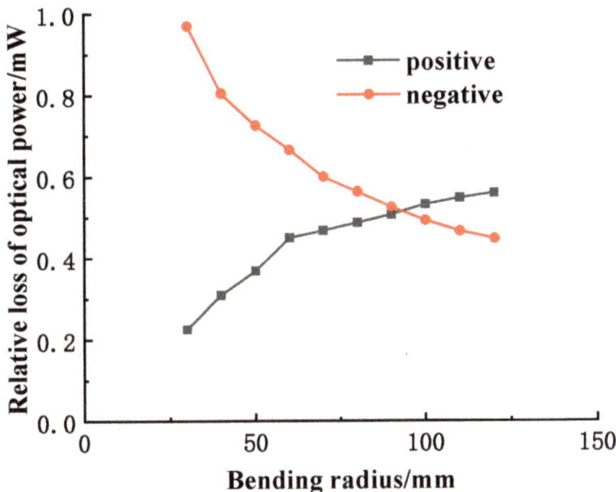

Figure 4. The relation curve between the relative loss of optical power and bending radius.

3. Sensor Signal Conditioning Circuit

The sensor conditioning circuit mainly includes functional module circuits such as amplification, filter, voltage conversion and so on, which can recognize the weak signal change of the sensor output. As shown in Figure 5, when the pulse signal drives the light emitter and the red light emitted by the light emitter is transmitted through the optical fiber, the bending information of the sensitive region of the fiber is converted into the voltage signal which is the output of the light receiver. Then the signal is amplified and filtered, finally being obtained by the microcontroller.

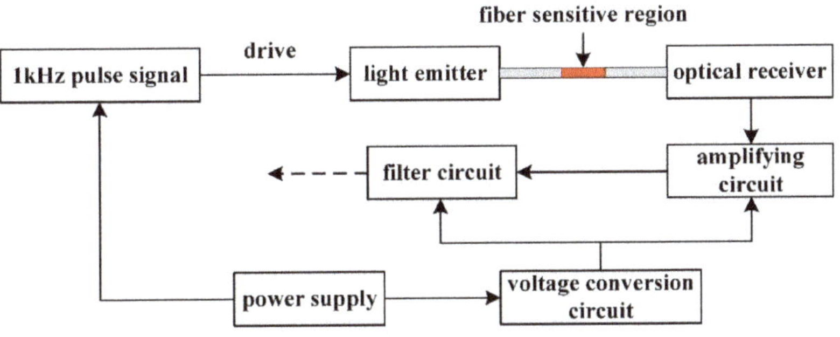

Figure 5. Sensor signal conditioning flow chart.

3.1. Amplifying Circuit

The input signal of the sensor is the bending information of the sensitive region of the fiber, and the output signal is a weak voltage signal. Generally, it is within 50 mV, and with the positive and negative bending of the sensitive region of the fiber, the dynamic range of the signal is wide, which, in the meantime, is easily affected by the working environment. Therefore, the input at the next stage must have large common-mode interference. In order to reduce the interference, the instrumentation amplifier AD620 is selected in the amplifier circuit for amplification. Compared to general-purpose amplifiers, AD620 has the advantages of the small operating current, high common-mode rejection ratio, low offset and drift, low noise and high closed-loop gain stability [18,19].

As shown in Figure 6, the amplifying circuit is a single-ended input: the inverting input of the AD620 is directly grounded, and the non-inverting input is connected to the sensor output signal. Meanwhile, this circuit can also play an important role in reducing common mode noise. Because the sensor output is also a single-ended signal and the other end is actually shared with the entire circuit system, it can be used as a double-ended input. In addition, a variable resistor R_g is connected across the circuit pins 1 and 8 to facilitate adjustment of the amplification factor G. Both dual power supply inputs are connected to the ground through a capacitor to ensure the stability of the supply voltage.

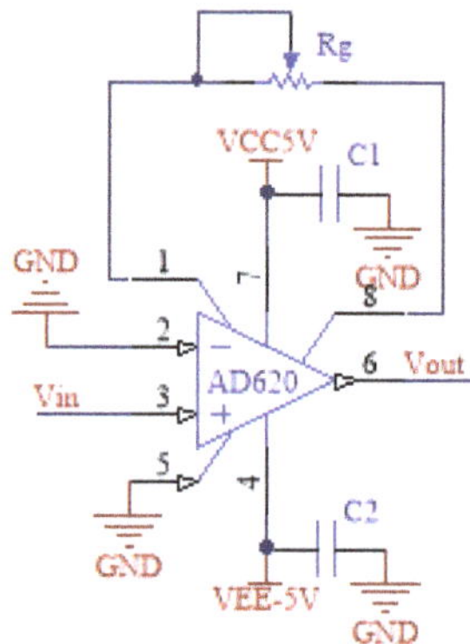

Figure 6. AD620 amplifier circuit.

3.2. Filter Circuit

The pulse frequency of the driving photodiode is 1 kHz, so the center frequency of the sensor output signal is about 1 kHz. The output signal of the sensor can be modulated by band-pass filtering. OP07 is chosen as the main chip [20]. The chip's high accuracy and extremely low input offset voltage eliminate the need for additional zeroing in many applications [21].

According to relevant knowledge of the analog circuit [22], the band-pass filter circuit can generally be formed by a low pass filter circuit and a high pass filter circuit connected in series. In order to reduce the output voltage at a faster rate outside the passband, improving the ability of the band-pass filter to remove noise, the second-order low-pass and second-order high-pass filter circuits are connected in series to form a fourth-order band-pass filter. As shown in Figure 7, the amplitude-frequency characteristic of the circuit is narrow, and the gain of the passband is 0 dB. At the frequency of 650 Hz or 1350 Hz, the amplitude attenuation is about 3 dB.

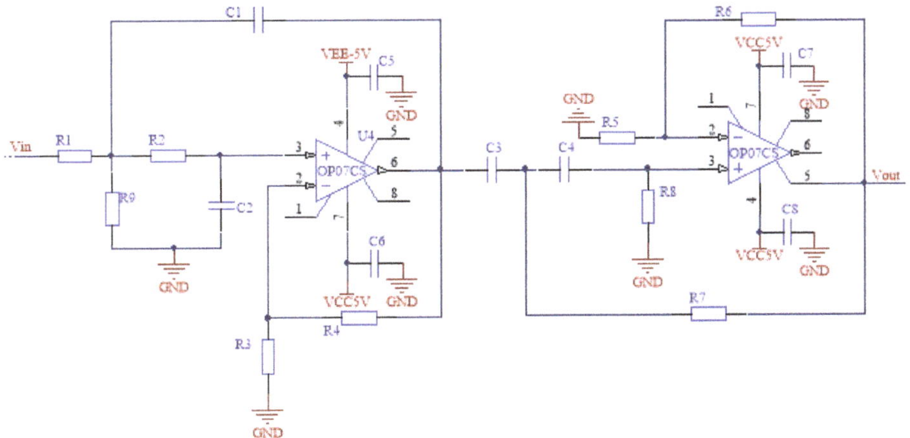

Figure 7. Band-pass filter circuit.

3.3. Summing Circuit

The micro signal output by the sensor has been amplified and filtered, and the final output signal is supposed to be a sine wave voltage signal [23]. If the sine wave signal is directly connected to the IO pin of the A/D input of the microcontroller, only the positive voltage in the range of 0–3.3 V could be measured, while the negative voltage would directly return to 0. Therefore, it is also necessary to carry out bias processing on the signal to raise the whole sine wave signal above the 0 level and ensure that the maximum amplitude of the signal is below 3.3 V. According to the principle of voltage divider, the designed circuit is shown in Figure 8. By adjusting the resistance of the slide rheostat to change the value of V_k, the sinusoidal signal can be moved up and down as a whole, and the sinusoidal signal waveform can be controlled between 0–3.3 V.

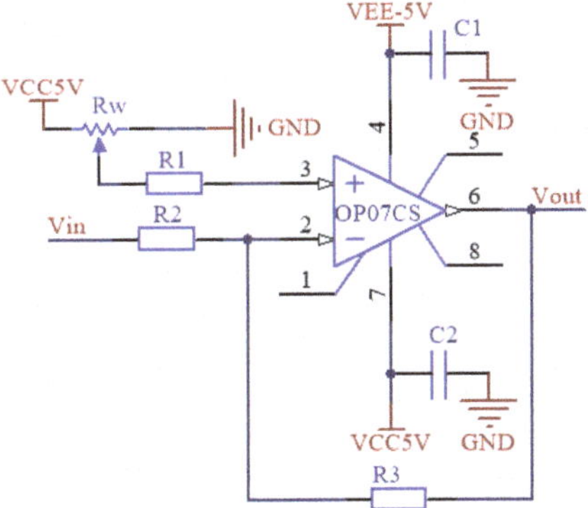

Figure 8. Band-pass filter circuit.

4. Sensor Characteristics Analysis

4.1. Linear Measurement Range Analysis

4.1.1. A Sensitive Region

As shown in Figure 9a, the sensitive region of the fiber is sequentially pressed against the surface of a series of cylinders with a radius of 10 mm, 20 mm, ... , 120 mm for positive and negative bending.

According to the output voltage of the sensor, the relationship curve between the output voltage of the sensor and the bending curvature can be obtained through linear fitting within a certain range, as shown in Figure 9b:

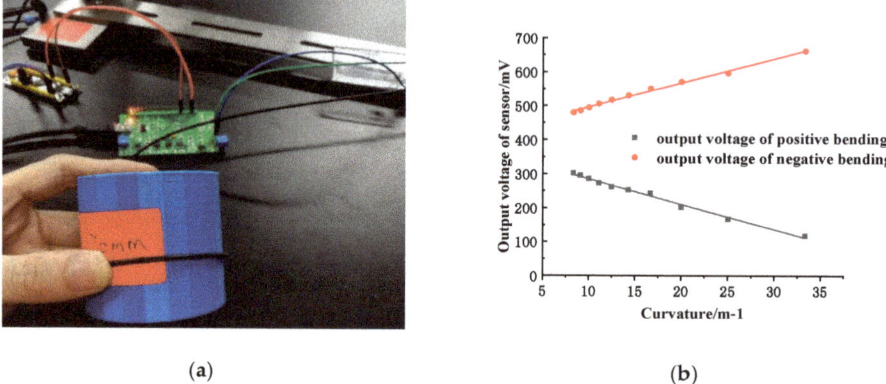

(a)   (b)

Figure 9. (**a**) bending on a cylinder; (**b**) linear fitting of positive and negative bending output voltage on a cylinder.

The linear equations of the sensor in positive and negative bending are obtained by linear fitting:

$$U_{postive} = 359.81715 - 7.45486C, \tag{9}$$

$$U_{negative} = 425.09514 + 7.18588C, \tag{10}$$

It can be seen from the above graph that the two fitting lines when the fiber is positively and negatively bent are basically symmetrical about the voltage value of 400 mV (that is, the output voltage value of the sensor when the fiber sensitive region is not bent). The measurement range of the fiber curvature sensor is defined as 8.33 to 33.3 m^{-1}. Within this range, the output voltage of the sensor is linearly related to the curvature.

4.1.2. Two Sensitive Regions

Two sensitive regions of the same length are processed on a fiber with a length of 1 m and a gap of 250 mm. When there is only one sensitive region, the voltage when the fiber is not bent is 400 mV, while after the second sensitive region is processed, the output voltage drops to 370 mV. The first sensitive region on the optical fiber was bent on a series of cylindrical cylinders, measuring the output voltage before and after processing the second sensitive region, then the relationship curve between the output voltage of the sensor and the bending curvature could be obtained, as shown in Figure 10a. The curve is fitted with a straight line within the range of linear measurement, as shown in Figure 10b.

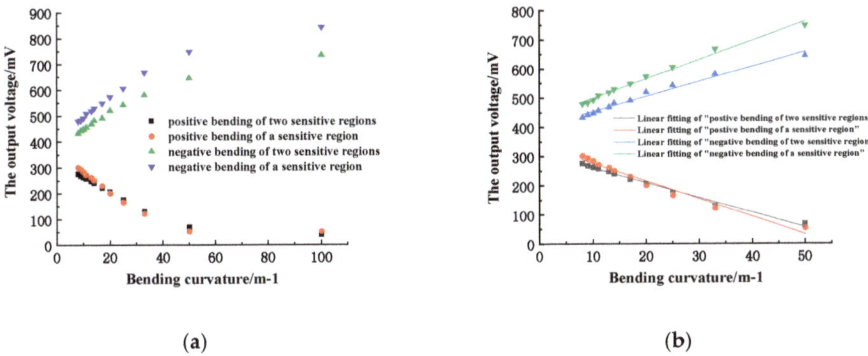

Figure 10. (**a**) relationship between output voltage of two sensitive regions and bending curvature; (**b**) linear fitting of two sensitive regions.

The curve shows that the linear measurement range of the sensor remains substantially unchanged after the second sensitive region is added. However, the output voltage will decrease when the fiber is bent, that is, the optical power loss will increase. According to the fitted line, it can be seen that after increasing the number of sensitive regions, the sensitivity of the sensor will decrease when bending positive or negative. Therefore, once the sensor's measuring sensitivity is guaranteed, multiple sensitive regions can be arranged on one optical fiber to realize the curvature measurement of multiple points.

4.2. Underwater Measurement Accuracy Analysis

4.2.1. Underwater Simple Beam Measurement

On land, the sensor can detect bending curvature within a certain range. Before applying it underwater, it is necessary to verify whether the sensor still meets some rules of underwater. Due to the limited experimental conditions, an experimental method similar to simply supported beams is used to roughly examine the characteristics of the sensor when it is used underwater.

As shown in Figure 11, the fiber sandwich is placed on a 50 mm wide water tank, and then directly traversed the sensitive area of the fiber interlayer through the thimble with the scale mark. The output voltage value of the sensor is recorded every 5 cm; until the thimble moves and the distance reaches 35 cm, the fiber interlayer is turned 180 degrees, the negative bending is measured according to the above operation, and finally the relationship between the sensor output voltage and the curvature is plotted, as shown in Figure 12.

Figure 11. Underwater simply supported beam measurement experiment.

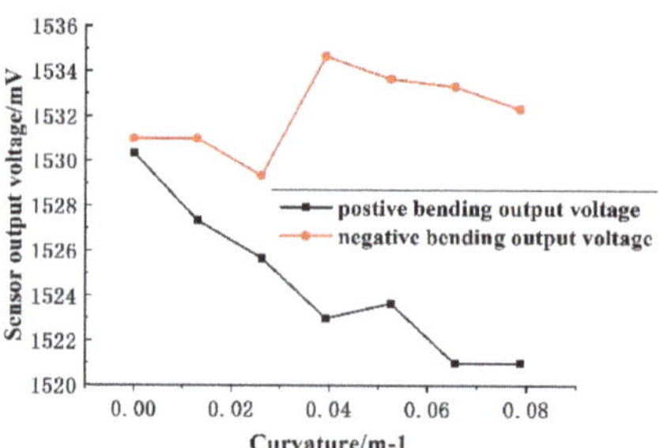

Figure 12. Output voltage versus curvature curve of underwater simply supported beam bending.

The curve in the figure shows: When the bending curvature of the fiber varies from 0–2.7 m^{-1}, the overall change trend of the output voltage of the sensor is consistent with the theory of power loss in the sensitive area of the fiber, that is, negative (positive) bending, as the bending curvature increases, the output voltage increases (decreases). However, from a partial perspective, there is no significant relationship between the output voltage of the sensor and the bending curvature of the fiber. Therefore, it can be concluded that the sensor is not suitable for measuring deformations with a curvature range of 0–2.7 m^{-1}; the power loss principle of the fiber sensitive area is still applicable underwater.

4.2.2. Underwater Dynamic Curvature Measurement

As shown in Figure 13, the optical fiber interlayer, is attached to the surface of a core with a bottom diameter of 30 cm and a height of 35 cm (taper of 0.86) underwater, moving slowly upward from the side of the cone near the bottom (the sensitive regions of the fiber are always kept close to the side of the cone in the process of moving), and the single-chip microcomputer is used to collect the real-time output voltage of the sensor.

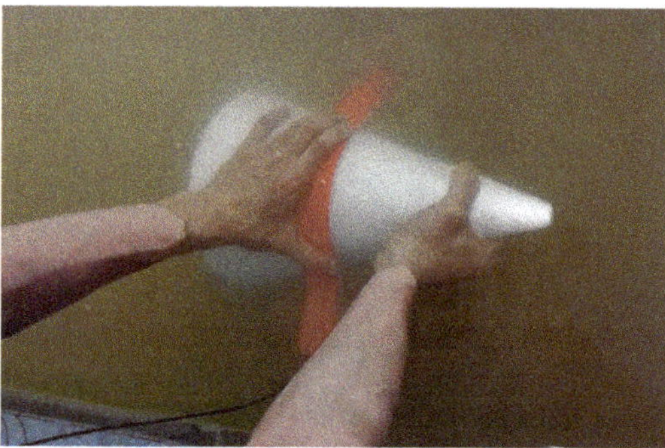

Figure 13. Measuring the curvature of the cone.

The relationship between the sensor output voltage and the equivalent bending curvature is plotted according to the measured data (the equivalent bending radius is the cross section radius parallel to the base of the cone at the contact point between the sensitive region and the cone's side, which is proportional to the actual bending radius, and the ratio coefficient is 0.92), as shown in Figure 14a.

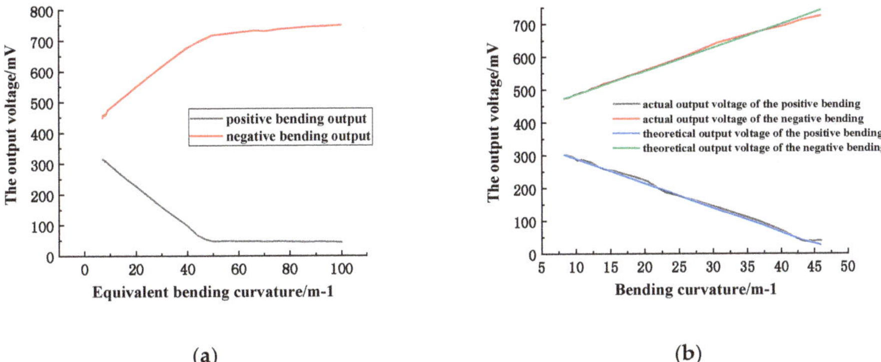

(**a**)   (**b**)

Figure 14. (**a**) relationship between output voltage and equivalent bending curvature; (**b**) relationship between the actual measured and theoretical output voltage and the bending curvature.

It can be seen from the above curve that when the equivalent bending curvature is in the range of 10–50 m^{-1} (the actual bending radius is about 30–120 mm), the output voltage is also substantially linear with the equivalent bending curvature. Within this range, compare the actual output voltage to the theoretical voltage (calculated by using the linear measurement equation of the sensor) and a curve varying with the curvature of a curve are plotted. As shown in Figure 14b, the relationship between the actual, theoretical output voltage and the curvature of curvature is substantially coincident, indicating that the measuring range of the sensor is also suitable for its underwater measurement.

4.2.3. Measurement of Underwater Static Curvature

As shown in Figure 15, the standard cylinder is regarded as the underwater deformation with known curvature. The sensitive region of the fiber is sequentially attached to the cylindrical surface with a radius of 120 mm, 100 mm, ... , 40 mm, 20 mm. The sensor's output voltage values of each time are correspondingly substituted into Equations (11) and (12) to obtain the measured curvature.

Figure 15. Measuring the curvature of a cylinder.

According to the Table 1 and the calculation, when the positive and negative bending radius of the fiber is 60 mm, the deviation between the measured curvature and the theoretical curvature is the maximum, but the deviation is different. $\Delta C_{positive\ max} = 0.76$ m^{-1} and $\Delta C_{negativemax} = 0.6$ m^{-1}. Therefore, the measurement accuracy of the sensor in the positive and negative linear range can be obtained as follows:

$$A_{postive} = \frac{\Delta C_{positivemax}}{C_{max} - C_{min}} \times 100\% = \frac{0.76}{50 - 8.33} \times 100\% = 1.82\%, \tag{11}$$

$$A_{negative} = \frac{\Delta C_{negativemax}}{C_{max} - C_{min}} \times 100\% = \frac{0.6}{50 - 8.33} \times 100\% = 1.44\%, \tag{12}$$

Table 1. Measuring standard cylindrical curvature data.

Bending Radius(mm)	Theoretical Curvature(m^{-1})	Forward Output Voltage (mV)	Measuring Curvature (mm)	Backward Output Voltage (mV)	Measuring Curvature (mm)
120	8.33	298	8.26	484	8.32
100	10.00	284	10.11	486	10.26
80	12.50	262	13.09	519	12.93
60	16.67	241	15.91	549	17.27
40	25.00	174	24.92	602	24.73
20	50.00	165	49.52	782	49.56

5. Conclusions

In this paper, the fiber-optic curvature sensor with strip type sensitive region is taken as the research subject. Through studying its sensing mechanism and underwater application, the following conclusions are obtained:

1. The accuracy of the model of bending power relative loss and bending radius is verified by measuring the optical power loss in the sensitive region of the fiber, and the characteristics of positive and negative bending are explained effectively.
2. The sensor conditioning circuit can effectively amplify and filter the weak voltage signal, and can simultaneously measure the curvature of different bending degrees with a good application range.
3. After adding a second sensitive region on one fiber, the output voltage of the fiber curvature sensor will decrease while the sensitivity will decrease, but the linear measurement range remains basically unchanged.
4. When the sensor is applied underwater, its linear measurement range can be defined as 8.3–50 m^{-1}, which is improved compared with the linear measurement range of 0–16.67 m^{-1} of the existing products.

Author Contributions: Methodology, Y.Z.; formal analysis, C.C., Y.H. and Y.G.; investigation, Y.H. and Y.Z.; resources, J.C.; data curation, Y.Z. and Y.G.; writing—original draft preparation, Y.H.; writing—review and editing, C.C.; visualization, C.C.; project administration, J.C.; funding acquisition, J.C.

Funding: This research was funded by the National Key Research and Development Program of China (2017YFC0307703), Natural Science Foundation of Zhejiang Province (Y17E090017), Special Project for Promoting Economic Development in Guangdong Province with grant NO. GDME-2018D004 and Key Research and Development Project of Zhejiang Province (2018C03SAA01010).

Conflicts of Interest: The authors declare no conflict of interest.

References

1. Shiming, L.; Ping, R.; Hongnian, M. Development and Countermeasures of Marine Underwater Engineering Technology. *Ocean Sci.* **1997**, *16*, 49–53.
2. Zhang, Q.; Liu, N.; Kong, W.X.; Pang, Y.C. Discussion on underwater deformation and leakage detection method for seismic shell equipment. *Ship Sci. Technol.* **2017**, *39*, 44–47.
3. Kocak, D.M.; Dalgleish, F.R.; Caimi, F.M.; Schechner, Y.Y. A Focus on Recent Developments and Trends in Underwater Imaging. *Mar. Technol. Soc. J.* **2008**, *42*, 52–67. [CrossRef]
4. Shen, Y. GPS Underwater Topographic survey without Tide Gauge. *Geotich. Investig. Surv.* **2002**, *28*, 37–40.
5. Liu, S.; Tian, J. Review on the Development of Underwater Topographic Measurement Technology. *Water Transp. Eng.* **2008**, *1*, 11–15.
6. Liu, Y. *Discussion on GPS Measurement Technology and Its Application in Measurement Engineering*; Heilongjiang Xilin Iron and Steel Group Co., Ltd. Equipment Engineering Department: Yichun, China, 2007.
7. Liu, J. *Principle and Method of GPS Satellite Navigation and Positioning*; Science Press: Beijing, China, 2003.
8. Yang, Q. Fiber Optic Sensor and Its Application. *J. Sens.* **1999**, *4*, 36–38.

9.   Liu, A.M.; Yu, Z.F.; Li, J.S. Automatic Monitoring Technology for Surface Settlement of Underwater Foundation. *Appl. Mech. Mater.* **2013**, *401–403*, 992–996. [CrossRef]

10. Zhou, M.; Yang, P. Establishment and Application of Deformation Monitoring System for Underwater Engineering Construction. *Railw. Technol. Innov.* **2012**, *3*, 101–103.

11. Di, H.; Liu, R. Research Progress in Measurement of Curvature by Optical Fiber Sensing Technology. *J. Sens. Microsyst.* **2011**, *30*, 5–7.

12. Liu, R.; Liu, P.; Fu, Z. Study on the Mechanism of Curvature Sensing Fiber Curvature Sensor. *Acta Opt. Sin.* **2007**, *27*, 807–812.

13. Di, H.; Fu, Y.; Liu, R. A Mezzanine Research System for Curvature Fiber Sensors. *Photoelectron. Laser* **2010**, *21*, 1597–1601.

14. Kuang, K.S.C.; Well, J.C.; Scully, P.J. An evaluation of a novel plastic optical fib re sensor f or axi al strain and bend measurements. *Meas. Sci. Technol.* **2002**, *13*, 1523–1534. [CrossRef]

15. Fu, Y.; Di, H. *Study on Simultaneous Measurement of Torsion and Bending Moment Using Curvature Fiber Sensor*; Institute of Robotics, Harbin Institute of Technology: Harbin, China, 2005.

16. Wang, Y.; Chen, J.; Rao, Y. Long period fiber grating sensors measuring bend-curvature and determining benddirection simultaneously. *J. Optoelectron. Laser* **2005**, *16*, 1139–1143.

17. Kovacevic, M.; Djordjevich, A.; Nikezic, D. Mont e Carlo simulation of curvature gauges by ray tracing. *Meas. Sci. Technol.* **2004**, *15*, 1756–1761.

18. Cui, G.L.; Che, X.L. Small Signal Acquisition System Based on STC12C5A60S2 and AD620. *Electron. Des. Eng.* **2012**, *20*, 112–114.

19. Liu, S.-L. Introduction of strumentation amplifier AD620 applied in precision data acguisition of high level switching power supply. *World Power Supply* **2005**, *1*, 59–62.

20. Chen, G.R. Development of digitally-controlled DC current source based on single-chip microcomputer. *Mod. Electron. Tech.* **2013**, *8*, 153–156.

21. Dan, H.; Gang, Z.; Jiming, S. Design and Implementation of a Practical Photoelectric Detection Circuit. *Instrum. Technol. Sens.* **2010**, *11*, 79–81.

22. Tong, S.; Hua, C. *The Basis of Analog Electronic Technology*, 4th ed.; Higher Education Press: Beijing, China, 2006; pp. 361–370.

23. Michalski, S.E.; Andrews, K.M.; Mieczkowski, D. Method and Apparatus for Low Power Offset Correction of Amplified Sensor Signals. U.S. Patent 4816752, 28 March 1989.

MDPI
St. Alban-Anlage 66
4052 Basel
Switzerland
Tel. +41 61 683 77 34
Fax +41 61 302 89 18
www.mdpi.com

Sensors Editorial Office
E-mail: sensors@mdpi.com
www.mdpi.com/journal/sensors